Advanced Security and Privacy for RFID Technologies

Ali Miri
Ryerson University, Canada

Managing Director: Lindsay Johnston
Editorial Director: Joel Gamon
Book Production Manager: Jennifer Yoder
Publishing Systems Analyst: Adrienne Freeland
Assistant Acquisitions Editor: Kayla Wolfe
Typesetter: Erin O'Dea
Cover Design: Jason Mull

Published in the United States of America by
Information Science Reference (an imprint of IGI Global)
701 E. Chocolate Avenue
Hershey PA 17033
Tel: 717-533-8845
Fax: 717-533-8661
E-mail: cust@igi-global.com
Web site: http://www.igi-global.com

Library of Congress Cataloging-in-Publication Data

Advanced security and privacy for RFID technologies / Ali Miri, editor.
p. cm.
Includes bibliographical references and index.
Summary: "This book addresses security risks involved with RFID technologies, and gives insight on some possible solutions and preventions in dealing with these developing technologies"-- Provided by publisher.
ISBN 978-1-4666-3685-9 (hardcover) -- ISBN 978-1-4666-3686-6 (ebook) -- ISBN 978-1-4666-3687-3 (print & perpetual access) 1. Sensor networks--Security measures. 2. Radio frequency identification systems--Security measures. I. Miri, Ali.
TK7872.D48A296 2013
621.3841'92--dc23
2012045191

British Cataloguing in Publication Data
A Cataloguing in Publication record for this book is available from the British Library.

To Heather, Setareh, and Bijan

Editorial Advisory Board

Table of Contents

Detailed Table of Contents

In this chapter, the author briefly reviews the various attacks on existing identification and authentication schemes and describe the challenges in their design for RFID systems. The chapter categorizes the RFID identification and authentication schemes into two general categories: cryptographic and non-cryptographic solutions. Cryptographic solutions are based on symmetric or asymmetric cryptography systems. Depending on the resource available on the RFID tags, algorithms based on standard cryptography cannot be utilized in an RFID system and new cryptographic algorithms must be designed. However, there remain security challenges in protecting the RFID systems that cannot be solved solely by relying on cryptographic solutions. The chapter also reviews these challenges and look at the countermeasures based on non-cryptographic solutions that would further protect RFID systems.

Section 4
Privacy

The chapter first investigates privacy issues in RFID systems, namely information privacy threats and location privacy threats. RFID systems should be able to resist tag information leakage and tag tracking attacks. Next, the author presents a few formal models in which the notion of privacy in RFID systems is defined. To measure the privacy level of various RFID protocols, a formal privacy definition is needed. Formal models for RFID systems are continually being developed. Here, the chapter describes definitions of RFID systems, adversaries, experiments, and privacy in the most popular models so far: the Avoine model, the Juels-Weis model, and the Vaudenay model.

In this chapter, the authors discuss the impact of providing tag privacy on the performance of an RFID system, in particular the complexity of identifying the tags being queried at the back-end server. A common technique to provide tag privacy is to use pseudonyms. That is, for each authentication session, a tag uses a temporary and random-looking identifier so that it is infeasible for attackers to relate two authentication sessions. A natural question which should arise here is how the server can identify a tag given that the tag's identity is changing all the time. This problem becomes even more serious when the shared secret key between a tag and the server is updated after every authentication session to provide forward privacy. In the first part of this chapter, the authors review different techniques to deal with this problem. They then point out that most of the existing techniques lead to vulnerability of the back-end server against Denial-of-Service (DoS) attacks. They illustrate some of these attacks by describing methods which attackers can use to abuse the server's computational resources in several popular RFID authentication protocols. Finally, the authors discuss some techniques to address the privacy vs. performance dilemma so that DoS attacks can be prevented while keeping tag identification efficient.

Preface

Radio-Frequency IDentification (RFID) devices and systems are fast becoming a ubiquitous tool in automatic identification technologies due to their relatively low cost and ease of deployment. These systems are made of three main components, *tags*, *readers*, and *back-end servers*. Common tags have a very simple circuitry, and do not have their own power sources. Tags receive their power through inductive coupling, electromagnetic backscatter and close coupling with readers, and typically engage in authentication and transfer of information once they are powered by any nearby reader. As increasing numbers of RFID devices are used in logistics, consumer applications, and business specific applications, security and user privacy concerns are becoming ever more important. However, the general architecture of these systems and make-up of their components, in particular resource constrained tags, pose considerable challenges in designing secure, privacy-preserving algorithms for these systems. Any possible solution not only has to be able to counter attacks against any typical computer systems, but also has to be able to address the unique vulnerabilities due to the use and set up of these systems. Any solution must also be extremely efficient and scalable.

This book attempts to bring together comprehensive coverage of security and privacy issues related to RFID systems from the formal definitions and standards used, to security modelling of existing threats and vulnerabilities, while detailing both general and specific solutions and countermeasures to these threats. Each chapter was contributed by expert authors in the field, and best attempts were made to make each chapter self contained. The book also contains all the required background information, and it is meant not only to be of value to those researchers who work in the security domain, but also to be accessible to those researchers and practitioners without any prior background in security and privacy technologies. This book is comprised of four major sections: *Background and Preliminaries*, *Attacks*, *Existing Solutions* and *Privacy*. Below, you will find a brief description of each chapter in each part, which can also be used as a map of how this book can be read.

SECTION 1: BACKGROUND AND PRELIMINARIES

Devising and implementation of cryptographic algorithms and protocols suitable for RFID systems represent a challenging area of research. This book has two chapters dedicated to covering fundamental cryptographic concepts and notions used in the remainder of the book and in the area as a whole. In the first chapter, titled "Security Terminology" by Li, basic definitions of security service requirements, notions of symmetric and public key cryptosystems, and those of computational security verses information theoretical security are described. This chapter is intended for researchers and practitioners without any background in security, and introduces them to the general necessary concepts and terminology without focusing on any specific algorithms or cryptosystems.

Standards are an essential part of today's modern manufacturing methods and they regulate product characteristics such as quality, reliability, and safety for consumers. They also enable inter-operability among the same kind of products from different manufacturers and increase competition by providing a wider market share for higher quality and better priced products. In the second chapter, titled "RFID Standards," Ilker *et al.* provide an overview of the RFID standardization bodies and their standards. The chapter includes a categorization of RFID systems and describes their operating principles from the standard's perspective.

SECTION 2: ATTACKS

RFIDs use wireless media, so all of the threats associated with wireless media usage exist in RFID systems. RFID systems are further challenged by these threats because of the tags' limited hardware and energy capacity for implementing robust security measures. The chapter titled "RFID Wireless Link Threats" by Ilker *et al.* discusses various different types of attacks that can be launched against low-end, mid-range and high-end RFID systems. These attacks include eavesdropping attacks, tampering, session hijacking, relay and replay attacks. Countermeasures against these types of attacks are also discussed.

Hardware attacks represent another class of important security threats to RFID systems, as they target architecture and the operations of the working components of these systems. Chapter four, "Hardware Attacks" by Kong *et al.* gives a survey of various kinds of hardware attacks and their countermeasures. Attacks discussed include cloning attacks, direct read attacks, and side channel attacks including power and timing analysis attacks.

Attacks on RFID systems can target different components and layers. In the chapter titled "Computer System Attacks," Ning gives a detailed account of some of these attacks on tags and readers in middleware, back-end systems, and wired links. These include buffer overflows, data flooding and spurious data, propagation of viruses and worms, application layer attacks and those on Object Naming Services, and network protocol attacks. Different countermeasures proposed in the literature to combat these attacks are also discussed.

SECTION 3: EXISTING SOLUTIONS

Chapter six, titled "An Overview of Cryptography," by Vahedi *et al.*, discusses some of the related number theoretic preliminaries; details some of key algorithms such as AES, RSA, and Elliptic Curve Cryptography; as well as hash functions, digital signature algorithms and identification and authentication protocols. The chapter concludes by highlighting some of the challenges in designing lightweight cryptographic techniques suitable for RFID applications.

Remote identification is one of the main tasks for RFID tags, and secure, practical authentication schemes represent the first line of defence against (wireless) attacks on RFID systems. In the seventh chapter, titled "Identification and Authentication for RFID Systems," Malek *et al.* first discuss the physical characteristics of typical RFID systems such as reading range, memory, and timing and their impact on any practical authentication scheme. The chapter also discusses different types of attacks such as eavesdropping, relay and replay attacks, and tampering against two general types of authentication that may be required: unilateral and mutual authentication. The chapter then gives a detailed treatment of both symmetric authentication solutions such as MAC/hash- and AES-based protocols, as well as asymmetric ones such as RabinXs, NTRU and ECC. The chapter concludes by providing coverage of some specially designed authentication schemes such as HB and HB# and non-cryptographic solutions.

SECTION 4: PRIVACY

In typical RFID systems, tags have unique identifiers that are readable without line-of-sight contact. Although this is one the key features that makes these systems usable in many different types of applications, it also can lead to serious breaches of privacy as it creates the possibility of user information leakage and location tracking. In the chapter titled "Privacy Issues in RFID," Song describes in detail some of the privacy requirements in terms of tag information and location, and introduces the Avoine, the Juels-Weis, and the Vaudenay formal models that have been proposed to study and measure privacy of these systems.

A common technique for dealing with privacy issues discussed in Chapter Eight is to use pseudonyms for RFID tags instead of their true identifiers. This technique, however, can pose challenges for readers and back-end servers trying to identify and authenticate tags in the system. Malicious attackers can also exploit such a set up in many of current protocols by exhausting back-end servers' computational resources. In Chapter nine, titled "DoS Attacks on RFID Systems: Privacy vs. Performance," Duc *et al.* highlight the key issues with regard to these type of Denial-of-Service (DoS) attacks, and discuss the DoS vulnerabilities of several privacy-preserving RFID authentication protocols including O-FRAP, Ryu-Takagi, and Burmester-Medeiros-Motta protocols. The chapter also includes a proposed countermeasure technique that takes advantage of a two-phase authentication protocol.

The scale and widespread use of RFID systems in various applications pose serious concerns about the effects of malware propagation and the impact on many systems that use RFIDs. The chapter "Malware Protection on RFID-Enabled Supply Chain Management Systems in the EPCglobal Network" by Yan *et al.* uses RFID-enabled supply chain management systems in the EPCglobal network as a case study to demonstrate the basic characteristics of RFID malware and key issues in RFID malware protection. The chapter introduces a demo security system, *RFscreen* and shows how it can be used to protect a given system at the tag, the reader and the back-end system layers. Countermeasures used in different phases of malware life cycle are also discussed in detail.

RFID technology plays an important role in automation and optimization of organizations' supply chains. However, it can enable adversaries to infer both spatial and temporal information about a given supply chain if they can create a covert channel by using tag tracking, tag duplication or modification and possible reader compromises. In the chapter "Addressing Covert Channel Attacks in RFID-Enabled Supply Chains," Chawla *et al.* model supply chains as network flow graphs, and introduce taint checking and verification algorithms that can be used as countermeasures against supply chain-based covert channels. Evaluation results based on a case study and other possible mitigating techniques including the use of authentication, pseudonym, re-encryption, and Physically Unclonable Functions (PUFs) are also discussed.

Chapter twelve, titled "Building Scalable, Private RFID Systems," by Lu, focuses on privacy-preservation for large scale RFID systems. After an overview of some of the existing protocols and highlighting some of their security vulnerabilities or lack of scalability, strong and weak formal security models are defined for this problem and it is argued that strong security models may not be acceptable for large scale RFID systems due to the lack of authentication efficiency. A weak formal privacy model called Refresh is introduced, which has looser constraints on the output of tags, such as randomization and unpredictability, while allowing RFID systems to achieve acceptable privacy protection as well as highly efficient authentication.

I hope that you find the content of this book beneficial to you, and that it helps all those who are interested and working in this very explosive area to make more secure and private systems.

Ali Miri
Ryerson University, Canada

Section 1
Background and Preliminaries

Chapter 1
Security Terminology

Ming Li
Shandong University, China & State Grid Corporation, China

ABSTRACT

The widespread use of RFID technology gives rise to security concerns. Cryptographic technology provides various valuable tools to enhance the security of RFID systems. In the literature, many cryptographic protocols have been proposed and designed for safeguarding RFID systems. In this chapter, the author describes some fundamental terminologies in information security and cryptology. More information on cryptography can be found in (Mao, 2003; Koblitz, 1994; Stinson, 2005; Stallings, 2006).

1.1 BACKGROUND

RFID (Radio-Frequency IDentification) systems are made up of readers and tags. The readers read the tags with non-contact communication. In order to complete this function, each tag comprises at least two parts. One is a microchip for storing and processing data, which is similar to that in a smart card or USB key, while the other is an antenna for receiving and transmitting radio waves. Because of its convenience and low cost, RFID technology has been widely used in industry to improve the efficiency of tracking and managing goods and production. For example, we are using RFID at the cashier of pharmacies to read the prices of different drugs. RFID has also been used for more complex applications, such as passport verification, employee cards and payment for bus tickets.

As with other new technology (such as computer networks), in the beginning, RFID systems designers were mainly focused on creating an available system without adequate consideration of security and privacy. Without access control, RFID tags could leak information and erode us of privacy. For example, an attacker could read the identifier in tags easily and track which items that an individual is using. If a RFID tags stores the Electronic Product Code (EPC) of an item,

DOI: 10.4018/978-1-4666-3685-9.ch001

we could identify the item by checking it on Object Name Service (ONS), which is provided by EPCglobal (Fabian et al. 2005).

Cryptography is the study of hiding information and protecting communications. In order to keep privacy in a RFID system, many cryptographic protocols have been designed to protect the sensitive information in tags, such as basic hash protocol and hash chain protocol. Cryptographic protocols hide tag ID by using cryptographic algorithms with secret keys. We give a simple example to show which aspects of security we need to consider transmitting a message protected by cryptography. Imagine that Alice is to send Bob a message "I am going to meet you at 19:00 tomorrow night." We list what services we will need to keep this information secure. That is to say, which aspects we should consider to implement a secure communication.

- **Confidentiality:** This message must be secret to others.
- **Authentication:** Alice knows for sure that she is communicating with Bob, and so does Bob.
- **Integrity:** The receiver, Bob, can verify if this message has been modified, such as if "tomorrow" has been changed to "today."
- **Non-repudiation:** Alice cannot deny that she has sent the message if she did send it, and Bob cannot deny receiving of the message if he does receive it.
- **Availability:** This message must be delivered in time, which means the communication channel have to always be in working order if needed.

The security requirements above can all be satisfied by cryptographic methods. If Alice wants to send the message secretly, she encrypts this message into a ciphertext using a cryptographic key. Then she sends the ciphertext of the message to Bob. Bob decrypts the cipher, firstly using a decryption algorithm with the right key, and reads the message. In general, encryption/decryption algorithms are either based on symmetric key cryptography or asymmetric key (public key) cryptography. In symmetric key cryptography, the encryption methods require that the sender and receiver share the same key. The encryption key and the decryption key are different in asymmetric cryptosystems. Whatever cryptographic algorithm Alice is using, she needs to share the right key or key pairs with Bob. We describe how to achieve key sharing in the next section. After that, some fundamental concepts of security requirements are explained. Lastly, we introduce the quantities that measure the security of an algorithm.

1.2 KEY SHARING

Before two parties start an encrypted communication, they have to share the right cryptographic keys. In symmetric cryptography, two parties must share the same key that is secret to all others. But in asymmetric key cryptography, the encryption key and the decryption key are different. Each user possesses a pair of keys, namely the public key and the private key. The public key (or a certificate) can be accessed by anyone, and may be stored, for example, in public servers managed by the so-called Certificate Authority (CA). Thus anyone else can send a ciphertext that is encrypted using the public key, but only the user who owns the private key can decrypt it. The foundation of public key (or asymmetric key) cryptography was a revolutionary change for key distribution. We have asymmetric key exchange protocols that can establish a secure communication channel in open networks.

For symmetric cryptography, it is more difficult to establish a shared cryptographic key. Two or more communication parties have to establish the shared key by some other means. A direct method is to establish the key using secure communication, such as mail, email or face-to-face.

One key establishment technique is key distribution, which requires a trusted dealer to delivers a key to two or more parties who want to build a secure communication channel. The disadvantages of this method are the bottleneck of the Trusted Third Party (TTP) and the inefficiency of the system for a large group of many communication parties.

Many more key establishment techniques are implemented using public key techniques. The Diffie-Hellman key exchange is a cryptographic protocol that enables two parties to establish a secret key through an insecure communication channel. A practical secure mechanism is a digital envelope technique combining symmetric cryptography and public key cryptography, which encrypts a secret symmetric key using a public key.

1.2.1 Symmetric Key

In symmetric key cryptography, the encryption methods require that the sender and receiver share the same key. Here are three possible means to achieve key distribution for symmetric key encryption between two parties, Alice and Bob.

1. Alice selects a key and physically deliver it to Bob.
2. If Alice and Bob have already shared a key, they can establish a new key under the protection of the old key.
3. If Alice and Bob are sharing different keys with a Key Distribution Center (KDC) separately, the KDC can deliver a key to Alice and Bob though secure channels.

In basic hash protocol (Weis et al., 2004), which is one of the first cryptographic protocol proposed for protecting RFID privacy, the keys of each tag are distributed physically before the use of these tags. Every tag has a secret key that is shared with the reader. When a tag is queried, it generates a random number n and sends $(n, H(k, n))$ to the reader as a response, where $H(k, n)$ is a hash function. The reader maintains a database that stores the keys of all tags. When the reader receives $(n, H(k, n))$, it tries all keys to hash n and checks which one matches the response. The reader verifies the ID of a tag when its hash value is found in the database. The key distribution here is similar to the first scenario above. The hash function is a one-way function that maps a message to a random string of bits. For example, Secure Hash Algorithm (SHA-1) (FIPS 2004), which is published by NIST as a Federal Information Processing Standard, produces a 160-bit digit from a message that is shorter than $2^{64} - 1$ bits.

In the third scenario, we make use of a trusted third party to help us distribute the keys. KDC (Stallings, 2006) is a trusted server that is responsible for managing all the keys in a system, and verifies the identity of every user. The KDC shares different keys, called user keys, with each user. When Alice and Bob want to establish a secret key to start a secure session, the KDC assigns a session key and sends it to Alice and Bob using their user keys separately. In general, a session key expires quickly, and a new session key can be created with the help of KDC before the expiration of the previous one. For reducing the burden of the KDC server, the new session key can also be established by the second method above. The Kerberos protocol is a successful commercial authentication protocol based on symmetric key cryptography. It realizes secure information exchange using a KDC. More information on Kerberos protocol can be found in (Miller et al., 1988) and (Kohl & Neuman, 1993).

1.2.2 Asymmetric Key

In a symmetric key cryptosystem, users who want to build a protected communication channel must have the same secret key prior to starting this channel. Before the appearance of asymmetric key cryptography, distribution or exchange of keys was extremely troublesome, because we needed a trusted channel to do the key exchange. In general, such a channel often had to be a physical channel. Thanks to the proposition of asymmetric key (or public key) cryptography, we can now exchange keys over an insecure communication channel. This feature encourages the development of electrical commerce. In asymmetric key cryptography, every party has a pair of keys. One is private and the other is public. We introduce the Diffie-Hellman key exchange protocol (Diffie & Hellman, 1976) to demonstrate how a public key protocol works (see Box 1).

In the Diffie-Hellman key exchange protocol, since $k = g^{ab} = g^{ba} (\bmod p)$, Alice and Bob possess the same group element k, which can serve as the shared secret key for symmetric cryptography. Others who do not know the private keys a and b, cannot compute the right secret key k. This is a basic version of the Diffie-Hellman protocol, which is not resistant to man-in-the-middle attack. That is to say, the basic Diffie-Hellman protocol cannot ensure the identity of the communicating parties. The security of the Diffie-Hellman protocol is based on the hardness of the discrete logarithm problem, which is believed to be intractable. This is called computational security that will be introduced in the next section. This protocol has been employed by a number of commercial products, such as Secure Sockets Layer (SSL) protocol (Rescorla et al., 2010) and IPSec protocol, which provide security for network communications.

We have other approaches to achieve key sharing without using a key exchange protocol. For example, in Pretty Good Privacy (PGP), a

Table 1. Huffman coding

a	*b*	*c*	*d*	*e*
0.10	0.12	0.15	0.20	0.43
0	1			
0.22		0.15	0.20	0.43
		0	1	
0.22		0.35		0.43
0		1		
0.57				0.43
0				1
1.0				

popular tool that provides cryptographic security, the session keys for a symmetric key algorithm are encrypted and distributed by the public key encryption. A cryptographic encryption of a message can also be completed by both symmetric key and asymmetric key encryption. Since public key encryption always consumes much more resources in computation, we use a combined cryptosystem to improve efficiency without losing security when we encrypt a long message. The digital envelope technique is a combination of symmetric key and asymmetric key cryptosystems. A digital envelope consists of a message encrypted by a symmetric key algorithm and a secret key that is encrypted using an asymmetric key algorithm with the receiver's public key. If the receiver wants to decrypt the message, he first decrypts the secret symmetric

Box 1.

Diffie-Hellman key exchange protocol

Input: A prime p, and a generator $g \in F_p^*$.

Output: A secret element $k \in F_p^*$ that is shared by Alice and Bob.

1. Alice chooses a number $a \in F_p^*$ randomly as her private key, computes the public key $g_a = g^a (\bmod p)$ and sends g_a to Bob.
2. Bob chooses a number $b \in F_p^*$ randomly as his private key, computes the public key $g_b = g^b (\bmod p)$ and sends g_b to Alice.
3. Alice computes $k = g_b^a (\bmod p)$.
4. Bob computes $k = g_a^b (\bmod p)$.

key using his private key and the asymmetric key algorithm. Then he decrypts the message using the secret symmetric key. This combination technique has been accepted by a number of applications, such as SSL protocol.

Although public keys can be accessed by anyone, the distribution of public keys also needs our attention in practice. We can publish it through public announcement for convenience, such as releasing the announcement on forums, newsgroups and by mailing list. But this approach has the weakness that anyone can forge a public announcement. To avoid forged information, a trusted third party CA is created to maintain a public key directory, which can be read by anyone but can only be modified by the CA. Public key infrastructure (PKI) (ITU-T 1993; Weise, 2001) is a perfect system that provide authentication and identification for each user, and distributes the public keys securely. In a PKI, public key certificates (Adams & Lloyd, 2002) are used to verify that a public key belongs to a specific user. A Public key certificate (or digital certificate) is an electronic document signed by a CA (Certificate Authority) to bind a public key to the owner's identity. The digital certificate includes the name of a person or an organization, their address and other useful information. The CA is in charge of verifying the truth of the user's information, and confirms it by signing the certificate. If we believe the CA, we believe that the certificate signed by the CA is valid. PKI has been widely applied in network communications and e-commerce. Several countries have passed legislation to give legal effect to digital certificates.

1.3 SECURITY REQUIREMENTS

Many network systems are typical open systems, such as computer internet, which means any terminal can join the networks without authentication. Most RFID systems can also be treated as open networks because the communications between tags and readers are not protected. The existence of attackers in this kind of system has to be of concern if we want to protect the privacy. Dolev and Yao (1981) suggested a model to describe the ability of an "active" saboteur in this kind of environment:

- He can obtain any message passing through the network.
- He is a legitimate user of the network, and thus, in particular, he can initiate a conversation with any other user.
- He will have the opportunity to be a receiver to any user.
- He will have the opportunity to masquerade as any sender to any user.

We should consider these attackers when we design a security protocol or construct a system. They not only damage the systems by sniffing, but also copy, forge, delete or modify sensitive information without authentication. Imagine that the users in a network want to send some messages that they do not want to let anyone else see, including the administrator of the network. Although the administrator has the right to access any data package that is being transmitted in the network, he cannot understand the data if the transmission is protected by cryptographic methods. Lot of research has been done on cryptology, which can provide specific services to protect the information against interception or modification, and help the systems work without interruption. Some of the main services that cryptography provides are confidentiality, integrity, authentication, non-repudiation and availability.

1.3.1 Confidentiality

Confidentiality means the information is intelligible only to specified users. This is one of the basic services provided by cryptographic techniques. Imagine that Alice wants to send Bob a secret message M that cannot be understand by others. She uses a key and encryption algorithm to encrypt this message to become a string which is referred to as a ciphertext, and sends it to Bob. Bob has the right key to decrypt the ciphertext. Even if an attacker intercepts the ciphertext, he cannot decrypt it because he does not have the encryption key used. Since Caesar's cipher, there have been many techniques to encrypt a message. Nowadays, we use symmetric encryption and asymmetric encryption to get confidentiality. AES, DES, 3DES, IDEA and RC5 are standard symmetric encryption algorithms. RSA, Elgamal, and NTRU are popular asymmetric encryption techniques.

1.3.2 Authentication

When Bob ascertains that the message M was indeed sent by Alice using some techniques, he authenticates the message. Authentication means that the receiver can verify the origin of a message. We can realize authentication using symmetric or asymmetric cryptosystems. In a symmetric system, each authenticated part keeps the same secret or cryptographic key. When one part shows that he/she possesses the secret or encrypts/decrypts a mesšage correctly, the other parts assure that he/she is authenticated. Message Authentication Code (MAC) is one such technique. A MAC can be produced by a keyed Hash function or block cipher. A hash function is a mathematical function that accepts a message as an input and outputs an almost random string. In other words, a good hash function ensures that it is hard for us to construct two different input messages that have the same output value. Hash functions also play an important role in protecting message integrity. An example of a keyed Hash function is HMAC Hash-based Message Authentication Code (HMAC) which is a secret key authentication algorithm that appears in RFC2104, and examples of those using block ciphers are OMAC and CBC-MAC.

Public key cryptography has made a profound impact in designing authentication algorithms. If someone signs a file using his private key, anyone else can verify this signature using the related public key. Most digital signature schemes are based on public key cryptography. For example, Digital Signature Standard (DSS) is suggested by National Institute of Standard and Technology (NIST). RSA, Elgamal and Schnorr schemes are also popular public key signature algorithms. Authentication is also a service that most RFID systems provide. There are many authentication protocols for RFID to help a reader identifying a tag without leaking any useful information to the attacker. Most of these protocols are based on hash functions or hash chains. For more information the reader is referred to Chapter Seven of this book.

1.3.3 Integrity

Integrity ensures that any illegitimate user's modification to the message can be detected. Modification includes insertion, deletion, substitution or changes to the status of a given message. For example, an attacker may change part of the ciphertext intended for Bob, and Bob may not detect these changes and accept the forged message as genuine. There are many schemes and protocols in cryptology that help with maintaining the integrity of messages or files. Hash functions are effective tools to help maintain integrity. Integrity is often concerned with authentication. In HMAC, DSS and Schnorr signature schemes, there are specific parts to ensure message integrity using hash functions. RSA-OAEP (Optimal Asymmetric Encryption Padding) is one helpful technique to ensure the integrity of a message without identifying the sender. RSA-OAEP was accepted by PKCS#1, IEEE P1363 and SET as a standard encryption scheme.

1.3.4 Non-Repudiation

If a protocol makes sure that neither the sender nor the receiver can deny the transmission, we say that it provides non-repudiation. This feature is necessary in many applications today such as e-banking and e-business transactions. Consider an internet transaction for which e-banking was used to pay for goods and services received. Clearly, there will be serious ramifications if the buyer denies making the purchase or the seller denies receiving the funds, despite doing so. PKI can be used to provide non-repudiation services by using a Trusted Third Party (TTP). A TTP is an organization that has the trust of communicating parties. When a transmission occurs, the TTP signs the message with some useful information (time stamp, sender, receiver ...) and passes it through the channel. Then no one can deny the message that he sends and receives because he cannot forge a valid signature. Non-repudiation service can therefore serve as an effective tool in dispute resolutions. Non-repudiation has not been widely considered in RFID systems yet, but it may be put into practice in RFID systems in the future when RFID is used for more complex applications.

1.3.5 Availability

When we implement a system, we should ensure that the service is available to authorized users whenever needed. There are various ways to break a system or make it unavailable or unusable. For example, one can cut the communication line or destroy one piece of hardware. Theses physical issues are of concern in the management of the systems, and solved by some sort of physical actions, such as setting up access controls to the hardware. There are also remote attacks that can interrupt the system, such as Denial of Service (DoS) attacks, Distributed DoS (DDoS) attacks and viruses (such as back door or trojan horses). DoS attacks are often used by hackers in computer networks by, for example, blocking a website or a payment gateway. They are also very effective attacks against RFID systems, and can be implemented against any part of these systems. For example, in a RFID system, if one forges enough (legitimate) requests to the reader to execute a DoS attack, the reader will not be able to accept requests from any other tags requiring normal service. Thus, the system will become unavailable to its users. The DoS attack can also be against the database, servers or communication lines in the system. Some DoS attacks are amendable automatically after enhancing the authentication of the system. For example, if we design communication protocols between readers and tags carefully to tell readers how to distinguish a normal legitimate user, then the readers can deny the requests that are part of a DoS attack. As to the protection of computer networks, we use firewalls, security protocols (IPSec and SSL), Intrusion Prevention System (IPS) or Intrusion Detection System (IDS). For more information on DoS attacks, readers can read (Stallings, 2006).

1.4 MEASURING SECURITY

In this section, we describe how to measure the security of a cryptography scheme or protocol. Information theory, which was founded by Shannon in 1948, provides a theoretical measure of information security (Shannon, 1949). Information theory was born for modeling communication between simple memoryless sources and channels, and it has since been applied in many other fields.

It should be noted that this theoretical measure of security does not outline the details of possible successful attacks against a given scheme or protocol. A more concrete, practical measure of security it that of computational security. In computational security paradigm, one measures time and resource complexities of attacks against a given scheme. If the resource and the time complexities of attack algorithms are beyond the capacity of existing computers and users, then one cannot realize these attacks in a "reasonable" time, and we consider the scheme to be computationally secure.

1.4.1 Information Theory

Information theory is based on probability theory, and has been widely applied in many areas (signal processing, data compressing, cryptography etc).

Entropy, which was proposed by Shannon in 1948, is a basic tool to measure information. That is to say, entropy quantifies the uncertainty when we get a random variable. We now introduce the notation and fundamental properties of entropy.

Assume that L is a language which has n different symbols $X = \{x_1, x_2, \cdots, x_n\}$. If a coding source S produces these symbols independently, and the probabilities that $x_1, x_2, \ldots, x_n$ appear are $p(x_1), p(x_2), \ldots, p(x_n)$ respectively, then the entropy of S is

$$H(S) = -\sum_{i=1}^{n} p(x_i) log_2(p(x_i)).$$

We define $0 log_2 0 = 0$ in the formula above. Entropy defined as above shows the number of bits that S outputs, which measures the amount of information that S can encode. The base of the logarithm can be any number. If the base is the natural logarithm e, the unit of entropy is called "nats." When we use base 2, the unit is "bits." We shall often use logarithms to base 2 for convenience here. If each x_i appears with the same probability, that is to say $p(x_i) = 1 / n$, we have $H(S) = log_2 n$. One important application of entropy is Huffman coding (Huffman, D.A. 1952), which is an entropy encoding algorithm for creating prefix codes.

Assume that we have an alphabet $\{x_1, x_2, \cdots, x_n\}$ and a weight set $p(x_1), p(x_2), \ldots, p(x_n)$. Huffman coding finds codewords $\{f(x_1), f(x_2), \cdots, f(x_n)\}$ for every symbol which satisfies the following: the average length $l(f(X)) = \sum_{i=1}^{n} p(x_i) \mid f(x_i) \mid$ is the shortest, where $\mid f(x_i) \mid$ is the length of the codeword of x_i. We give an example to show how Huffman coding works. Assume that the symbol set is $\{a,b,c,d,e\}$, and the probabilities that each symbol appears are

$$p(a) = 0.10, p(b) = 0.12,$$
$$p(c) = 0.15, p(d) = 0.2, p(e) = 0.43.$$

We first choose the two symbols a,b with the lowest probabilities. Then we set $a = 0$ and $b = 1$ and delete them from the set. After that, we insert a new symbol with probability $p(a) + p(b) = 0.22$. We repeat these steps until there is just one symbol in the set with probability 1. Then we get the code of each symbol. The whole process constructs a tree, which is shown in Table 1. The last row, 1.0, is the root and the first row is the leaves. At last, we get the code of each symbol from the root to each leaf. Thus, we get the codes

$$f(a) = 000, f(b) = 001,$$
$$f(c) = 010, f(d) = 011, f(e) = 1.$$

The average length of the codewords is

$$(0.10 + 0.12 + 0.15 + 0.20) \times$$
$$3 + 0.43 \times 1 = 2.14.$$

The entropy of variable X is

$$H(X) = 0.10 \times 3.322 + 0.12 \times 3.059 + 0.15 \times$$
$$2.737 + 0.20 \times 2.322 + 0.43 \times 1.218 = 2.10,$$

which is close to the average length. In fact, we have the proposition

$$H(X) \leq l(f(X)) \leq H(X + 1).$$

As entropy measures the quantities of uncertainty, and we want an encrypted cipher to be as random as possible, the encryption algorithm should output ciphers with high entropy. Shannon proposed a perfect security scheme based on entropy in his 1949 paper. But in that perfect secure encryption scheme, the key space is as large as the message space, which makes the scheme infeasible to implement practically.

Entropy is also useful in cryptanalysis, which is to guess messages or keys given some ciphertexts. The redundancy of language (Shannon,

1949; Cover et al., 1978) is an application of entropy in cryptanalysis. In English, if every letter appears with the same probability, we have the entropy of this language $H_L = log_2 26 = 4.7$. But in everyday life, some letters are used more than others, such as E, A and T. Moreover, we often encounter some specific strings in English, such as "the," "ing" and "ed." These circumstances diminish the entropy of English. The entropyofEnglishisbelievedtobe $1.0 < H_L < 1.5$. If we take 1.5, we get the redundancy of English $R_L = 4.7 - 1.5 = 3.2$. The redundancy rate is $3.2 / 4.7 = 68\%$. The redundancy of language is used to estimate unicity distance, which is a measure of the length of ciphertext that is needed to break the cipher. Unicity distance tells us how many spurious keys we get in one analysis given a number of ciphertexts. Although the unicity distance is a useful measure in cryptanalysis, it does not mean that we can find the right key in practice. The unicity distance may be too large for us to complete the analysis using existing computers. This is called computational security, which is introduced in the next section. More information on cryptanalysis using entropy can be found in (Hellman, M.E. 1977) and (Stinson, D.R. 2005) for interested readers.

1.4.2 Computational Security

In modern cryptographic schemes, security relies on the assumption that adversaries have limited computational capacity. This is in accordance with practical requirements. In such cases, security cannot be guaranteed if the adversaries have enough (or unlimited) computational power, which is impossible in real-life. We take RSA (Rivest et al., 1978) for an example, which is a popular public key cryptographic algorithm that has been widely used in business. RSA Laboratories gave the following description of the security of RSA in the standard PKCS#6 (Public Key Cryptography Standards) (PKCS 2010).

The most important result thus far is the factorization of RSA-155 (a number with 155 digits), which was completed in August 1999 after seven months. A group consisting of, among several others, Arjen K. Lenstra and Herman te Riele performed the necessary computations on 300 workstations and PCs.... The result indicates that a well-organized group of users such as distributed.net might be able to break a 512-bit key in just a couple of days.

RSA Laboratories also recommended that we should use RSA with at least 1024 bits keys for corporate use.

Although security here is relative to our computational ability, we still have ways to measure it. If an successful attacker against one system can also break another system that is believed to be secure, or solve a mathematical problem that is widely believed to be hard (e.g. the integer factorization problem or discrete logarithm problem), we say that this system is computationally secure. The process that a problem is transformed into another problem is called "reduction" in complexity theory. A mathematical problem is hard or intractable if there is no efficient algorithm for solving it, or if it can be reduced to another hard problem that has been proven to be hard. We first show how to measure the complexity of an algorithm. Complexity includes time, space complexity, the number of gates in a circuit, the amount of communication, etc. Assume that we have a complexity function $T(n)$, where n is the length of input. We consider the bound on the growth rate of $T(n)$. When we say $T(n) = O(f(n))$, we mean there are exist a constant $c>0$ and a natural number N satisfying $T(n) \leq cf(n)$ for all $n>N$, where $O(f(n))$ is called big O notation. For example, if the complexity function $T(n) = 3n^2 + 2n + 1$, we get $T(n) \leq 4n^2$ where $N = 2$ and $c = 4$. We say that $T(n) = O(n^2)$ here. In fact, $T(n) = O(f(n))$ indicates the upper bound on the growth rate of $T(n)$. Because the term in $T(n)$ with the highest growth rate is the one with the largest exponent, we omit the other terms except $3n^2$ when $n \to \infty$. We omit the constant factor 3 for convenience because 3 does not depend on n. Thus, we get $T(n) = O(n^2)$ in this example. In general, big O notation make use of the term in $T(n)$ that grows fastest. For example,

- If $T(n) = 5n^4 + n^2 + 6$, then $O(T(n)) = O(5n^4) = O(n^4)$.
- If $T(n) = 2^n + 5n^{19}$, then $O(T(n)) = O(2^n)$.
- If $T(n) = n! + n^{20}$, then $O(T(n)) = O(n!)$.

When $f(n)$ is a polynomial, we say that the complexity is polynomial. If the order of $f(n)$ is exponential, we have an exponential algorithm. We also have a lower bound on the grow rate of $T(n)$, which is written as $T(n) \in \Omega(n)$. Like the definition of $O(f(n))$, $T(n) \in \Omega(n)$ means there exists a constant c to make $T(n) \geq cf(n)$ for all n greater than a number N. We always use $O(f(n))$ to represent the complexity of an algorithm.

How can we prove computational security? Let us inspect the security of RSA as an example, which is a popular public key algorithm nowadays (Box 2).

The function $\phi(N)$ in RSA algorithm is Euler's totient function, which is the number of positive integers that is less than N and relatively prime to N. Because

$$m = c^d = m^{ed} = m^{k(p-1)(q-1)+1} = m(\text{mod } N),$$

we can decrypt the cipher successfully if we have the private key d. The security of the RSA algorithm relies on the integer factorization problem.

Integer Factorization Problem

- **Input:** An odd composite number N.
- **Output:** A prime p that satisfies p|N.

N can be factored if we calculate the private key e successfully. That is to say, we could solve the integer factorization problem if we totally broke the RSA algorithm. The details of the reduction from calculating e to integer factorization are shown in (Koblitz, N. 1994). We emphasize that the integer factorization problem is hard when we choose appropriate variables. For example, for constructing secure RSA cryptosystems, we choose a composite number that is the product of two large primes. These two primes must be randomly chosen and about the same size (but not too close). It is widely believed that the integer factorization problem is a hard problem, which means we have no polynomial algorithm to solve it. The best published asymptotic running time for the factorization problem is

$$exp((\frac{64}{9}b)^{\frac{1}{3}}(logb)^{\frac{2}{3}}),$$

which is the time complexity of the general number field sieve (GNFS) algorithm, where n is a composite number that is b – bits long. This is a sub-exponential time algorithm. In 2009, several researchers factored a 232 digit number (768 bits) utilizing hundreds of machines by number field sieves for two years. This makes us consider 1024 bit-RSA at least in commercial use. Shor (1994) showed that a quantum computer could factor a composite number in polynomial time, which can be used to break RSA. But fortunately, quantum computers are not practical in the foreseeable future.

The security of all public key cryptosystems rely on hard problems, such as Elgamal encryption and Diffie-Hellman key agreement (on discrete logarithm problem), or the Rabin cryptosystem (on integer factorization).

The security of modern symmetric algorithms is also measured by computational security. For example, DES was shown to not be secure in 1998. A custom DES-cracker was built by the Electronic

Table 2. Security bits for some cryptographic algorithms (bits)

symmetric key	ECC keys	RSA keys
80	160	1024
112	224	2048
128	256	3072

Frontier Foundation (EFF) at the cost of approximately US $250,000, which cracked DES using a brute-force attack in 2 days. COPACOBANA is another more efficient DES cracker built for approximately $10,000. This machine breaks DES in less than one day. The substitution for DES is AES, which is believed secure because all the known attack algorithms are not practical.

1.4.3 Security Bits

Since the computational security of a cryptographic scheme is measured by the computational complexity of the efficient attack algorithms, we quantify security here using security bits, which is the binary length of the complexity. For example, triple-DES with a key size of 168 bits provides 112 bits of security, because the complexity of the best brute force attack is 2^{112} up to now. Note that the key length is distinct from the security bits of an algorithm, and the key length is definitely not smaller than the security bit length, because an algorithm can always be cracked by brute force. If we execute exhaustive search to attack a symmetric encryption algorithm with key size 128 bits, we must try at most 2^{128} keys to find the right key. Most symmetric key algorithms are designed to have security equal to their key length, which means that there is no faster attack on them than brute force.

The security of public key algorithms is often compared with symmetric key algorithms that can only be cracked by brute force attacks. For example, the security of 1024-bit RSA is often compared with 80-bit symmetric key algorithms in many standards documents, such as triple-DES (112-bit keys) and AES (128-bit keys). This comparison helps us to select appropriate encryption algorithms and key sizes in applications. If a system requires security that 80-bit symmetric key cryptographic algorithm can provide, then AES (128-bit keys), 1024-bit RSA and 160-bit ECC (Elliptic Curve Cryptosystem) are all appropriate choices. The next security level is 112 bit, which matches the security of triple-DES. We list the security that several common public key cryptographic algorithms provide in Table 2.

SUMMARY

This chapter gave a brief introduction to some fundamental terminology in information security

Box 2.

RSA Algorithm

Key Generation: 1. Choose p and q randomly, where p and q are both large primes.
2. Calculate $N = pq$.
3. Calculate $\phi(N) = (p-1)(q-1)$.
4. Choose e randomly, where $1 < e < \phi(N)$ and $gcd(\phi(N), 1) = 1$.
5. Calculate d, where $ed \equiv 1 (\bmod\, \phi(N))$.
6. Publish the public key (e, N) and keep (d, N) as the private key.

Encryption: $c = m^e (\bmod\, N)$, where m is a message and c is the encrypted cipher.

Decryption: $m = c^d (\bmod\, N)$.

and cryptology. Its purpose was to describe the basic aspects of security and tools for security. Specific techniques and algorithms were avoided in this chapter.

We first introduced symmetric key and asymmetric key cryptography, which are important tools for protecting communications. Then we described the aspects of concern fin information security. Finally, we showed how to measure security using entropy and computational complexity.

REFERENCES

Adams, C., & Lloyd, S. (2002). *Understanding PKI: Concepts, standards, and deployment considerations* (2nd ed.). Addison-Wesley.

Cover, T. M., & King, R. C. (1978). A convergent gambling estimate of the entropy of English. *IEEE Transactions on Information Theory*, *24*(4). doi:10.1109/TIT.1978.1055912

Diffie, W., & Hellman, M. E. (1976). New directions in cryptography. *IEEE Transactions on Information Theory*, *22*(6), 644–654. doi:10.1109/TIT.1976.1055638

Dolev, D., & Yao, A. C. (1981). On the security of public key protocols. In *Proceedings of IEEE 22nd Annual Symposium on Foundations of Computer Science*, (pp. 350-357).

Fabian, B., Guenther, O., & Spiekermann, S. (2005). Security analysis of the object name service for RFID. In *International Workshop on Security, Privacy and Trust in Pervasive and Ubiquitous Computing*, July 2005.

Garfinkel, S., & Spafford, G. (1996). *Practical Unix & internet security. O'Reilly*. Associates.

Hellman, M. E. (1977). An extension of the Shannon theory approach to cryptography. *IEEE Transactions on Information Theory*, *23*(3). doi:10.1109/TIT.1977.1055709

Huffman, D. A. (1952). A method for the construction of minimum-redundancy codes. *Institute of Radio Engineers*, *40*(9), 1098–1101. doi:10.1109/JRPROC.1952.273898

Koblitz, N. (1994). *A course in number theory and cryptography*. Springer Verlag. doi:10.1007/978-1-4419-8592-7

Kohl, J., & Neuman, C. (1993). The Kerberos network authentication service (v5). In *The Internet Engineering Task Force Request For Comments (IETF RFC) 1510*. Retrieved from http://www.ietf.org/rfc/rfc1510.txt.

Mao, W. B. (2004). *Modern cryptography: Theory and practice*. Prentice Hall.

McClure, S., Scambray, S., & Kurtz, G. (1999). *Hacking exposed: Network security secrets and solutions*. McGraw-Hill.

Miller, S., Neuman, B., Schiller, J., & Saltzer, J. (1988). Kerberos authentication and authorization system. In *Section E.2.1, Project Athena Technical Plan, M.I.T. Project Athena*. Cambridge, MA: MIT.

Rec, I. T. U.-T. X.509 (revised). (1993). *The directory-authentication framework*. Geneva, Switzerland: International Telecommunication Union.

Rescorla, E., Ray, M., Dispensa, S., & Oskov, N. (2010). Transport layer security (TLS) renegotiation indication extension. In *The Internet Engineering Task Force Request For Comments (IETF RFC) 5746*. Retrieved from http://tools.ietf.org/html/rfc5746

Rivest, R. L., Shamir, A., & Adleman, L. (1978). A method for obtaining digital signature and public-key cryptosystems. *Communications of the ACM*, *21*(2), 120–126. doi:10.1145/359340.359342

Sawyer, S., & Tapia, A. (2005). The sociotechnical nature of mobile computing work: Evidence from a study of policing in the United States. *International Journal of Technology and Human Interaction*, *1*(3), 1–14. doi:10.4018/jthi.2005070101

Shannon, C. E. (1948). A mathematical theory of communication. *The Bell System Technical Journal*, *27*(3), 379–423.

Shannon, C. E. (1949). Communications theory of secrecy systems. *The Bell System Technical Journal*, *28*, 656–715.

Shor, P. W. (1994). Algorithms for quantum computation: Discrete log and factoring. In S. Goldwasser (Ed.), *Proceedings of the 35th Annual Symposium on the Foundations of Computer Science*, (pp. 124-134).

Stallings, W. (2006). *Cryptography and network security: Principles and practice* (4th ed.). New Jersey: Prentice-Hall.

Stinson, D. R. (2005). *Cryptography: Theory and practice* (3rd ed.). Chapman & Hall/CRC.

Weis, S. A., Sarma, S. E., Rivest, R. L., & Engels, D. W. (2004). Security and privacy aspects of low-cost radio frequency identification systems. In Goos, G., Hartmanis, J., & van Leeuwen, J. (Eds.), *Security in Pervasive Computing, LNCS* (*Vol. 2802*, pp. 55–59). Berlin, Germany: Springer. doi:10.1007/978-3-540-39881-3_18

Weise, J. (2001). Public key infrastructure overview. In *Global Security Practice, Sun BluePrint™ OnLine*. Retrieved from http://www.sun.com/blueprints/0801/publickey.pdf

Chapter 2
RFID Standards

Ilker Onat
University of Ottawa, Canada

Ali Miri
Ryerson University, Toronto, Canada

ABSTRACT

There are many RFID standards defined by different standardization bodies and organizations. These evolving standards are often overlapping may be confusing to the practitioners. In this chapter, a summary of the RFID technology is given with the relevant standardization bodies and their RFID standards.

2.1 INTRODUCTION

Standards are important in regulating many parts of our lives. By providing regulations for interoperability for almost all modern human activity, they are now an essential part of today's businesses and governments. Standards also regulate product characteristics such as quality, reliability, and safety for consumers. Standards are an essential part of today's modern manufacturing methods and one of the major driving forces in technological advancement. They provide interoperability among the same kinds of products from different manufacturers and increase competition by providing a wider market share for higher quality and better priced products. Standards also define the practices and relevant metrics and assimilate new technologies to practical widespread use. General industrial adoption plans of RFID technology has increased interest in standardization. In this chapter, we describe the standardization bodies and summarize their RFID standards. We also emphasize some supply chain management considerations. RFID technology is poised to change the way many companies do business and will allow businesses to share information effectively. Different businesses can use the same information only if they agree on the content and format.

DOI: 10.4018/978-1-4666-3685-9.ch002

We first categorize RFID systems and describe their operating principles from the standards perspective. We then give an overview of the RFID standardization bodies and their standards.

2.2 AN OVERVIEW OF RFID SYSTEMS

An RFID system consists of two components: a transponder, or a tag and a reader, or an interrogator. The transponder carries the actual data and is attached to the object to be identified. RFID systems are categorized according to fundamental operating principles, tag complexity, operating frequency, range and powering methods.

As a major classification, according to the power source, tags are classified into three categories:

- **Passive tags:** Tags with no built-in power source.
- **Semi-passive tags:** Tags that use batteries for some of their operations.
- **Active tags:** Tags that use batteries as their sole source of energy.

2.2.1 Operating Principles of RFID Systems

There are three main types of physical operation for RFID tags: inductive coupling, electromagnetic backscatter and close coupling.

Inductively coupled transponders receive energy from the reader generated strong electromagnetic field which passes through the transponder's coil area. This electromagnetic field induces a current proportional to its strength (decreasing with distance), the coil area and the number of windings, providing energy to the transponder. The majority of inductively coupled systems use either the 30-300 kHz low frequency or the 3-30 MHz high frequency ranges. They constitute about 90% of today's RFID systems (Finkenzeller (2003)). Their range is less than 1m. Inductively coupled systems use *load modulation* to transfer data from the transponder to the reader. A resonant transponder is a transponder with a self-resonant frequency that is the same as the transmission frequency of the reader (Finkenzeller (2003)). In load modulation, a resistor on a resonant transponder is switched on and off according to data which effects the voltage across the reader and transmits data.

Electromagnetic backscatter transponders reflect back the electromagnetic waves created by the reader. The radiation power of the waves decreases with the square of the distance from the source, therefore a much weaker signal is returned to the reader by the passive backscatter transponders. Increasing frequency increases reflectivity hence these transponders use the ultra high frequency (UHF) range at 900 MHz or 2.4 GHz. Short wavelengths at these frequencies enable the use of smaller antennae than with inductively coupled system coils. UHF backscatter systems are also called long-range systems since they can transmit from up to 5 m away from the reader.

Close coupling systems are powered through the magnetic field generated on the transponder coil when it is placed between the two windings of the reader carrying a high frequency alternating current. The transponder can be coupled up to 1cm away from the reader. The frequency used is usually less than 30 MHz. Close coupling is used in contactless smart cards common in secure identification systems.

Because of the very limited useful energy that can be converted and used at the passive transponders, they can send information only very short distances. *Active transponders* on the other hand, are very similar to other battery powered wireless devices, differing only in their activation methods. Active backscatter transponders have significantly higher ranges but their use is limited because of the maintenance and cost issues associated with battery use. The last transponder type is the *semi-passive transponder* which uses batteries to retain memory contents or to process data; their radio functions use power supplied through the reader as in passive readers.

2.3 RFID STANDARDS ORGANIZATIONS

2.3.1 International Standards Organizations

There are three standardization bodies at the international level: ISO/IEC, EPCglobal and ITU-T.

- **ISO (International Organization for Standardization):** ISO is a network of the national standards institutes of more than 150 countries. ISO develops standards for a broad range of products. It specified RFID standards for various purposes at different frequencies and with different operating principles. IEC (International Electrotechnical Commission) is the charter specializing in all electrotechnologies including electronics, multimedia, and telecommunications. ISO/IEC JTC1 (Joint Technical Committee 1 of ISO and IEC) produces standards in the field of Information Technology which includes capture, representation, processing, security and management of information. A JTC1 standard starts with an "ISO/IEC" before the number. The JTC1 standards are similar to those developed by ISO.
- **EPCglobal Standards:** EPCglobal is a joint venture of GS1 and GS1 US. Its aim is to achieve worldwide adoption and standardization of Electronic Product Code (EPC) technology. The main focus of this body is to create a worldwide standard for RFID. As an independent body, the Auto-ID Center was set up in 1999 to develop new low-cost disposable RFID device standards for global supply chains. UHF was selected as the operating frequency because of the range and data rate requirements. The tasks of the Auto-ID Center were taken over in 2003 by EPCglobal, which administers and develops new RFID standards. With the establishment of EPCglobal, sponsored by many of the world's leading corporations, investment and research began to gain pace. The former Auto-ID Center still exists as seven academic research around the globe engaged in research on RFID.
- **ITU-T (International Telecommunication Union):** The ITU is the United Nations agency for information and communication technologies and it provides global telecommunication standards. Its publications are called ITU-T Recommendations.

2.3.2 European Standards Organizations

- **CEN (European Committee for Standardization):** CEN works in a large number of areas. Its members are 30 European countries. It produces European standards, technical specifications and reports. CENELEC (European Committee for Electrotechnical Standardization) develops electrotechnical standards. Only one organization per country may be a member of CENELEC. It delivers European Standards (EN), harmonization documents, technical reports and workshop agreements.
- **ETSI (European Telecommunications Standards Institute):** ETSI produces standards for wireless mobile, wireline and internet communications technologies. Membership of ETSI is open to any company or organization interested in the

creation of telecommunications standards and standards in other electronic communications networks and related services. It delivers European Standard (EN), ETSI Standard and guidelines.

2.4 MAJOR RFID STANDARDS

Before the development of RFID standards, users of RFID systems had to buy complete systems from the same manufacturer as the manufacturers rarely produced components compatible with each other. Early RFID systems were hence proprietary systems. The lack of interoperability was an obstacle to the proliferation of the technology. Incentives for investment and research were few. The picture began to change when the ISO started to develop RFID standards. The following is a list of major ISO RFID standards based on the frequency.

- **ISO 11784/11785:** Radio frequency identification of animals
- **ISO/IEC 14443:** Contactless integrated circuit cards, proximity cards
- **ISO/IEC 15693:** Contactless integrated circuit cards, vicinity cards
- **ISO/IEC 18000:** RFID for Item Management
- **ISO/IEC 24730:** Real Time Locating Systems

Contact cards have replaced the ISO 10536 based close coupling smart cards, hence devices based on this standard are not popular today. ISO 14443 describes contactless proximity coupling smart cards. This standard is for smart cards operating at 13.56 MHz. It has two versions, A and B. They are incompatible since they use different modulation and anti-collision schemes. ISO standard 15693 defines the standard for smart cards with a maximum range of 1m. RFID systems based on this standard can be found as part of access control systems.

The ISO 18000 series of standards are important because of the area it targets. It was developed to determine the use of the same air interface protocols for most of the widely used frequencies for a wide range of item management tasks. The series includes the following:

- **ISO/IEC 18000-1:** Generic Parameters for the Air Interface for Globally Accepted Frequencies
- **ISO/IEC 18000-2:** Parameters for Air Interface Communications below 135 KHz
- **ISO/IEC 18000-3:** Parameters for Air Interface Communications at 13.56 MHz
- **ISO/IEC 18000-4:** Parameters for Air Interface Communications at 2.45 GHz
- **ISO/IEC 18000-6:** Parameters for Air Interface Communications from 860 to 960 MHz
- **ISO/IEC 18000-7:** Parameters for Air Interface Communications at 433 MHz

Since each part operates at a different frequency range, their operation principles and target applications are different. This framework allows developers to select options suitable for their application requirements. Among those standards, ISO/IEC 18000-6 is widely used for supply chain management because of its capabilities which satisfy general supply chain requirements such as a medium-range reading distance, multi-read capabilities, and the capacity for high-speed item identification. Supply chain operations cover the life cycle of products from their raw material phase to disposal or recycling phases. It is widely predicted that supply chain operations will be one of the major applications of RFID technology.

EPCglobal standards are currently dealing with the UHF Class 1 Gen 2 air interface protocol. These conformance standards address the EPCglobal equivalent of ISO/IEC 18000-6C. EPCglobal also specifies interoperability test requirements. EPCglobal operates a certification programme for UHF Class 1 Gen 2 devices. Products can be certified as compliant with the conformance standards. The devices are interoperable if they satisfy basic requirements and are tested against other system components that are compliance certified. The collaboration of ISO and EPCglobal as two independent standardization bodies was complicated by the fact that EPCglobal has decided to create its own UHF air interface protocol. EPCglobal accepted an incremental improvement of classes of tags. The original EPCglobal classes are listed below.

Class 1: A passive read-only backscatter tag with one-time programmable memory.
Class 2: A passive backscatter tag with read-write memory.
Class 3: A semi-passive backscatter tag with read-write memory.
Class 4: An active tag that uses a built-in battery to power the tag circuitry including the transmitter.
Class 5: An active networked RFID tag that can communicate with other peer tags and devices.

EPCglobal submitted the UHF Gen 2 Class 1 air-interface protocol to ISO in 2005. ISO has approved the EPC Gen 2 Class 1 UHF standard, publishing it as an amendment to its 18000-6 standard, as 18000-6C. RFID standards from ISO and EPCglobal with their corresponding frequencies are also given in Table 1.

2.5 CLASSIFICATION OF RFID RELATED STANDARDS

In order to describe RFID standards, it is important to classify the target areas of standards in RFID systems. The following are the main areas of standardization (Chartier, Consultants & van den Akker (2008)).

1. Frequency
2. Security and privacy
3. Middleware
4. Data
5. Applications
6. Real-time locating

2.5.1 Frequency (Air Interface) Standards

RFID frequency regulations govern the use of the frequency spectrum for RFID devices. Since RFID systems operate with electromagnetic waves, the interference to and from other devices should be minimized. Because of the strong RF signals required to power passive tags, available operating

Table 1. Major RFID standards as applied to frequency

Frequency Spectrum			
2*	**LF** **125/134.2kHz**	**HF** **13.56 MHz, 433 MHz**	**UHF** **900 MHz, 2.45 GHz**
5*ISO	ISO 11784 ISO.IEC 18000-2A ISO/IEC 18000-2B	ISO/IEC 14443 (13.56 MHz) ISO/IEC 15693 (13.56 MHz) ISO 18000-3 (13.56 MHz) ISO 18000-7 (433 MHz)	ISO 18000-6A (900 MHz) ISO 18000-6B (900 MHz) ISO 18000-6C (900 MHz) ISO 18000-4 (2.45 GHz) ISO/IEC 24730-2 (2.45 GHz)
3*EPCglobal			Class 0 Class 1 Class 1 Gen 2

frequencies for RFID systems are significantly restricted. For this reason, RFID devices, most of the time, can only use industrial, scientific or medical (ISM) application frequencies.

The air interface standards specify the communication rules between the reader and the tag. The specification is done for all relevant network layers. At the physical layer, modulation and bit encoding rules are defined. At the MAC layer, the anti-collision algorithm is defined. Upper layer commands and responses, and physical memory architectures are also a part of the air interface standards.

The ISO/IEC 18000 series of air interface standards, currently published in seven parts, detail diverse RFID technologies using different frequency ranges. There are also other major standards relating to livestock tracking systems (ISO 11785), proximity cards (ISO 14443) and vicinity cards (ISO 15693).

EPCglobal also introduced a separate air interface standard for UHF frequencies. EPCglobal has defined classes of tags with increasing sophistication levels. As originally developed, Class 0 and Class 1 tags were not compatible with each other. They were not compatible with the ISO's air interface standard either. EPCglobal then developed second generation protocols named Gen 2, merging the old Class 0 and Class 1. ISO has approved the EPC Gen 2 Class 1 UHF standard, publishing it as an amendment to its 18000-6 standard.

2.5.2 Security and Privacy Standards

Security and privacy standards for RFID systems aim to achieve confidentiality, integrity, availability and non-repudiation. Confidentiality is the assurance that all the data in the system is protected from unauthorized access. This includes the protection of communications channels as well as devices. Integrity is the assurance that data and system resources are protected against modification. Availability provides protection against any DoS (Denial of Service)-type attacks. Non-repudiation is the assurance that a reader or tag cannot deny data alterations. This requires authentication between RFID layers.

There are many proposals introducing privacy enhancing and security techniques to RFID technology. Only a few of these techniques are included in RFID standards so far. There are four separate areas in an RFID system where security features can be applied:

- **The RFID tag:** Permanent encoding, password protection;
- **The air interface protocol:** Password protection, but most RFID applications provide open access to the reading of data;
- **The RFID interrogator:** Protection against unauthorized emulation;
- **The network:** Networks containing RFID devices must take into account the capabilities and characteristics of RFID devices.

Since passive RFID tags are resource constrained, security proposals for RFID systems should take into account this limitation. There are various security proposals from academia (Juels (2004), Chae, Yeager, Smith & Fu (2007), Peris-Lopez, Hernandez-Castro, Estevez-Tapiador & Ribagorda (2006*a*), Peris-Lopez, Hernandez-Castro, Estevez-Tapiador & Ribagorda (2006*b*), Benoit, Canard, Girault & Sibert (2006)). However, so far standardization bodies were only able to include a smaller and simpler subset of those proposals into RFID standards. In (Phillips, Karygiannis & Kuhn (2005)), current EPC and ISO/IEC standards are analyzed from the security perspective. Most of the protocols analyzed use CRC error detection for integrity. The EPC Class 1 Gen 2 standard uses one-time pad stream cipher reader-to-tag communications for confidentiality. The ISO/IEC 18000-3 protocol uses 48-bit password protection on read commands. Smart cards, on the other hand, have ample resources to implement strong security measures when compared to EPC tags. They receive energy from much smaller distances and they are built with higher computational resources. They are mostly used in payment systems and their security features are addressed from the beginning. High-end smart-cards based on ISO 14443 and 15693 implement strong cryptographic algorithms with AES, triple-DES and SHA protocols (Phillips, Karygiannis & Kuhn (2005)). In general, RFID smartcards are built with basic standard suites and expanded with proprietary security algorithms by the vendors according to their security requirements.

EPCglobal has defined a secure reader protocol between RFID readers and application software called Reader Protocol v1.1 (*Reader Protocol (RP) Standard, EPCglobal Inc.* (n.d.)). The protocol defines three layers: the reader layer, the messaging layer and the transport layer. The reader layer deals with message content. At the messaging layer security rules related to framing and connections are detailed. The network layer defines networking facilities. ISO/IEC published the report TR24729-4 as an implementation guideline for tag data security. Security advice for backend systems is provided in the standard ISO/IEC 24791-5.

2.5.3 Middleware Standards

RFID middleware standards define the data encoding and protocol rules between the RFID reader and the application software. The widespread adoption of RFID requires an efficient supporting infrastructure that manages readers without dealing with application details. In (Floerkemeier & Lampe (2005)), authors describe the main features an RFID middleware must provide to readers and applications. RFID systems use limited communication bandwidth. In order to manage a large number of tags, multiple readers must coordinate their reading process. Readers transmitting at close proximity might interfere with each other. High power wireless transmissions by readers in order to give energy to and read back data from passive tags can be obstructed by wireless transmission factors such as multipath fading, or absorption by the nearby objects. False readings caused by collisions or transmission errors must also be taken into account when designing RFID middleware.

The EPCglobal middleware standards are the Low Level Reader Protocol (LLRP) and Application Level Events (ALE). LLRP (*Low Level Reader Protocol (LLRP) Standard, EPCglobal Inc.* (n.d.)) specifies an interface between RFID readers and clients. It provides the formats and procedures of communications in terms of messages. The client messages are to discover the readers, and to get and set reader configurations. The client messages serve also to access the inventory through readers. Reader messages to the client reports the reader status, inventory or RF survey. Being an application layer protocol, LLRP does not provide retransmission, or reordering facilities. Client and reader consistency is provided with update messages.

ALE (*Application Layer Events (ALE) Standard, EPCglobal Inc.* (n.d.)) provides methods for filtering and grouping captured RFID tag data. This standard was previously known as Savant. It was developed to provide middleware between RFID readers and databases. It is located between readers and applications, and manages retrieved tag information. The protocol filters and stores data in order not to overflow underlying communication networks.

The ISO protocols for RFID middleware are the ISO/IEC 15961, 15962, 15963 and 24791. The purpose of these standards is to provide a common data protocol that has no restrictions on frequencies used in applications and air interface protocols and to describe numbering systems for tags. ISO/IEC 15961 provides the data protocol application interface. It allows data and commands to be specified in a standardized way. ISO/IEC 15962 provides data encoding rules and logical memory functions such as processing of the data and its presentation to the RF tag. ISO/IEC 15963 provides unique identification of tags and a numbering system. ISO/IEC 24791-2 is written for data management which provides operations on tag data.

2.5.4 Data Standards

Data standards deal with the structure of RFID data, data dictionaries and encoding methods. EPCglobal defined the EPC Tag Data Standard (TDS) (*EPC Tag Data Standard (TDS), EPCglobal Inc.* (n.d.)). This standard defines the data encoding on the tags and on the information system. ISO/IEC 15459 defines unique identifiers. ISO/IEC 15418 defines data dictionaries and ISO/IEC 15434 defines a syntax for high-capacity automatic data capture media.

2.5.5 Application Standards

Major standards cover sectors such as automotive, baggage handling, airline/defence items and libraries. Application standards specify a set of requirements which result in a single air interface protocol. Standards from ISO 17363 to ISO 17367 define generic application standards for different industry sectors:

- ISO/IEC 17363, Supply Chain Applications of RFID, Freight containers
- ISO/IEC 17364, Supply Chain Applications of RFID, Returnable Transport Items
- ISO/IEC 17365, Supply Chain Applications of RFID, Transport Units
- ISO/IEC 17366, Supply Chain Applications of RFID, Product Packaging
- ISO/IEC 17367, Supply Chain Applications of RFID, Product Tagging

2.5.6 Real Time Locating Standards

Real-time locating systems (RTLS) are wireless systems helping to locate the position of an item in a defined space. Position is derived with various triangulation and multilateration methods using radio link. RTLS in asset management tries to establish interoperability of products for the growing RTLS market. ISO/IEC 24730 defines two air interface protocols and a single application program interface (API) for RTLSs.

CONCLUSION

In this chapter we have given an overview of wireless RFID standards. Practitioners and researchers of RFID technology must pay attention to the works of the main standardization bodies as they continue to evolve. An EPCglobal subscription might be beneficial to supply chain operations. Ensuring compliance with ISO and EPCglobal standards must be the first step before selecting a vendor.

REFERENCES

Benoit, C., Canard, S., Girault, M., & Sibert, H. (2006). Low-cost cryptography for privacy in RFID systems. In *The Proceedings of the International Conference on Smart Card Research and Advanced Applications, CARDIS '06.*

Chae, H., Yeager, D., Smith, J., & Fu, K. (2007), Maximalist cryptography and computation on the WISP UHF RFID tag. In *The Proceedings of the Conference on RFID Security.*

Chartier, P., Consultants, P., & van den Akker, G. (2008). *GRIFS, global RFID forum for standards, RFID standardisation state of the art report. Technical report.* CEN.

EPCglobal Inc. (n.d.). *Application layer events (ALE) standard.* Retrieved from http://www.epcglobalinc.org/standards/ale/ale_1_1_1-standard-core-20090313.pdf

EPCglobal Inc. (n.d.). *EPC tag data standard* (TDS). Retrieved from http://www.epcglobalinc.org/standards/tds/tds_1_6-RatifiedStd-20110922.pdf

EPCglobal Inc. (n.d.). *Low level reader protocol (LLRP) standard*. Retrieved from http://www.epcglobalinc.org/standards/llrp/llrp_1_1-standard-20101013.pdf

EPCglobal Inc. (n.d.). *Reader protocol (RP) standard*. Retrieved from http://autoid.mit.edu/CS/files/11/download.aspx

Finkenzeller, K. (2003). *RFID handbook: Fundamentals and applications in contactless smart cards and identification*. John Wiley & Sons, Inc.

Floerkemeier, C., & Lampe, M. (2005). RFID middleware design: Addressing application requirements and RFID constraints. In the *Proceedings of the 2005 Joint Conference on Smart Objects and Ambient Intelligence, sOc-EUSAI '05*.

Juels, A. (2004). Minimalist cryptography for low-cost RFID tags. In *The Proceedings of the International Conference on Security in Communication Networks, SCN '04, Vol. 3352 of Lecture Notes in Computer Science,* (pp. 149–164).

Peris-Lopez, P., Hernandez-Castro, J. C., Estevez-Tapiador, J., & Ribagorda, A. (2006b). M2AP: A minimalist mutual-authentication protocol for low-cost RFID tags. In *The Proceedings of the International Conference on Ubiquitous Intelligence and Computing, UIC '06, Vol. 4159*, (pp. 912–923).

Peris-Lopez, P., Hernandez-Castro, J. C., Estevez-Tapiador, J. M., & Ribagorda, A. (2006a). EMAP: An efficient mutual authentication protocol for low-cost RFID tags. In *The Proceedings of the OTM Federated Conferences and Workshop: IS Workshop, IS '06, Vol. 4277*, (pp. 352–361).

Phillips, T., Karygiannis, T., & Kuhn, R. (2005). Security standards for the RFID market. *IEEE Security Privacy, 3*(6), 85–89. doi:10.1109/MSP.2005.157

Section 2
Attacks

Chapter 3
RFID Wireless Link Threats

Ilker Onat
University of Ottawa, Canada

Ali Miri
Ryerson University, Toronto, Canada

ABSTRACT

This chapter gives an overview of wireless link threats against RFID systems. A major portion of the RFID tags are passive devices without their own power source and they can be easily attacked. It is difficult to implement countermeasures in RFID tags due to major resource constraints. In this chapter, major attack types against RFID systems are described. The vulnerabilities of RFID systems are explained along with the proposed solutions and design methods against the attacks.

3.1 INTRODUCTION

RFID devices are used in logistics, consumer applications and different business specific applications. With the increasing number of RFID devices used in daily life, security and user privacy concerns are also becoming critical. Since RFID uses a wireless medium, all the threats associated with wireless media usage exist in RFID systems. RFID systems are further challenged by these threats because of the tags' limited hardware and energy capacity for implementing robust security measures.

Tags can be passive or active. A passive tag obtains all of its energy for communications and data processing from the electric or magnetic field of the reader. An active tag on the other hand includes a battery. The general operation of a passive tag is given in Figure 1. In this chapter we will overview the possible wireless link based attacks against RFID systems using passive tags and summarize the current protection algorithms against them.

According to hardware complexity, RFID tags can be low-end, mid-range and high-end systems. Low-end systems are mostly low-cost, low-

DOI: 10.4018/978-1-4666-3685-9.ch003

Figure 1. Passive tag communications

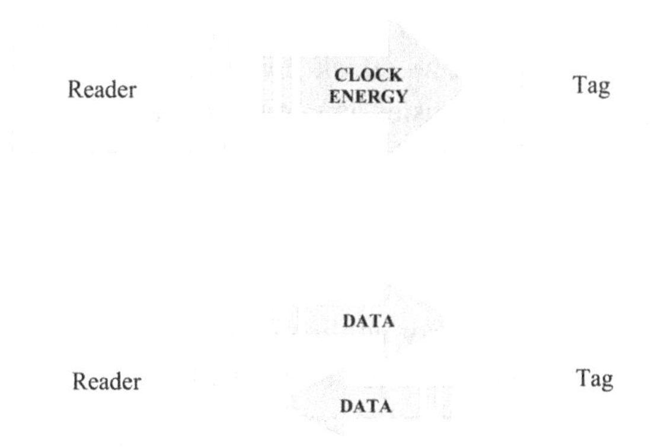

power one-bit transponders with no medium access control. Mid-range tags allow reading and writing on their memory. They can be addressed and they can support cryptographic security primitives. High-end tags such as smartcards allow complex authentication algorithms. Top-end smartcards carry cryptographic co-processors allowing complex calculations.

3.2 EAVESDROPPING ATTACKS

Eavesdropping is the interception of the communication between the reader and the tag or the unauthorized listening to and storing of information in the open wireless medium. A similar attack is the skimming attack where the attacker actually provides power to the tag and acts as a false reader. In an eavesdropping attack, the attacker passively listens and records to the bits over the air. Therefore he has to be in the vicinity of the reader and the tag in order to perform the attack. The attacker also has to have the right RF equipment before recovering and storing the wireless data. The attacker with better wireless equipment with higher sensitivity RF devices will be able to eavesdrop correctly further away from the reader and the tag than the attacker with simple RF equipment. As long as the communication standard and the frequency is known, obtaining the bits from the wireless medium is an easy task for the passive attacker listening to the medium. The severity of the this attack against different RFID systems is mainly determined by the range of the wireless communications, i.e., the distance between the tags and the reader. This range is determined by the operating frequency of the reader, the physical coupling method and the transmit power of the active elements. The recovery of useful data by eavesdropping can be prevented by application layer encryption of the transmitted data. Many HF RFID systems have components that can implement application layer encryption however these attacks are still important since many application layer algorithms are designed based on the assumption that the HF RFID communication range was small. In (Hancke, 2011) the vulnerabilities of such systems are discussed.

It is also important to make a distinction between channels in this attack. The reader-to-tag channel, also called the forward channel, carries a higher RF power signal since the power emitted will be used to power the tag and its RF logic. The tag-to-reader channel, or the backward channel, has much lower RF power since the tag is passive and using the reader transmitted signal. Depending on the distance, the attacker may only sense the communications but cannot recover the data, or can recover only the forward channel's or both the forward and backward channels' data.

3.2.1 Near-Field RFID Systems

Systems with communication ranges of up to 1m are labeled as *near-field communications* or *remote-coupling systems*. Almost all remotely coupled systems are based upon an inductive (magnetic) coupling (Finkenzeller, 2003). These systems are also called high-frequency (HF) systems. There are three major standards for passive near-field devices operating at a frequency of 13.56 MHz: ISO 14443A, ISO 14443B and ISO 15693. Because of their limited range these systems are often deemed secure against eavesdropping attacks. The advertised operational range of these channels is less than 10 cm and therefore several implemented systems assume that the communication channel is limited and therefore secure. However, there are practical attacks against these systems. Eavesdropping attacks are mentioned regularly in the literature. In (Hancke, 2011), it is shown through experiments that for inductively coupled RFID systems, forward channel data can be recovered up to10m away whereas backward channel data can be recovered up to 1m away with high quality RF equipment and data analysis tools. These results differ slightly when the RF propagation environment changes from open spaces to corridors.

3.2.2 Long-Range RFID Systems

Long-range RFID systems operate at the UHF or microwave frequency. Their range is between 1m and 15m. The majority of these systems use backscattering as their physical operating principle and operate at around 900 MHz.

3.2.3 Measures against Eavesdropping Attacks

Eavesdropping attacks are mainly dependent on the attacker's monitoring equipment and the RF propagation environment. Even near-field communication systems are not immune to eavesdropping attacks. An attacker can recover data beyond the advertised operating range of the RFID system. From the attacker's point of view, there are many technical improvements available to better capture and analyze data like the improved tuning techniques. The designer's of RFID systems should make few assumptions on the technical capabilities of the attackers and devise and implement strong application layer security measures. The most effective protective measure against an attack involving eavesdropping is to not transfer any content but the ID of the tag itself (*EPCglobal Inc.,* n.d.). This way, all the necessary information about the tag will be securely stored at the secure backend servers and the ID itself will not reveal any useful information to the attackers. For applications where relevant content has to be stored on the tags themselves, only strong encryption procedures can provide reliable protection against eavesdropping (Kim, Kim, Han, & Choi, 2006).

3.3 TAMPERING

A major threat to RFID based information systems is the alteration of the data in the tag memory or while the information is being retrieved. In this section, we will give an overview of these data tampering attacks on RFID systems. The characteristics and the effects of data tampering attacks on RFID systems are summarized along with the approaches proposed to defend against them.

In all information systems, the malicious alteration of entities is called tampering. The wireless and mostly unattended nature of RFID systems exposes tags to physical and RF transmission access. In the context of an RFID system, the part of the system most vulnerable to tempering attacks is the tag. Tampering of the tag can happen in two ways. Either the information transmitted or received over the air can be altered, or the data inside the tag itself can be modified. In general there are two major protection mechanisms against tampering attacks. Using *tamper-evidence* measures provides quantifiable evidence for the existence of tampering. *Tamper-resistance* measures provide the system under attack the means to protect themselves against tampering. As a result of a successful tampering attack, the tag under attack might become unusable, or it might become a malicious or false tag itself.

The tampering problem in information technology is generally studied in the context of software tampering. Various measures exist to detect and act upon once the software alterations. However, since the processing part of most RFID systems consists only of simple logic, software tampering is not a serious threat for these systems (Gandino, Montrucchio & Rebaudengo, 2010). Hardware tampering is a more serious threat for RFID systems. If the RFID tag is accessed and reprogrammed, insider attacks are possible over the tampered tag. Using tamper resistant hardware is the only prevention mechanism against such attacks. However, the high cost of tamper resistance schemes make them infeasible to be implemented on low-cost RFID tags.

Data can be tampered with by impairing part of it, by inserting completely different data or by swapping it with data from other locations.

3.3.1 Measures against Tampering

Tamper Detection Methods

1. **Watermarking:** In general, watermarking consists of embedding information into original data. When the original data changes, embedded data and original data generated become incoherent. Watermarking is proposed for RFID tag tamper detection in (Potdar & Chang, 2006). This system detects tampering on the memory of RFID tags compliant with the EPC96 standard. The authors propose embedding a watermark into the serial number. The watermark is generated by performing 3 one-way functions on the three fields set by the standard. This system allows detection and correction of tampering on three fields, of an affected tag. This method can detect tampering by the weak attackers employing random modifications, however it is weak against more organized attackers that perform modifications that take into account this watermarking measure.
2. **Cryptography:** Public or private key cryptography can be used to authenticate tag IDs or tag data. Any tampering on the ID or data will result in different decrypted data that will reveal the tampering. However, in order to implement cryptographic measures, especially public-key cryptography, tags need to have a large memory, which most of the low-end tags do not possess. In (Bernardi, Gandino, Lamberti, Montrucchio, Rebaudengo & Sanchez, 2008), an RSA based authentication approach is developed. The ID of the tag is encrypted and written to the user memory. The encrypted data is later decrypted and compared to the tag ID. However, if the whole data from the tag is copied into another tag, this false tag cannot be detected with cryptographic schemes. Another approach is to encrypt tag ID and data at the backend and use the encrypted values in the tag. Any attacker trying to modify these values will create tags that do not decrypt meaningfully which will make the tampering evident.
3. **Write Record Keeping:** In (Yamamoto et al., 2008), the authors propose a method to detect tag memory tampering. They use a private memory region in the tag, which is only readable and not writable by the reader, but writable by the tag itself. Any write to the user memory automatically adds an entry into the tag's private memory region. The first part of the special memory area represents the pointer to the area for the next insertion, and the number of recorded writing operations. Tampering is detected when there is an overlap in the written memory regions.

Tamper Resistance Methods

1. **Protected Memory:** There are two classes of memory that provide this kind of protection: read-only memory and permanently lockable memory are types do not allow any party to

write data once the information is burned into the memory. Permanently lockable memory is more flexible than read-only memory but requires stricter management.

2. **Passwords:** A weak protection against tampering is the use of passwords. Passwords do not provide strong security since the attacker who can eavesdrop on both way directions of communication can recover the password from tag reader exchanges.
3. **Cryptography:** Many challenge-response protocols can be used for RFID authentication. In private key based approach, both ends use a shared secret. In order to alter the content of data that use such protocols, the encryption scheme itself has to be broken. This requires that the attacker to obtain the keys and the pseudo-random generation logic. The only drawback of this strong tamper protection mechanism is the requirement that the tags have crypto-processors, which do not exist in low-end RFID systems.

3.4 SESSION HIJACKING

In general, session hijacking refers to the exploitation of a valid session to gain unauthorized access to the system under attack. In RFID systems, session hijacking attacks are possible during authentication sessions between the tags and the reader. In these attacks, an attacker reads the tags after a user is authenticated. For example, the attacker reads the protected data while the authenticated user is reading the tags. In (Zhou & Huang, n.d.), a secure RFID access control mechanism is described. To counter hijacking attacks this paper proposes an RFID activity watchdog that detects the number of working readers. If the number is larger than the known authenticated reader count, the RFID reading activity is suspended and an alarm is raised. The authenticity of the reader is also periodically checked through a challenge-response algorithm. In (Weis, Sarma, Rivest, & Engels, 2004), frequency hopping by the reader is suggested as a measure against session hijacking. Passive tags may be designed such that their operating frequency is determined by the reader.

3.5 REPLAY ATTACK

In a replay attack, the attacker records valid credentials and data from authenticated users and replays them at a later time of his choosing. Therefore the replay attack is an impersonation attack. The main method against the replay attacks is to bind information exchanged to its correct context (Aura (1997)). This way, once a message is replayed, it will be determined that it is out of context. Practical examples of relay attacks are shown by the successful attacks on the Texas Instrument Digital Signature Transponder (Bono, Green, Stubblefield, Juels, Rubin & Szydlo (2005)) and during credit card transactions (Hancke (2006)). This interception and retransmission of RFID queries can be used against contactless payment systems, access control systems or passport readers.

3.5.1 Measures against Replay Attacks

An extensive study of protection mechanisms against replay attacks for cryptographic protocols was done in (Aura, 1997). Strong cryptographic measures defeat most of the other attacks; however, they can be bypassed with elaborate replay attacks. In order protect against replay attacks in RFID systems, session tokens, timestamps or challenge-response authentication can be used.

A session token in the RFID systems identifies a reading round between the reader and the tag. It can consist of single or multiple exchanged messages. In general, RFID reading sessions are stateful and session state and history is kept at the reader. Session tokens are usually created with pseudo-random number generators. A session number should be unique and non-predictable. When a session token is replayed by the attacker, it will not be the same as the expected session token.

Time-stamping is another way of preventing a replay attack. This method requires time synchronization between the reader and the tag. Timestamps are appended to exchanged messages and parties accept the messages as valid only if the response is received within a given tolerance. In (Deursen & Radomirović, 2009), the authors analyze the replay attacks targeting the challenge-response mechanism in authentication protocols. They choose the notion of *recent aliveness* as the most appropriate authentication requirement for RFID protocols.

3.6 RELAY ATTACKS

High-end smartcards can perform public and private key calculations. They are also tamper resistant and carry large-enough memory to hold an operating system and perform cryptographic calculations (Anderson, Bond, Clulow, & Skorobogatov, 2006). Relay attacks bypass the authentication step of the RFID communication protocols. Since non-trivial authentication can only be applied in high-end tags, these attacks generally target high-end contactless (ISO 14443) or contact (ISO 7816) smartcards. The acquired identity and credentials of the legitimate device are relayed to a distant location to be used to gain unauthorized access. Relay attacks pose a serious threat to the security of RFID systems. They operate at the physical layer and application-layer encryption protocols are easily bypassed. They are simple and can be mounted with simple low-cost equipment. The relay attack is different from the replay attack, since the attacker in the relay attack does not store or replay the previous message. In general, relay attacks can be implemented with different tools and techniques. In a relay attack, the attacker needs to build false tags and false readers both in hardware, firmware and radio circuitry. The setup is rather simpler to build than other complex attacks such as power-analysis or hardware tampering and cryptanalysis. In addition, current smartcard standards are delay tolerant and they open the doors for attacks on the cards implementing the protocol without additional security measures against this attack. In Figure 2 we give the general scenario and the components of the relay attack. Next, we explain the relay attack and describe prevention mechanisms.

A detailed analysis of RFID relay attacks is given in (Lima, Miri, & Nevins, 2008). In a relay attack, the attacker acts as a middle-man between the tag and the reader. In order pose as a legitimate reader to the actual tag, and a legitimate tag to the actual reader, the only constraint the attacker has to consider is the increased delay occurring because of the increased signal propagation path and the processing delay at the relay. Once the tag-reader communication's delay constraints of the system under attack are tackled, information can be relayed to distant fake nodes that use the signals obtained from the relay attack. Smartcards are used for authentication and user and account verification. A smartcard's secure non-volatile memory holds all the necessary data for these tasks. There are many successful relay attack implementations in the literature.

3.6.1 Relay Attacks against Proximity Smartcards

Contactless smartcards (ISO 14443) are also called proximity cards. They have a very short range (less than 10cm) which is sometimes falsely viewed as a security feature. They use inductive coupling with load modulation. ISO 14443 A and B only differ in their modulation, coding and initialization schemes. This allows for interoperability of contactless smartcard products. Readers in this standard are called Proximity Coupling Device

Figure 2. General description of a relay attack system

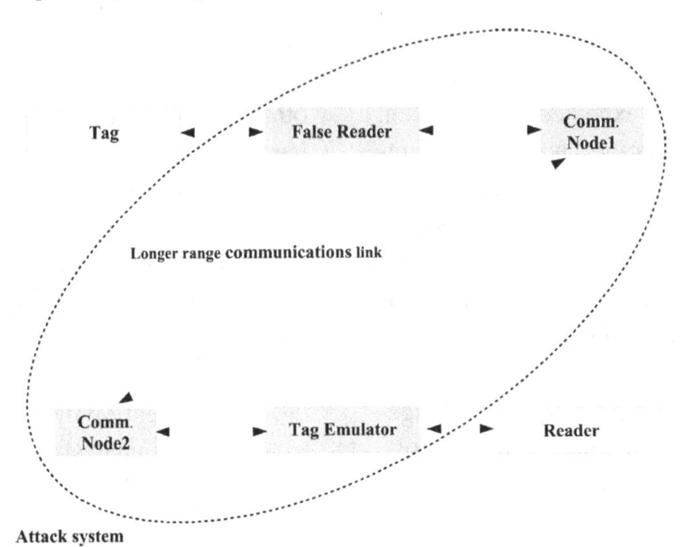

(PCD) and tags are called Proximity Integrated Circuit Cards (PICCs). They serve in many security sensitive payment applications, e-passports and cellular devices. They can support secure messaging, cryptographic tokens and authentication mechanisms. They can be built tamper resistant with hardware memory firewalls and sensors. Specific security algorithms are not specified and can be proprietary to manufacturers. Those using crypto co-processors can be used with complex cryptographic algorithms. In Hancke (2006) and Kfir & Wool (2005), the authors describe successful relay attack implementations against ISO 14443 A type proximity tags with practical details of false tag and false reader construction.

3.6.2 Relay Attacks against Contact Smartcards

Relay attacks can also be mounted without a wireless link. Contact smartcards (ISO 7816) embedded with a computer chip are widely used in healthcare, banking, entertainment and transportation. The standard does not specify any particular on-chip security algorithm. It is not possible to mount wireless link dependent attacks such as man in the middle, eavesdropping or replay against contact smartcards. Although they are more secure than contactless smartcards, invasive attacks such as side-channel attacks or relay attacks are possible. Customers with smartcards authorize transactions with the card and PIN without guarantees of the amount charged and who is to be paid. From the customer's point of view, it is not possible to guarantee the authenticity of the payment terminal. Measures at the smartcard cannot protect customers from the relaying of authentication information to false tags far away. In (Drimer & Murdoch, 2007), a practical relay attack against contact smartcards is detailed.

3.6.3 Measures against Relay Attacks

Cryptographic measures at the application layer can protect against skimming attacks but cannot protect against relay attacks. A relay attack is invisible to application layer security and therefore new protection measures should focus on the physical layer. For contactless smartcards, shielding the tag can prevent against relay attacks performed without the user's knowledge. Additional physical controls at the reader side to help ensure that the reader is in fact reading the tag that is presented to it and not some remote victim tag can also be implemented.

An effective measure against relay attacks is to measure the signal propagation time where it is ensured that the verifier is involved in the exchange. The first distance-bounding protocol based on the single-bit roundtrip delay was introduced in (Brands & Chaum, 1993). The verifier transmits challenges and records the response timing. The prover then transmits a message authentication code (MAC). Premature requests made from a false verifier to the prover can only succeed with a small probability without being detected with the guessed challenges. In (Hancke & Kuhn, 2005), the authors introduce a distance-bounding protocol for contactless smartcards. High-resolution timing information about the arrival of individual data bits is used to measure the round-trip delay. With this method, an upper bound for the distance between the reader and tag is established and if the message exchanges take more time than this bound, the existence of a relay attack is revealed. The authors of (Drimer & Murdoch, 2007) also describe a distance-bounding security scheme for contact smartcards with a similar round-trip delay measurement method.

CONCLUSION

In this chapter, we have given an overview of wireless link threats against RFID systems. Since most of RFID tags are passive devices without their own power sources, attacks against RFID systems are relatively easy to mount and implementing countermeasures is more difficult. We have also summarized proposed solutions and design methods against these vulnerabilities. For further proliferation of RFID technology, efficient solutions must be adopted in RFID standards.

REFERENCES

Anderson, R., Bond, M., Clulow, J., & Skorobogatov, S. (2006, February). Cryptographic processors - A survey. *Proceedings of the IEEE, 94*(2), 357–369. doi:10.1109/JPROC.2005.862423

Aura, T. (1997). Strategies against replay attacks. In *The Proceedings of the 10th IEEE Workshop on Computer Security Foundations, CSFW '97*, (pp. 59-68).

Bernardi, P., Gandino, F., Lamberti, F., Montrucchio, B., Rebaudengo, M., & Sanchez, E. (2008). An anti-counterfeit mechanism for the application layer in low-cost RFID devices. In *The Proceedings of the 4th European Conference on Circuits and Systems for Communications, ECCSC*, (pp. 227–231).

Bono, S. C., Green, M., Stubblefield, A., Juels, A., Rubin, A. D., & Szydlo, M. (2005). Security analysis of a cryptographically-enabled RFID device. In *The Proceedings of the 14th Conference on USENIX Security Symposium, SSYM '05*, (pp. 1–15).

Brands, S., & Chaum, D. (1993). Distance-bounding protocols. In *Advances in Cryptology EUROCRYPT '93* (pp. 344–359). Berlin, Germany: Springer-Verlag.

Deursen, T., & Radomirović, S. (2009). Algebraic attacks on RFID protocols. In *The Proceedings of the 3rd IFIP WG 11.2 International Workshop on Information Security Theory and Practice. Smart Devices, Pervasive Systems, and Ubiquitous Networks, WISTP'09*, (pp. 38–51). Berlin, Germany: Springer-Verlag.

Drimer, S., & Murdoch, S. J. (2007), Keep your enemies close: distance bounding against smartcard relay attacks. In *The Proceedings of 16th USENIX Security Symposium on USENIX Security Symposium, SS '07*, (pp. 1–16).

EPCglobal Inc. (n.d.). Retrieved from http://www.epcglobalinc.org

Finkenzeller, K. (2003). *RFID handbook: Fundamentals and applications in contactless smart cards and identification*. John Wiley & Sons, Inc.

Gandino, F., Montrucchio, B., & Rebaudengo, M. (2010). Tampering in RFID: A survey on risks and defenses. *Mobile Networks and Applications*, *15*(4), 502–516. doi:10.1007/s11036-009-0209-y

Hancke, G. (2006). Practical attacks on proximity identification systems. In *The Proceedings of the IEEE Symposium on Security and Privacy*.

Hancke, G. (2010). Practical eavesdropping and skimming attacks on high-frequency RFID tokens. *Journal of Computer Security - Special Issue on RFID System Security, 19*(2), 259–288.

Hancke, G. P., & Kuhn, M. G. (2005). An RFID distance bounding protocol. In *The Proceedings of the First International Conference on Security and Privacy for Emerging Areas in Communications Networks, SECURECOMM '05*, (pp. 67–73).

Kfir, Z., & Wool, A. (2005). Picking virtual pockets using relay attacks on contactless smartcard. In *The Proceedings of the First International Conference on Security and Privacy for Emerging Areas in Communications Networks, SECURECOMM '05*, (pp. 47–58).

Kim, H. S., Kim, I. G., Han, K. H., & Choi, J. Y. (2006). Security and privacy analysis of RFID systems using model checking. *High Performance Computing and Communications, Vol. 4208 of Lecture Notes in Computer Science,* (pp. 495–504). Berlin, Germany: Springer.

Lima, A., Miri, A., & Nevins, M. (2008). RFID relay attacks: System analysis, modelling, and implementation. In *Security in RFID and sensor networks*, (pp. 49–75). Auerbach Publications, Taylor & Francis Group.

Potdar, V., & Chang, E. (2006). Tamper detection in RFID tags using fragile watermarking. *IEEE International Conference on Industrial Technology, ICIT'06* (pp. 2846–2852).

Weis, S. A., Sarma, S. E., Rivest, R. L., & Engels, D. W. (2004). Security and privacy aspects of low-cost radio frequency identification systems. *Security in Pervasive Computing, Vol. 2802 of Lecture Notes in Computer Science,* (pp. 50–59). Berlin, Germany: Springer.

Yamamoto, A., Suzuki, S., Hada, H., Mitsugi, J., Teraoka, F., & Nakamura, O. (2008). A tamper detection method for RFID tag data. In *The Proceedings of the IEEE International Conference on RFID*, (pp. 51–57).

Zhou, Z., & Huang, D. (n.d.). RFID keeper: An RFID data access control mechanism. In *The Proceedings of the IEEE Global Telecommunications Conference, GLOBECOM '07*, (pp. 4570–4574).

Chapter 4
Hardware Attacks

Fanyu Kong
Institute of Network Security, Shandong University, China

Ming Li
Shandong University, China & State Grid Corporation, China

ABSTRACT

In a secure system, the algorithms, protocols, and digital data are finally implemented and stored on hardware, such as chips, DSP, and registers. Knowledge of the implementation may be used to carry out attacks against the system without attacking the algorithms and protocols directly. The hardware which implements the system deserves much attention and scrutiny. Several hardware attacks are shown in this chapter, which is helpful in designing a secure RFID system.

4.1 BACKGROUND

Radio Frequency Identification (RFID) is one of the essential pervasive computing technologies and widely used in the Internet of Things. A typical RFID system consists of three main components: a RFID tag, a RFID reader and a back-end database. The RFID tag's core component is an integrated circuit microchip with a CPU, a memory and an antenna. The RFID's security and privacy have been important issues while the RFID's applications increase rapidly.

The RFID's security is a complex problem concerning not only technological issues (mathematical, software and hardware) but also information system management issues (authorization, social engineering and so on). While IC chips and embedded systems have been widely applied, hardware attacks play a more important role in breaking the security and privacy of an information system. The electromagnetic emission attack, timing attack, simple or differential power analysis attacks and data tampering are several hardware attacks that are often used. For RFID tags, besides these attacks, the cloning attack and data copying attack are important attacking methods via physical access to the RFID devices.

In this chapter, we give a survey of various kinds of hardware attacks and their countermeasures. Side channel attacks (timing attack, power analysis

DOI: 10.4018/978-1-4666-3685-9.ch004

attack and so on) are applicable for various implementations of cryptographic algorithms, including software implementations, ASIC/FPGA, smart card and RFID devices. (Oren, Yossi, & Shamir, 2007) proposed a practical remote power analysis attack on passive RFID tags without the help of physical access to the RFID device under attack. The cloning attack and direct read attack are direct physical attacks on the RFID devices. The tampering attack is doing destruction to the integrity of the information stored in the RFID memory. In 2009, (Gandino et al., 2009) presented a good survey on tampering attacks in RFID devices.

4.2 CLONING ATTACK

The main function of RFID systems is to identify different items. The identification is implemented by attaching a tag to each item. Each tag has a unique serial number (such as an EPC Code), and can be accepted by the legal reader by wireless communication. If we clone a tag, we copy the identifier of the item. This is similar to forging the signature of a file. The cloning attack is a severe challenge to many RFID applications, such as access control, ticketing, payment, passport verification and supply chains. There are also many countermeasure techniques against the cloning attack.

(Goodin, 2009) showed that Chris Paget had demonstrated how to clone the unique electronic identifier used in US passport cards and next generation drivers' licenses using inexpensive devices. Most of the existing RFID standards and systems are vulnerable to cloning attacks. But with the application of RFID tags in more sensitive fields, such as access control, electronic passports and payment, the security against cloning attack becomes more and more significant.

Cloning of RFID tags is concealed because of the wireless signal transmission between the tags and readers. Cryptographic algorithms can be designed for some schemes to detect illegal cloning or reading. (Lehtonen et al., 2009) proposed a novel method to protect the RFID applications by detecting the different cloned RFID tags with the same ID number without the use of cryptographic methods.

As there is a lack of security consideration on existing RFID systems and standards, we should design new schemes with or without cryptographic techniques to protect RFID tags. For example, a Gen2 tag will transmit its EPC to any reader query, and an adversary can easily get it by reading the tag remotely. Since some Gen2 tags are field-programmable, they are vulnerable to cloning attacks. Therefore, authenticity for tags is required to prevent cloning attacks. Providing the authentication of readers to tags is a common way to defeat cloning attacks. Many proposed cryptographic protocols using a shared secret key to authenticate readers, such as in (Juels, 2005; 2006; Abawajy, 2009).

Electronic passports, an important application of RFID technology, are susceptible to the tag cloning attack. International Civil Aviation - (ICAO 2006) introduced RFID tags to passports, visas and other travel documents, which have been endorsed by ISO as Standard 7501 1 & 2. Basic access control (BAC) is an optional function that helps electronic passports - ePassports be secure against skimming and eavesdropping attacks. But it has been shown by (Juels, A. 2005) that BAC keys have low entropy and are not secure in practice. (Avoine, Kalach, & Quisquater, 2008) analyzed that the entropy of the worst BAC key in Belgian ePassports is only 23 bits. (Witteman, 2005) proposed an attack against the BAC key of the ePassport. He found that the entropy of the BAC key on Dutch passports is around 35 bits, which means that the ePassport can be broken in two hours on a personal computer. Active authentication (AA) is often used to enhance the security of ePassports. In tags with active authentication, a secret private key stored in the tag is used to help establish a secure communication between the tag and reader.

4.3 DIRECT READ

Because there is some secret information stored in RFID tags, we must prevent the illegal reading and tampering of the tags. Some examples of physical methods for protecting privacy include kill commands, clipped tags, Faraday cages, active interference and block tags.

Kill commands are used to make the information stored in the tags unavailable. According to EPCGlobal standards, when a tag is killed, it is disabled permanently. It is a nonreversible operation. One often uses this command to keep the privacy of items after the payment of these products in stores (drug stores, for example). But sometimes, consumers do not always want their tags to be killed, such as for goods returns and exchange.

A Faraday cage is a mesh made by conducting material. An external static electrical field will redistribute the electrical charges in the conducting material to cancel the field's effects in the cage. Thus, external radio waves are also blocked by such a cage. As a result, passive tags are unable to receive signals to get power, and active tags cannot transmit signals. Putting tags into a Faraday Cage Wallet is a good idea to block the illegal read and tampering. But note that this technique can also be used to block the regular read and tracing of the tags for illegal purpose. The seller is unable to detect one item that has been stolen if its tag is torn or in a Faraday cage.

The concept of the clipped tag, which is seen in (Moskowitz & Karjoth, 2005; Moskowitz, Lauris, & Morris, 2007), was first introduced by Paul Moskowitz and Guenter Karjoth in 2005. In their papers, the authors suggested that the read range of the tag may be reduced from many meters to a few centimeters by shortening the antenna. One way to achieve that is to allow the tearing off of a portion of the antenna. Another proposed method is to scratch off the exposed conducting lines which are connected to the tag antenna. After the payment of products, a consumer may tear off a portion of the tag. Then this tag can only be read within a few centimeters. This prevents remote tracing and reading, while the tags may still be used later for goods returns or recalls.

Active interference shields tag using special equipment that transmits noise. This method is suitable in transport. One can block all the illegal reading using equipment that actively transmits signals. The disadvantage of this method is that the interference may influence normal wireless communications or wireless devices.

Sometimes people wish to possess the live RFID tags after payment for goods returns. At the same time, they do not want the RFID tags to be scanned illegally. The RSA blocker tags, which were proposed by (Juels, Rivest, & Szydlo, 2003), help us protect live RFID tags against unauthorized reading. An RSA blocker tag is an RFID tag that responds to all requests, but blocks illegal scanners.

4.4 SIDE CHANNEL ATTACK

For a long time, the security of cryptographic algorithms has been viewed as a nearly pure mathematical problem. While the security of symmetric cryptographic algorithms is based on s-box construction, the security of public-key cryptographic algorithms is related to hard/intractable mathematical problems such as the Integer Factoring Problem and the Discrete Log Problem. With a tighter formal definition, the concept of provable security, which originated in (Goldwasser & Micali, 1984) and became a practical paradigm from (Bellare & Rogaway, 1993), means that the security of cryptographic algorithms can be proved or argued mathematically or reduced to a hard/intractable problem.

However, a cryptographic primitive with a mathematical security proof may suffer from so-called side channel attacks, which can retrieve a part of the secret key by analyzing the leaked information in the concrete implementation of cryptographic algorithms such as timing information, power consumption and so on. The first kind of side channel attack, namely the Timing Attack, was proposed by (Kocher, 1996) at CRYPTO 1996. Other well-known powerful side channel attacks include the Fault Attack, which was proposed by (Boneh, DeMillo, & Lipton, 1997) at EUROCRYPT 1997, and the Power Analysis attack (Simple Power Analysis-SPA and Differential Power Analysis-DPA), which was proposed by (Kocher, Jaffe, & Jun, 1999) at CRYPTO 1999.

Side channel attacks may not be able to obtain all the bits of the whole secret key. However, for many cryptographic algorithms, a part of the secret key is sufficient to break the whole cryptosystem. The so-called partial key exposure attacks aim to recover the whole secret cryptographic key by using the critical part of the secret cryptographic key. An important research problem is how to implement a practical and efficient attack method by combining side channel attack and mathematical algorithms including some computational number theory and algebraic algorithms.

4.4.1 Power Analysis

At CRYPTO 1999, (Kocher, Jaffe, & Jun, 1999) proposed the Simple Power Analysis and Differential Power Analysis attacks on DES algorithm and RSA. Cryptographic primitives are implemented on various kinds of hardware devices such as the CPU of a personal computer, the micro-controller of an RFID card or smart card. The computational process of a cryptographic primitive can leak information about the secret key because different operations consume different amount of electric power and therefore have different power consumption traces.

In the simple power analysis (SPA) attack, the attacker aims to retrieve useful information about the secret key by using a single power consumption trace. For either symmetric algorithms (DES, AES and so on) or public-key algorithms (RSA, ECC and so on), the execution sequence of cryptographic operations, which produces a specific power trace, is related to the secret key. For example, in RSA or ECC, a basic implementation algorithm is the square-and-multiply method (see Box 1) for $y = x^d \bmod n$ or double-and-add method (see Box 2) for $Q = kP$, which can be seen in the excellent survey by (Gordon, 1998). The '1' bit or '0' bit in the secret integer d or k means a different branch of the execution procedure, which can be seen in the power consumption trace. Let d, k be $m-$bit integers and d_i, k_i denote the $i-$th least significant bit. It is noted that there are some methods such as the window method, signed-digit recoding algorithms and so on to improve the efficiency of exponentiation in RSA and point multiplication on elliptic curves.

Box 1.

The Square-and-multiply algorithm for $y = x^d \bmod n$,
which is seen in the survey by (Gordon, D.M. 1998).

Input: $m-$bit integers x, d, and n.

Output: $y = x^d \bmod n$.

1. $y = x$;
2. For $i = m-2$ to 0 do
2.1 $\{y = y^2 \bmod n$;
2.2 If $(d_i == 1)$ then
2.3 $y = y \cdot x \bmod n;\}$
3. Return y.

They have a similar power analysis strategy as the basic square-and-multiply and double-and-add algorithms.

Some power consumption measure errors or physical noise may make it infeasible to implement a successful Simple Power Analysis attack. The Differential Power Analysis attack combines the statistical differential analysis functions and power consumption analysis to mount a much more powerful attack. In the DPA attack, the attacker first executes m encryption operations and measures many power consumption traces to determine whether a guess of a part of the secret key is correct.

At CRYPTO' 1999, (Kocher, Jaffe, & Jun, 1999) gave a practical example on the Data Encryption Standard-DES algorithm. The DES encryption/decryption algorithm consists of 16 rounds of operations and each round performs an s-box substitution, permutation and so on. The s-box is the most critical security component of DES.

The DPA attack proposed by (Kocher, Jaffe, & Jun, 1999) is shown as follows. Let C be a ciphertext and K_s be the 6 key bits. The function $D(C, b, K_s)$ computes the value of b which is the DES intermediate value. The attacker first performs m encryption operations and obtains power consumption traces $T_{1...m}[1...k]$. Let $\Delta_D[j]$, which was given by (Kocher, Jaffe, & Jun, 1999), be defined in Box 3.

If the guessed K_s is incorrect, one has

$$\lim_{m \to \infty} \Delta_D[j] \approx 0.$$

The high-order DPA attack uses a joint statistical analysis on the power consumption traces and is a generalization of the first-order DPA attack.

Box 2.

The double-and-add algorithm for $Q = kP$.
which is seen in the survey by (Gordon, D.M. 1998).

Input: The $m-$bit integer k and the point P on an elliptic curve.
Output: $Q = kP$.
1. $Q = P$;
2. For $i = m-2$ to 0 do
2.1 $\{Q = 2Q$;
2.2 If $(k_i = 1)$ then
2.3 $Q = Q + P;\}$
3. Return Q.

Box 3.

$$\Delta_D[j] = \frac{\sum_{i=1}^{m} D(C_i, b, K_s) T_i[j]}{\sum_{i=1}^{m} D(C_i, b, K_s)} - \frac{\sum_{i=1}^{m} (1 - D(C_i, b, K_s)) T_i[j]}{\sum_{i=1}^{m} (1 - D(C_i, b, K_s))}$$

$$\approx 2 \left(\frac{\sum_{i=1}^{m} D(C_i, b, K_s) T_i[j]}{\sum_{i=1}^{m} D(C_i, b, K_s)} - \frac{\sum_{i=1}^{m} T_i[j]}{m} \right)$$

Elliptic curve cryptosystems, which were introduced independently by (Miller, 1985) and (Koblitz, 1987), are the public key cryptographic algorithms defined on the abelian group $E(F_q)$ of rational points over an elliptic curve. Typically, the finite fields F_p, where p is a large prime, or F_{2^n} are widely used. For example, an elliptic curve over F_p is defined as the form

$$y^2 = x^3 + ax + b$$

with $a, b \in F_p$. The point $P = (x, y)$, where x and y satisfy the above equation, is called a point on the elliptic curve. Then the point addition formula $P + Q$ and point doubling formula $2P$ are given to construct an abelian group of points. The main operation of elliptic curve cryptosystems is scalar multiplication $Q = kP = P + \ldots + P$, which is also called point multiplication. The basic scalar multiplication algorithms are the double-and-add method and signed-digit binary method, in which the point addition and point doubling are computed alternately.

The idea of the simple power analysis attack on elliptic curve cryptosystems is to retrieve the useful information of the secret value k by distinguish point addition from point doubling by measuring the power consumption trace of computing the scalar multiplication $Q = kP$. Similar to the differential power analysis attack on symmetric cryptographic algorithms, the DPA attack on elliptic curve cryptosystems uses the statistical differential analysis of many power consumption traces of ECC cryptographic operations to improve the ability to attack successfully.

Since point addition or doubling on elliptic curves are not simple operations and consist of finite field computations such as addition, multiplication and inversion of integers or polynomials, there are a few special power analysis attacks on elliptic curve cryptosystems. In 2003, several new power analysis attacks such as the refined power analysis, the zero-value point attack and the doubling attack were proposed. (Goubin, 2003) proposed the refined power analysis attack on elliptic curve cryptosystems, which improved the differential power analysis by using special points $(x, 0)$ or $(0, y)$. The Zero-value Point Attack, which was proposed by (Akishita & Takagi, 2003), can work for more elliptic curves without zero-value coordinates by using the zero-value registers which may appear in the computation process of scalar multiplication. (Fouque & Valette, 2003) introduced the doubling attack, which can be applicable for the left-to-right binary double-and-add algorithm for point multiplication on elliptic curves.

For RFID devices, the attacker can measure power traces in the communication period between the RFID tag and the reader when the RFID tag implements cryptographic algorithm operations. (Oren, Yossi, & Shamir, 2006) proposed the first power analysis attack on passive RFID tags in the thesis 'Remote Password Extraction from RFID Tags'. The proposed power analysis attack can work for passive UHF tags adhering to the EPCGlobal standard (Electronic Product Code), when the ultra high frequency band ranges roughly from 300MHz to 3GHz. The power analysis can recover the kill passwords from both EPC Gen-1 and EPC Gen-2 tags which operate in the 900 MHz frequency range.

4.4.2 Timing Analysis

At CRYPTO 1996, (Kocher, 1996) introduced the Timing Attack on implementations of some public-key cryptographic algorithms such as the Diffie-Hellman key exchange protocol, the RSA algorithm and the DSS signature algorithm. These primitives consist of time-consuming exponentiations of large integers and suffer from timing attacks more easily than symmetric algorithms. Kocher's timing attack was also shown by experimental results using the RSAREF toolkit.

The integer exponentiation $y = g^x \bmod p$ can be implemented by using the basic square-and-multiply method or the more efficient window method. Indeed, the computation time is related to the secret integer x. Therefore, one can guess the bit values of x by measuring the computation time of exponentiation $y = g^x \bmod p$.

At CARDIS 1998, (Dhem et al., 1998) proposed some improvements to show how to implement a practical timing attack, which is able to break a 512-bit key RSA in a few minutes by collecting 300, 000 timing measurements. The timing attack can also be applied to elliptic curves and hyper-elliptic curves, since the basic scalar multiplication operation $Q = nP$ on elliptic curves is analogous to the exponentiation $y = g^x \bmod p$.

At Usenixsec 2003, (Brumley & Boneh, 2003) proposed and implemented a practical timing attack against OpenSSL servers. Their experiments showed that one can obtain the secret 1024-bit RSA private key in several hours by using the timing information of an SSL server.

4.5 PHYSICAL TAMPERING

RFID tampering is a kind of physical attack method, which can modify the data, information or programming codes stored in RFID devices or damage the RFID devices. Generally, the identification code and other application data are stored in programmable memory (such as EEPROM, Flash and so on) in the RFID tag. Therefore, if the attacker has physical access to the RFID tag memory, it follows that he can modify the data in the RFID tag memory. A tampering attack maybe produce some dangers such as disabling the availability of the RFID system. A tamper-evident method means that the tampering attack can be detected. A tamper-resistant method means that the device is protected from tampering attacks.

There are some software or hardware countermeasures to defeat tampering attacks on RFID systems. In cryptography, it is important to provide data integrity for detecting or preventing illegal modifications of the data by malicious attackers. Hash algorithms, Message Authentication Codes (MAC), and other cryptographic algorithms are applicable to protect data integrity. Other kinds of methods of protecting data integrity or transferring secret information is information hiding such as digital watermarking. Therefore, cryptographic algorithms (HASH, MAC and so on) and information hiding are important techniques for defeating tampering attacks. (Gandino, Montrucchio, & Rebaudengo, 2009) presented an excellent survey on the tampering attack and countermeasures for RFID systems.

4.6 HARDWARE PROTECTION

These hardware attacks, including the cloning attacks, direct reads, tampering attacks and side channel attacks, have be serious threats not only to RFID systems but also to other kinds of information systems. Therefore, many countermeasures have been proposed to defeat these hardware attacks.

4.6.1 Tamper-Proof Hardware

There have been some consideration and attempts to design hardware countermeasures against the above hardware attacks. (Gennaro et al., 2004) proposed the concept of tamper-proof hardware, which prevents the data stored in the device being modified illegally. (Tiri, Akmal, & Verbauwhede, 2002) had proposed the idea for designing specific hardware with constant power consumption, which can also protect a device from tampering attack.

However, as noted by (Gandino, Montrucchio, & Rebaudengo, 2009), the tamper-resistant hardware is too expensive to be applied in RFID tags.

4.6.2 Algorithm Level

There are many algorithmic countermeasures for side channel attacks. At CARDIS 1998, (Dhem et al., 1998) gave some countermeasures to protect RSA cryptosystem from the timing attack. A natural method, which was proposed by (Kocher, 1996) at CRYPTO 1996, is data blinding before the cryptographic operations. For example, in RSA cryptosystem, the public exponent e and the private exponent d satisfy the equation

$$ed \equiv 1 \bmod \phi(n)$$

where $\phi(n) = (p-1)(q-1)$ and $n = pq$ is the product of two primes p and q. One can choose a pair of integers (v_i, v_f) randomly, which satisfies

$$v_f^{-1} = (v_i)^e.$$

The RSA encryption/decryption operations are $c = m^e \bmod n$ and $m = c^d \bmod n$ respectively. The improved RSA operations against timing analysis attack are $c = (m \cdot v_i)^e \bmod n$ and $m = (c \cdot v_f)^d \bmod n$ respectively. The idea of this countermeasure method comes from the design of RSA blind signatures proposed by (Chaum, 1983).

The Montgomery multiplication algorithm is one of the most efficient modular multiplication algorithms and widely used in RSA, ElGamal, DSA and Elliptic Curve Cryptosystem. The ordinary Montgomery multiplication algorithm has a final subtraction which is executed conditionally. Thus the countermeasure proposed by (Dhem et al., 1998) is to use the Montgomery multiplication algorithm with a fixed subtraction or without subtraction, which can be implemented by appropriate modification to the algorithm.

At CHES 1999, (Coron, 1999) proposed several efficient countermeasure algorithms against simple and differential power analysis attacks for elliptic curve cryptosystems. The fundamental idea of preventing Simple Power Analysis attacks is to remove the data-dependent conditional branches and to avoid the SPA characteristics of the secret cryptographic key. (Coron, 1999) proposed a direct method for preventing simple power analysis attacks which implements a fixed double-and-add-always computation sequence of scalar multiplication. The drawback of this method is the requirement of more computational cost. When the basic double-and-add algorithm needs about $\frac{3}{2}m$ point additions or doublings, the SPA-resistant algorithm (see Box 4) requires $2m$ point additions or doublings.

Therefore, many novel algorithms have been designed to improve the efficiency of SPA-resistant algorithms for elliptic curve cryptosystems. Point addition and doubling consist of many basic finite field operations such as addition/subtraction, multiplication and so on. In 2004, (Chevallier-Mames, Ciet, & Joye, 2004) proposed the notion of side-channel atomicity and some practical and efficient SPA-resistant methods for elliptic curve cryptosystems which split point addition or doubling into small side-channel equivalent basic operation blocks. Since the difference in power consumption between point addition and doubling leads to the SPA attack, another method is to design a united formula for point addition and doubling to defeat SPA.

An interesting type of SPA-resistant methods is the randomized signed-digit recoding algorithms which have good computational efficiency. However, most of them have been broken and proven to be unable to resist the SPA attack. At CHES 2003, (Karlof & Wagner, 2003) proposed the hidden Markov model cryptanalysis which is used to

Box 4.

(Coron, 1999) modified binary algorithm against SPA.
Input: The $m-$bit integer k and the point P.
Output: The point multiplication $Q[0]=kP$.
1. $Q[0]=P$;
2. For $i=m-2$ to 0 do
2.1 $\{Q[0]=2Q[0]$;
2.2 $Q[1]=Q[0]+P$;
2.3 $Q[0]=Q[k_i];\}$
3. Return $Q[0]$.

Box 5.

(Coron, 1999) first algorithm against DPA.
Input: The $m-$bit integer k and the point P.
Output: $Q=kP$.
1. Select an integer r of size l bits randomly, where l is tens.
2. Compute $k'=k+r\cdot\#E(F_q)$.
3. Using scalar multiplication algorithm, compute $Q=k'\cdot P$ $=(k+r\cdot\#E(F_q))\cdot P=kP$.
4. Return Q.

Box 6.

(Coron, 1999) second algorithm against DPA.
Input: The $m-$bit integer k and the point P.
Output: $Q=kP$.
1. Selecting a point R randomly and computing $S=kR$.
2. Computing scalar multiplication $Q'=k(P+R)$.
3. Computing $Q=Q'-S=kP$.
4. Return Q.

Box 7.

(Mamiya, Miyaji, & Morimoto, 2004) BRIP algorithm.
Input: The point $P\in E(F_q)$ and the $k-$bit integer n.
Output: $Q=nP$.
1. Select randomly a point $R\in E(F_q)$;
2. $T[0]=R$; $T[1]=-R$; $T[2]=P-R$;
3. For $i=k-1$ downto 0 do
3.1 $\{T[0]=2T[0]$;
3.2 If $n_i==0$ then
3.3 $T[0]=T[0]+T[1]$;
3.4 Else
3.5 $T[0]=T[0]+T[2];\}$
4. Return $Q=T[0]+T[1]$.

analyze randomized countermeasure algorithms with a probabilistic finite state machine.

The goal is to defeat the differential power analysis by blinding the intermediate data in the cryptographic operations. A simpler method is adding random physical noise to the device which only increases the time and space costs of the attackers for differential power analysis attack. For example, the S-box of the AES algorithm is a finite field computation and has a good algebraic property. Therefore, multiplicative masking method is an efficient countermeasure against differential power analysis attack.

(Coron, 1999) proposed three countermeasure algorithms against differential power analysis attacks for elliptic curve cryptosystems. The common technique of the three countermeasures is to blind or randomize of the secret scalar k or the point P for the point multiplication computation $Q = k{\cdot}P$. Let $\# E(F_q)$ be the number of rational points over an elliptic curve. Thus it follows that $\# E(F_q){\cdot}P = O$, where the point O is the point at infinity. Coron's first method is to randomize the secret integer k by using the order $\# E(F_q)$. See Box 5 for the algorithm.

Coron's second method is to randomize the point P by using a random point R, which is a similar method as Chaum's RSA blind signature. See Box 6 for the algorithm.

Coron's third method is to randomize the point P with projective coordinates. The affine coordinates (x, y) of the point P correspond to the projective coordinates (X, Y, Z) with the relation $x = \frac{X}{Z}, y = \frac{Y}{Z}$. See Box 7 for the algorithm.

The projective coordinates (X, Y, Z) are equivalent to $(\lambda X, \lambda Y, \lambda Z)$, where $\lambda \neq 0$ is an element in the finite field F_q. Therefore, one can blinding the point P by using λ with projective coordinates.

At CHES 2004, (Mamiya, Miyaji, & Morimoto, 2004) proposed a countermeasure (called BRIP) against new power analysis attacks such as RPA and so on, which uses a random initial point (RIP). (Avanzi, 2005) presented an excellent survey of side channel attacks and the countermeasures of elliptic curve cryptosystems.

As noted by (Oren, Yossi, & Shamir, 2006; Kasper, Oswald, & Paar, 2009), the power consumption measuring method of RFID tags is different from other devices since RFID tags obtain electric power from the antenna by receiving the electromagnetic signal generated by the reader. (Oren, Yossi, & Shamir, 2006) proposed a practical power analysis attack on RFID tags to guess the Kill password successfully.

CONCLUSION

In practice, malicious attackers often launch an attack by combining various kinds of attack methods such as hardware attacks and mathematical analysis tools. For example, an attacker tries to retrieve secret information by using the timing attack, simple power analysis and differential power analysis on the device. It is possible that the attacker only recover a part of the secret key, not the whole cryptographic key. Then the attacker can continue the attack further by adopting some mathematical analysis.

Therefore, it is interesting to find the relations and conjunctions of various hardware attacks or mathematical analysis algorithms.

REFERENCES

Abawajy, J. (2009). Enhancing RFID tag resistance against cloning attack. In *Third International Conference on Network and System Security*, October 19-October 21, 2009, (pp. 18–23). IEEE Computer Society.

Akishita, T., & Takagi, T. (2003). Zero-value point attacks on elliptic curve cryptosystem. In *Information Security Conference - ISC '03, LNCS 2851*, (pp. 218-233). Springer-Verlag.

Avanzi, R. M. (2005). *Side channel attacks on implementations of curve-based cryptographic primitive*. Retrieved from http://eprint.iacr.org/2005/017.pdf

Avoine, G., Kalach, K., & Quisquater, J.-J. (2008). ePassport: Securing international contacts with contactless chips. In *Financial Cryptography and Data Security, FC 2008*, (pp.141-155). Cozumel, Mexico.

Bellare, M., & Rogaway, P. (1993). Random oracles are practical: A paradigm for designing efficient protocols. In *First ACM Conference on Computer and Communications Security*, (pp. 62-73). ACM Press.

Boneh, D., DeMillo, R. A., & Lipton, R. J. (1997). On the importance of checking cryptographic protocols for faults. In *EUROCRYPT' 1997, LNCS 1233* (pp. 37–51). Berlin, Germany: Springer.

Brumley, D., & Boneh, D. (2003). Remote timing attacks are practical. In *Proceedings of the 12th conference on USENIX Security Symposium* - Volume 12, (p. 1).

Chaum, D. (1983). Blind signatures for untraceable payments. *Advances in Cryptology- Proceedings of Crypto 82*, (pp. 199-203).

Chevallier-Mames, B., Ciet, M., & Joye, M. (2004). Low-cost solutions for preventing simple side-channel analysis: Side-channel atomicity. *IEEE Transactions on Computers*, *53*, 760–768.

Coron, J.-S. (1999). Resistance against differential power analysis for elliptic curve cryptosystems. In *CHES' 1999, LNCS 1717* (pp. 292–302). Berlin, Germany: Springer.

Dhem, J.-F., Koeune, F., Leroux, P.-A., Mestré, P., Quisquater, J.-J., & Willems, J.-L. (1998). A practical implementation of the timing attack. In *Third Smart Card Research and Advanced Application Conference - CARDIS 98, volume 1820 of Lecture Notes in Computer Science*, (pp. 167-182). Berlin, Germany: Springer-Verlag.

Fouque, P.-A., & Valette, F. (2003). The doubling attack - Why upwards is better than downwards. In *Cryptographic Hardware and Embedded Systems - CHES '03, LNCS 2779*, (pp. 269-280). Springer-Verlag.

Gandino, F., Montrucchio, B., & Rebaudengo, M. (2009). Tampering in RFID: A survey on risks and defenses. [Springer Netherlands.]. *Mobile Networks and Applications*, 1–15.

Gennaro, R., Lysyanskaya, A., Malkin, T., Micali, S., & Rabin, T. (2004). Algorithmic tamper-proof (ATP) security: Theoretical foundations for security against hardware tampering. In *TCC 2004*, (pp. 258-277).

Goldwasser, S., & Micali, S. (1984). Probabilistic encryption. *Journal of Computer and System Sciences*, *28*, 270–299.

Goodin, D. (2009, 2nd February). Passport RFIDs cloned wholesale by $250 eBay auction spree. *The Register*. Situation Publishing Limited. Retrieved from http://www.theregister.co.uk/2009/02/02/low_cost_rfid_cloner/

Gordon, D. M. (1998). A survey of fast exponentiation methods. *Journal of Algorithms*, *27*, 129–146.

Goubin, L. (2003). A refined power-analysis attack on elliptic curve cryptosystems. In *Public Key Cryptography - PKC'03, LNCS 2567*, (pp. 199-210). Springer-Verlag.

ICAO. (n.d.). *Machine readable travel documents: ICAO Doc 9303*. Retrieved from www.icao.int/mrtd/

Juels, A. (2005). Strengthening EPC tags against cloning. In M. Jakobsson & R. Poovendran (Eds.), *ACM Workshop on Wireless Security (WiSe)* (pp. 67-76).

Juels, A. (2006). RFID security and privacy: A research survey. *IEEE Journal on Selected Areas in Communications*, *24*(2), 381–395.

Juels, A., Molnar, D., & Wagner, D. (2005). Security and privacy issues in e-passports. In M. Jakobsson & R. Poovendran (Eds.), *Proceedings of the First International Conference on Security and Privacy for Emerging Areas in Communications Networks (SECURECOMM)* (pp. 74-88). Washington, DC: IEEE Computer Society.

Juels, A., Rivest, R. L., & Szydlo, M. (2003). The blocker tag: Selective blocking of RFID tags for consumer privacy. In V. Atluri (Ed.), *8th ACM Conference on Computer and Communications Security*, (pp. 103-111). ACM Press.

Karlof, C., & Wagner, D. (2003). Hidden MARKOV model cryptanalysis. In *Cryptographic Hardware and Embedded Systems - CHES 2003, LNCS 2779*, (pp. 17- 34). Springer-Verlag.

Kasper, T., Oswald, D., & Paar, C. (2009). *New methods for cost-effective side-channel attacks on cryptographic RFIDs*. In RFIDSec'09, July 2009.

Koblitz, N. (1987). Elliptic curve cryptosystems. *Mathematics of Computation*, *48*, 203–209.

Kocher, P. (1996). Timing attacks on implementations of Diffie-Hellman, RSA, DSS, and other systems. In *CRYPTOŸÏ 1996, LNCS 1109* (pp. 104–113). Berlin, Germany: Springer.

Kocher, P., Jaffe, J., & Jun, B. (1999). Differential power analysis. In *CRYPTO 1999, LNCS 1666* (pp. 388–397). Berlin, Germany: Springer.

Lehtonen, M., Michahelles, F., & Fleisch, E. (2009). How to detect cloned tags in a reliable way from imcomplete RFID traces. In *2009 IEEE International Conference on RFID* (pp. 257-264). Piscataway, NJ: IEEE.

Mamiya, H., Miyaji, A., & Morimoto, H. (2004). Efficient countermeasure against RPA, DPA, and SPA. In *CHES' 2004, LNCS 3156* (pp. 343–356). Berlin, Germany: Springer.

Miller, V. S. (1985). Use of elliptic curves in cryptography. In *Proceedings of Crypto 85, LNCS 218*, (pp. 417-426).

Moskowitz, P., & Karjoth, G. (2005, November 7). IBM proposes privacy-protecting tag. *RFID Journal*.

Moskowitz, P., Lauris, A., & Morris, S. S. (2007). A privacy-enhancing radio frequency identification tag: implementation of the clipped tag. In *Fifth IEEE International Conference on Pervasive Computing and Communications Workshops (PerComW'07)*, March 19-March 23 2007, White Plains, New York.

Oren, Y., & Shamir, A. (2006). *Power analysis of RFID tags*. Advances in Cryptology - CRYPTO 2006. Retrieved from http://www.wisdom.weizmann.ac.il/yossio/rfid/

Oren, Y., & Shamir, A. (2007). Remote password extraction from RFID tags. *IEEE Transactions on Computers*, *56*(9), 1292–1296.

Tiri, K., Akmal, M., & Verbauwhede, I. (2002). A dynamic and differential CMOS logic with signal independent power consumption to withstand differential power analysis on smart cards. In *28th European Solid-State Circuits Conference(ESSCIRC 2002)*, (pp. 403-406).

Witteman, M. (2005). Attacks on digital passports. *What the Hack*. Retrieved from http://wiki.whatthehack.org/index.php/Track:Attacks_on_Digital_Passports

Chapter 5
Computer System Attacks

Zhang Ning
XiDian University, People's Republic of China

ABSTRACT

The study of computer system attacks is an important part RFID security and privacy. This chapter provides a general overview of computer system attacks organized by target. Attacks on EPC entities - tags, readers, middleware, and back-end systems - are categorized and discussed, as well as wired link attacks. Countermeasures to the attacks are summarized and evaluated based on the discussion. The Denial of Services (DoS) attack is highlighted in the discussion.

5.1 INTRODUCTION

Security and privacy in RFID systems is a topic that deserves careful consideration. In this chapter, attacks can be various, especially, since RFID systems have a computer-driven back-end. Different sorts of computer system attacks are presented within the scope of RFID.

Throughout this chapter we will be considering the EPC Global network. EPC Global is an organization set up to achieve worldwide adoption and standardization of Electronic Product Code (EPC) technology. The main focus of the group is currently to create both a worldwide standard for RFID and the use of the internet to share data via the EPC Global Network.

According to the EPC Global framework, as in Figure 1, an RFID system consists of tags, readers, middleware and back-end. Any of these four entities or the communication paths between them can be the target of an attacker. We will do a comprehensive analysis of the computer system attacks on each entity and the wired link between middleware and the back-end. EPC Global network, by design, is also susceptible to DoS attacks. Our objective is to provide a reference for readers that acquaint them with computer system attacks on RFID systems.

DOI: 10.4018/978-1-4666-3685-9.ch005

5.2 ATTACKS ON TAGS

With wider usage of RFID, for instance, in many countries, new passports contain an RFID tag with an encrypted form of the data that is written in clear text on the passport, tag data security becomes our first consideration for security purposes. Generally, low-cost RFID tags (such as EPC Class-1 Generation-2 tags) have very limited resources, and may, therefore, not be able to support sophisticated security procedures based on encryption. This problem is exacerbated by the constant pressure from industry to develop ever cheaper tags. Surprisingly, these limitations may actually be an advantage to the security architect. Thus in RFID deployment, the most effective attacks are those on the tags and the ones resulting from the communications channel between tags and readers (wireless link attacks, which was discussed in Chapter 4). We will discuss attacks on tag data in this section.

5.2.1 Data Integrity

A powerful attacker that is able to modify data on an RFID tag can be very dangerous, possibly compromising the integrity of tag data. Here, modification means deletion, change or insertion; it can corrupt the data integrity.

Consider the following example - a known terrorist modifies the biographic data (name, address, etc.) on his passport tag and is able to cross borders into any country since his modified data does not appear on any existing watch list. The amount of impact that this attack may have will, of course, depend on the application in which the tags are used, as well as the degree to which tag data are modified. Data might be modified in such a way that the ID of the tag and any security related information (i.e. keys, credentials) remain unaltered. Thus the inconsistency between data stored on the RFID tag and the corresponding tagged object/human may have serious implications (i.e. in health care applications, tags are used that may contain critical information about a patient's health or a medicine's recommended dosage). Hence the reader can be fooled into thinking that it is communicating with an unmodified tag, while critical information might have been falsified.

There are several open-source libraries for reading/manipulating data on RFID tags.

RFIDIOt is an open-source Python library for exploring RFID devices written by Adam Laurie. RFIDIOt caused a huge furor in the market with its release. Attackers can use it to manipulate RFID devices. It provides support like READ, WRITE, DEBIT, LOGIN, etc. while working with external readers. It supports ISO standards such as 14443A and 14443B in a variety of 13.56 MHz and 125/135 KHz RFID bands. RFIDIOt facilitates the scripting of malicious RFID queries. For example, Rieback (2008) stated that, using RFIDIOt API, Pieter Siekermann and Maurits van der Schee from the University of Amsterdam successfully attacked the Dutch RFID public transportation (OV Chipkaart) system, manipulating the data on single-use MIFARE Ultra light cards to exploit a hole in the back-end RFID middleware, allowing free travel. For more information about RFIDOt, the reader is refereed to http://www.rfidiot.org/.

RFDump is a similar open-source software toolkit created by Lukas Grunwald and Christian Bottger for the purpose of security auditing of RFID tags. It is periodically updated to emerging RFID standards such as e-passport and Mifare encryption currently found on many pay as you go systems. RFDump is a back-end GPL tool to directly inter-operate with any RFID ISO-reader to make the contents stored on RFID tags accessible. RFDump can be used to detect RFID tags and show their data information: tag ID, tag type, manufacturer, etc. The user data of a tag can be displayed and modified using either a Hex or an ASCII editor. In addition, the integrated cookie feature demonstrates how easy it is for a company

Figure 1. The EPC global framework (Traub et al (2010))

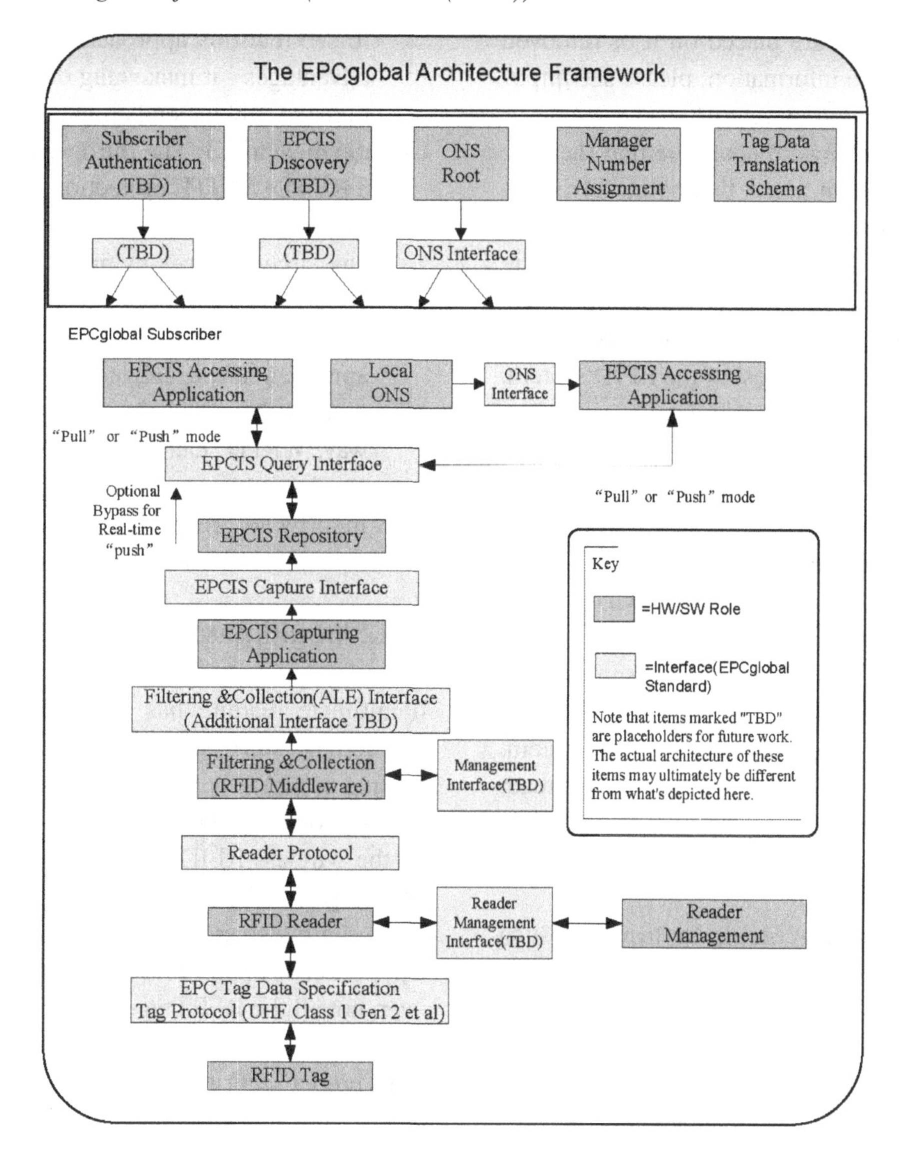

to abuse RFID technology to spy on their customers. For more information, please see http://www.rfdump.org/.

LibRFID is an open-source library written by Harald Welte. It implements the PCD (reader) side protocol stack of ISO 14443 A, ISO 14443 B, ISO 15693, Mifare Ultralight and Mifare Classic. It is designed to support iCODE*1 and other 13.56MHz based transponders. It works mainly with OpenPCD and Omnikey Cardman 5121/5321. It directly drives the readers and so does not use PCSCD which must be stopped. For more information, please see http://openmrtd.org/projects/librfid/.

TagEventor is a Linux open-source client for Touchatag (Tikitag). It uses the PCSC-Lite daemon and can be run in foreground or daemon mode to make tag events available to user-space applications. The software is currently a daemon that monitors the presence of one or more RFID tags on a connected reader and generates "system events" when tags are placed on it or removed from it. For more information, please see http://code.google.com/p/tageventor/.

However, we have to note here that the ease with which such an attack can be performed is highly dependent on the RFID standard used and the READ/WRITE protection employed. The more sophisticated the protection employed, the more difficult it is for the attacker to modify the tag, and the greater the cost of the RFID system.

After modifying the tag data by injecting malicious code into the RFID tag, the tags can be used to propagate hostile code that could subsequently infect other entities in the RFID network (readers, middleware, back-end and connecting networks) according to Rieback, Crispo and Tanenbaum (2006). In this scenario, an attacker uses the memory space of RFID tags in order to store the infecting viruses or other RFID malware and spread them through the back-end system. Although this type of attack is not widespread, laboratory experiments have proven that they are feasible.

Countermeasure 1: A simple way to prevent such attacks is to sign all data on the tag with the private key of the tag issuer (for example, the passport office that supplied the ePassport). This means that we would employ digital signature schemes to guarantee the data integrity of each tag. However, with such an approach, it is critical that all verifiers, by which we mean reader and back-end system are aware of every tag issuer's public key (for signature verification purposes), and the data can only be modified by the tag issuer (the only one who has the private key to sign the data on the tag) when it needs to be updated.

Countermeasure 2: An alternate approach is for tags to have read only registers which make it impossible to write over any data. For low cost EPC systems, such as EPC Generation 1, this is very efficient. However, it is easy to see that this approach has obvious disadvantages - it makes tag data corrections, updates, and appends impossible.

Countermeasure 3: The third way is to employ a READ/WRITE protection system. While the READ command can be execute under any circumstances by anyone, the WRITE command can only be executed using special circumstances by dedicated equipment. This approach can be realized either in a physical way or through software. In the physical way, WRITE can only be executed over a special frequency using special equipment. In software, group signature and proxy group signature schemes can be used.

5.2.2 Illegitimate Reading

Illegitimate reading has other names such as skimming or sniffing, they all mean unauthorized reading of tags. In Chapter 4, we saw that RFID tag data could also be collected by eavesdropping on the wireless RFID channels, and that eavesdropping is also a kind of illegitimate reading.

In contrast to most electronic products, RFID tags are not equipped with an on/off switch. Moreover, not all the RFID tags support protocols for authenticated read operations. In general, if a tag lacks proper security measures, it answers to any reader. Thus attackers may easily read the contents of RFID tags without leaving a trace.

Even the small amount of data stored on an RFID tag, like manufacturer or product type, can lead to privacy violations. It is convenient to thieves that an attacker can read tags and, with the data on them, identify items that are worth stealing. If additional data is stored on tags, as in the electronic passport scenario, the problems become even more severe.

The scope of such attacks remains small, however, since the attacker requires close proximity to the RFID tag. Ten meters is the upper limit for inductively coupled systems, while the construction of special readers with longer than normal radio ranges requires additional expense. Illegitimate reading of data must be prevented in RFID systems because the data must be treated confidentially since it may be privacy sensitive. A "good" RFID system must be able to cope with the threats of illegitimate reading of data. There are two categories of countermeasures to prevent illegitimate reading of data.

Countermeasure 1: Physical measures are dedicated to deactivating RFID tags or blocking the communication around tags to avoid illegitimate reading. There are several physical countermeasures to illegitimate reading.

RFID standards and product specifications generally indicate the read ranges at which they intend tags to operate. These ranges, called the Nominal read range, represent the maximum distances at which a normally operating reader with an ordinary antenna and power output can reliably scan tag data. ISO 14443, for example, specifies a nominal range of 10cm for contactless smartcards.

The "Kill" approach is promoted by EPC to protect the privacy of consumers. In practice, each tag contains a 16 bit (in other literature 24 bit is stated) kill password. If a kill message with this password is sent to the tag, it can no longer be queried. The Kill command deactivates the tag permanently. As an alternative approach with the same result, there are tags where the antenna can be removed manually so that the tags can no longer be queried.

A blocker tag, which was proposed by Juels, Rivest and Szydlo (2003), is a passive RFID device that uses a sophisticated algorithm to simulate many ordinary RFID tags simultaneously. The blocker tag is very similar to a regular RFID tag, except that it has the ability to block the singulation algorithm used by the reader to singulate tags. By sending two different UIDs to the reader, the blocker tag simulates a collision. If this is done every time a reader broadcasts a Select command, the reader is tricked into believing that all possible tags are in its interrogation zone. Blocker tags may thus be used to establish a safe zone around the tag, preventing the reading of tags within the zone. However, this approach gives individuals a lot of control. In addition, a blocker tag may be used maliciously to circumvent RFID reader protocols by simulating multiple tag identifiers.

An RFID tag may be shielded from scrutiny using what is known as a Faraday Cage-a container made of metal mesh or foil that is impenetrable by radio signals (of certain frequencies). At best, Faraday cages represent a very partial countermeasure to consumer privacy attacks. In some sensitive applications such as passports or pharmaceutical pedigrees, the Faraday Cage can block illegitimate reading efficiently. One practical proposal was the use of aluminum lined wallets to protect RFID payment cards and passports against unauthorized reading. Many companies embraced this countermeasure and sell these types of products. However, since the sniffing of confidential data can nevertheless be performed at the time of actual use, the approach does not seem to be very effective. Still, the shape and the cost of the Faraday cage are the main problems to using this approach.

Active jamming of RF signals is another, related physical means of shielding tags from view. Active jamming of RF signals refers to the use of a device that actively broadcasts radio signals in order to disrupt the operation of any nearby RFID readers. The consumer could carry a device that actively broadcasts radio signals so as to block and/or disrupt the operation of any nearby RFID readers. However, this physical means of shielding may disrupt nearby RFID systems and may be illegal - at least if the broadcast power is too high - and is a crude, sledgehammer approach. It could cause severe disruption of all nearby RFID systems, even those in legitimate applications where privacy is not a concern.

Countermeasure 2: Rieback, Crispo and Tanenbaum (2005) and Juels, Rivest and Bailey (2005) propose very similar devices, respectively called an "RFID Guardian" and an "RFID Enhancer Proxy" (REP). A Guardian acts as a personal RFID firewall. It intermediates reader requests to tags, selectively simulating tags under its control. As a high-powered device with substantive computing power, a Guardian can implement sophisticated privacy policies, and can use channels other than RFID (e.g., GPS or Internet connections) to supplement ambient data. For example, a Guardian might implement a policy like: "My tags should only be subject to scanning within 30m of my home (as determined by GPS), or in shops that compensate consumer tag-scanning with coupons for a 10% discount." The logistical questions of how a Guardian should acquire and release control of tags and their associated PINs or keys are challenging problems that merit further research.

Countermeasure 3: The software option countermeasure is mainly a cryptographic measure to prevent illegitimate reading. Cryptographers have created a lot of new low-power algorithms for RFID tags, including stream and block ciphers, and public-key cryptographic primitives or even no cryptographic primitive such as pseudonyms and error correction codes. Tags and readers may mutually authenticate each other with keys, using well understood protocols. Tags may even encrypt their contents using random nonce to prevent tracking. Unfortunately, supporting strong public key cryptography is beyond the resources of low cost tags, although countermeasures do exist for more expensive ones. However, an important limitation on employing these schemes in RFID systems is that the latter have inherent vulnerabilities such as possible power interruptions or the disruption of wireless and wired channels. Moreover, we have to keep in mind that employing all these encryption techniques even in non-critical applications such as RFID on underwear or chewing gum is definitely not worthwhile. This topic will be further discussed in Chapter 7 and 8.

5.2.3 Spoofing, Counterfeiting, and Mimicking

Attackers can create authentic RFID tags, by writing appropriately formatted data on blank RFID tags. For example, thieves could re-tag items in a supermarket identifying them as similar, but cheaper, products. This is known as a cloning attack. While tag cloning is a kind of physical attack, spoofing is a variant of cloning that does not physically replicate an RFID tag.

Spoofing attacks supply false information that looks valid and that the system accepts. In this type of attacks, an adversary impersonates a valid RFID tag to gain its privileges. Typically, spoofing attacks involve a fake domain name, Internet Protocol (IP) address, or Media Access Code (MAC). In order to achieve spoofing, the attackers employ special devices with increased functionality that are able to emulate RFID tags given some data content. This impersonation requires full access to the same communication channels as the original tag, which includes knowledge of the protocols and secrets used in any authentication that is going to take place. An example of such an attack was performed by researchers from John Hopkins University and RSA Laboratories who succeeded in unlocking a vehicle immobilizer system by reverse engineering and cracking the system and subsequently spoofing the reader using the data obtained (New York Times, "*Graduate Cryptographers Unlock Code of 'Thiefproof' Car Key*" by John Schwartz. 29 January 2005).

Countermeasure 1: Spoofing attacks are generally prevented by restricting access to the "correct" information. Without this information, the attack cannot be performed. A secret key, needed as part of an authentication procedure, may be introduced as part of the "correct" information. This key is then stored in a restricted area of memory that cannot be read and is never transmitted by the tag as plaintext. This way, adversaries cannot hold the complete "correct" information, and will never pass as an authentic tag.

However, many systems rely on the secrecy of the algorithms and protocols to enhance the security provided by cryptography, and hence settle for short key lengths. This was the case with the immobilizer system spoofed by the researchers from John Hopkins University and RSA Laboratories. This violates Kerchoff's law which states that a system should be secure even if everything except the key is known. Thus, spoofing attacks are best prevented by proper cryptographic protocols with sufficiently long keys.

Countermeasure 2: Spoofing could be combated by using authentication protocols or a second form of authentication such as one-time passwords, PINs or biometrics. In 2007, engineering researchers at the University of Arkansas developed a unique, robust method to prevent counterfeiting of passive RFID tags by 'Fingerprinting' RFID tags' keys. For details, please refer to Chinnappa (2007).

Tag counterfeiting is a similar problem to that of credit card fraud where a card is duplicated and possibly used in multiple places at the same time. Since the incorporation of RFID technology in sensitive applications such as passports or pharmaceutical pedigrees, the possibility of creating counterfeiting tags has created some concerns. If RFID tags are used for access control systems, for example, in the form of an ID card, counterfeiting is also a severe problem: If it is possible to copy tags, tags no longer provide protection and become a bad security device. Unwanted copying of tags is also a problem when they are used to combat product counterfeiting, e.g. for pharmaceutical items.

Mimicking of tags is equivalent to counterfeiting: The attacker becomes able to pretend that a tag is there when it no longer is. This way, an attacker could remove a high valued item with an RFID tag affixed by replacing the item (including the original tag) with a cheaper mimicking device.

Mimicking is easier than counterfeiting a tag. Only a subset of the tag's complete functionality is required to mimic it. Hence mimicking can be done with less information and hardware equipment than required for counterfeiting a tag.

Countermeasures to counterfeiting and mimicking: For low cost tags, the current proposal promoted by GS1 for combating counterfeiting is "track and trace". The idea is to maintain an item's history. This means that read-outs by readers are centrally recorded. Based on this, one assumes that an item is genuine if it can produce a valid, i.e. complete and reproducible, item history. Thus, "track and trace" can be regarded as a plausibility check. Counterfeiting is combated with the "track and trace" approach for supply chain applications, but counterfeiting cannot be effectively prevented by this means. As counterfeited tags cannot be operatively distinguished, the back-end database should detect rare conditions. An example of a rare condition is the following: a tag cannot be in the toll gate on one motor way and fifteen minutes later in the toll gate of another 500 miles away. The design of back-end databases should be considered case by case. If RFID tags only emit unique numbers for identification, they can be copied or mimicked easily. But with RFID tags that can prove their authenticity, counterfeiting can be prevented much more effectively.

5.3 ATTACKS ON READERS

Known system attacks on readers are mainly impersonation and counterfeiting attacks. There are four scenarios.

Scenario 1: In some cases, RFID readers are installed in locations without adequate physical protection. Unauthorized intruders may set up hidden readers of a similar nature nearby to gain access to the information being transmitted by the readers, thus threatening the privacy of the RFID system.

Scenario 2: Attackers can even compromise the readers themselves, thus affecting their integrity, then counterfeiting the identity of the reader and performing unauthorized writing to the tag. Unauthorized readers may also compromise privacy by accessing tags without adequate access controls.

Scenario 3: In some sensitive deployments of RFID systems such as passports and ePay cards, illegitimate readers may elicit the confidential information stored in tags. However, if things are more complicated, the reader needs to access the back-end to retrieve the necessary credentials.

Scenario 4: Information collected by counterfeited readers and passed to the RFID application may have already been tampered with, changed or stolen by an adversary. Malicious code can be injected into the middleware and other back-end systems.

The feasibility of these attacks depends on the security measures employed for authenticating the RFID reader and varies from "very easy" to "practically impossible".

An RFID reader can also be a target for viruses. Viruses can be spread by counterfeited tags with malicious code, and then sent by the readers to the middleware and back-end systems. In 2006, researchers demonstrated that an RFID virus was possible. We will cover this topic in the next section in detail.

When an RFID reader scans a tag, it expects to receive information in a predetermined format. However, an attacker could write carefully crafted data on an RFID tag, which is so unexpected that its processing corrupts the reader's back-end software.

Countermeasures:

Reject abnormal tag information: Readers can reject tag replies with anomalies in response times, signal power levels or data format which do not match the physical properties of the tags. If passive tags are used, this can be a way of preventing spoofing attempts.

Random transmit frequency: Readers can also use random frequencies with tags designed to follow a frequency dictated by the reader. Readers can change frequencies randomly so that unauthorized users cannot easily detect and eavesdrop on traffic.

Reader detectors: RFID environments can be equipped with special devices to detect unauthorized read attempts or transmissions on tag frequencies. These read detectors may be used to detect unauthorized read/update attempts on tags if they are used together with specially designed tags that can transmit signals over a reserved frequency, indicating any attempts to kill or modify tags.

Using authentication protocol: In addition, data transmitted between the reader and middleware could require verification of the reader's identity. Authentication mechanisms can be implemented between the reader and the back-end application to ensure that information is passed to a valid processor.

5.5 ATTACKS ON MIDDLEWARE

Middleware in RFID systems refers to the software technology designed to manage and transfer information to avoid overloading public and corporate networks. EPC middleware uses a distributed architecture that works on different computers throughout an organization. The middleware and back-end systems are the real "brains" of any RFID deployment. Thus the middleware itself is a common target of attackers.

5.4.1 Buffer Overflow

Buffer overflow is one of the most frequent sources of security vulnerabilities in software.

Found in both legacy and modern software, buffer overflows cost the software industry hundreds of millions of dollars per year. Buffer overflows have also played a prominent part in events of hacker legend and lore. Programming languages such as C or C++ are not memory safe. C library functions such as strcpy (), strcat (), sprintf () and vsprintf () operate on null terminated strings and perform no boundary checking. The function gets () is another function that reads user input (into a buffer) from stdin until a terminating newline or EOF is found. The scanf () family of functions may also result in a buffer overflow. Hence the best way to deal with buffer overflow problems is to not allow them to occur in the first place. Developers should be educated about how to minimize the use of these vulnerable functions. Buffer overflow stores data or code beyond the bounds of a fixed-length buffer.

Middleware systems are designed to accept tag data of a certain size. Adversaries may use RFID tags to launch buffer overflow attacks on the middleware and back-end. Although this might not be trivial, considering the memory storage of RFID tags, there are still commands that allow an RFID tag to send the same data block repeatedly in order to overflow a buffer in the back-end RFID middleware. Other options include the use of other devices with more resources such as smart cards or devices that are able to emulate multiple RFID tags. For example, consider a middleware system written in C or C++ code, which reads tag data into a predefined memory size. If an intruder brings a tag with more capacity, it may force the back-end system to have a buffer overflow, thus leading to a system crash.

RFID tags are limited to 1024 bits or less. However, commands like ‘write multiple blocks’ from ISO-15693 can allow a low cost RFID tag to repeatedly send the same data block, with the net result of filling up an application-level buffer by meticulous formatting of the repeatedly sent data. An attacker can also use contactless smart cards, which have a larger amount of available storage space.

Countermeasures: The usual countermeasures to buffer overflow attacks can be used in RFID systems. The general idea is to make bounds check, which can prevent buffer overflow attacks by detecting whether an index lies within the limits of an array. Static or dynamic source code analyzers should be employed to check the code for buffer overflow problems.

5.4.2 Malware

RFID tags can be used in order to propagate hostile code that could subsequently infect other entities in the RFID network (readers, middleware and back-end systems). In this scenario, an adversary uses the memory space of RFID tags in order to store the infecting viruses or other RFID malware and then spread them through the back-end system.

Middleware can be infected by viral tags. Considering the fact that middleware applications use multiple scripting languages such as JavaScript, PHP, XML etc; an adversary may exploit this and inject malicious code in order to compromise the middleware systems. More specifically, RFID tags can be employed in order to perform code insertions in RFID applications that use web protocols and intercept scripting languages. In the same way, SQL injections can also be performed, a special code insertion attack based on unexpectedly executing SQL statements that may lead to unauthorized access to back-end databases and subsequently reveal or even modify data stored in the back-end RFID.

In March 2006, Melanie R. Rieback of Vrije University, Amsterdam released a paper (Rieback, Crispo and Tanenbaum (2006)) regarding the possibility of using tags and their data to attack middleware and back-end databases. The paper proposed that there were vulnerabilities in middleware applications that left room for tags to be written with malicious payloads that could affect back-end database systems, and possibly lead to a virus. At the core of the paper was the idea that even though RFID tags did not have a lot of storage space, it may still be possible to perform certain attacks through special data written to them. Rieback et al. have done significant work in this area while providing a proof-of-concept for both Linux and Windows based systems. In their paper, they target Oracle with Server-Side Includes (SSI) performing SQL injection and script based attacks. They used PHP along with SSI to achieve the above result.

Example 1 (SQL injection attack): SQL is an interactive query language which can access a database. SQL injection attacks employ this interaction to access the external database interface, insert user data to the actual database operating language, form new SQL sentences, and modify the function of the original operation to invade the database. SQL injection is a type of traditional "hacking" attack that tricks a database into running SQL code that was not intended. Attackers may use those special SQL sentences to read, modify or delete the data in the database, and may also gain users' names and passwords from the database and other important information, even obtaining the database administrator's privileges. In the case of RFID systems, the vulnerability occurs mostly in middleware.

Because of the limitations of tag data storage, it was believed that the injection attack was impossible on RFID systems. But many experiments have showed that SQL injection attacks can be disastrous to RFID systems. Such as when used in an airport, by injecting a very small amount of SQL, such as the command: ;shutdown- which shuts down an SQL server instantly, and has only 12 characters of input, a great deal of harm can be done. By injecting another command: drop table <tablename>, the specified database table will be deleted. When used in an access control system, an unauthorized person can use a counterfeited tag to access legitimate logs. The middleware authenticates the tag with the following SQL query: SELECT * FROM Table WHERE ID = $ID AND Name = '$Name'. The SQL injection attack is made by a malicious string in the Name field of the tag: a' OR 'a' = 'a. The resulting SQL query during execution is: SELECT * FROM Table WHERE ID = $ID AND Name = 'a' OR 'a = a', thus the query when executed will always return a value, which in turn grants the person access to the door.

Example 2 (Malicious file): Another research group has used malware to attack middleware systems. Lukas Grunwald, a security consultant with DN-Systems in Germany and an RFID expert, started cracking RFID systems in 2006 (see RFDump in section 6.1). In 2007, he showed how to construct a malicious JPEG2000 image file that contains an e-passport photo to crash RFID middleware by exploiting the buffer overflow vulnerability in an off-the-shelf JPEG library. Although the JPEG data is protected by a digital signature, an attacker is still able to crash the system by loading malicious data.

Example 3 (Code insertion): Code insertion can also target web-based components, such as remote management interfaces and web-based database front-ends (like Oracle iSQL*Plus). An attacker might inject malicious code into an application, using any script language (i.e. VBScript, CGI, Java, JavaScript, PHP, and Perl, etc.). HTML insertion and Cross-Site Scripting (XSS) are common code insertions. Usually, the presence of the following special characters in input data: < > ? ' % ;) (& + - is a sign of this kind of attack.

RFID tags with data written in a scripting language can perform code insertion attacks on RFID middleware and back-end systems. If the RFID applications use web protocols to query back-end databases (as EPC Global does), there is a chance that RFID middleware clients can interpret the scripting languages. If this is the case, then the RFID middleware will be susceptible to the same code insertion problems as web browsers. Client-side scripting exploits generally have limited consequences because web browsers have limited access to the host. However, an RFID-based JavaScript exploit could still compromise a machine by directing the client's browser to a page containing malicious content, like an image containing the recently discovered WMF-bug:

```
<script> document.location= ' http://
ip/malicious/code.wmf '; </script>.
```

Server-side scripting, on the other hand, has obvious far-reaching consequences; it can execute payloads with the web server's permissions. Server-Side Includes (SSIs) can execute system commands like:

```
<!--execcmd="rm"—R/">.
```

These scripting-language payloads are activated when they are viewed by a web client (i.e. the WWW Management Interface).

Rieback (2008) summarized all of these attacks such as code insertion, SQL insertion, buffer overflow, Quines and Command attacks in her PhD thesis. Since none of the attacks exclusively target middleware, we will introduce other attacks, such as "worms" and "viruses" in the next section.

Rieback's research group created a modular test platform to imitate RFID middleware architecture, and showed that this platform can be used to attack multiple databases. Their RFID Reader Interface consisted of a Philips MIFARE/I.Code RFID reader running on Windows XP. The RFID Reader Interface communicated with both ISO-15693 compatible Philips I.Code SLI tags and Philips MIFARE contactless smart cards. The WWW-based Management Interface ran Apache, Perl, and PHP, and the DB Gateway connected to MySQL, Postgres, Oracle and SQL Server databases. A variety of different platforms were tested. These efforts were met with moderate but not unqualified success. By experiment, it was shown that some RFID middleware components were more susceptible to attacks than others. The WWW management interface was a large source of vulnerabilities; upon script exploitation, the compromised Apache web server allowed unauthorized system commands and manipulation of the back-end.

To prevent malware in middleware, the suggested countermeasures are as follows:

Sanitize the input: Code injection attacks are easily prevented by sanitizing input data. In RFID systems, tag data should be sanitize before it is processed by middleware systems. It is much easier to only accept data that contains the standard alphanumeric characters (0-9,a-z,A-Z) than to explicitly strip off the special characters (such as "',*,..."). However, it is not always possible to eliminate all special characters. So sanitizing rules should be designed very carefully to leverage security and readability rates.

Isolate the RFID middleware server: In case the middleware server is compromised, the RFID middleware server should not automatically grant full access to the rest of the back-end infrastructure. Isolating the RFID middleware server so that access to the rest of the network will not be provided is a simple measure to further promote system integrity.

Finally, review of the source code is necessary. It is well known that a large complex software system cannot be bug free, but RFID middleware source code is less likely to contain malware if it is reviewed frequently.

There are other considerations regarding middleware.

Since middleware is the main provider of scalable security infrastructure to thwart RFID attacks, communications protocols with cryptographic algorithms are embedded, thus some credentials are stored in it. The data security of the middleware system is also a concern. Moreover, since RFID middleware includes networking devices, an adversary may take advantage of the system's limited resources and cause a denial of service in the RFID middleware. For instance, a stream of packets may be sent to the middleware, so that the network or processing capacity is swamped and subsequently denies access to regular clients.

Finally, used to authenticate the handshake between the reader and the back-end, RFID middleware is the key enabling infrastructure that leverages existing investments and security levels of robust RFID systems. The security and design of secure communication in middleware must be given the utmost consideration.

5.5 ATTACKS ON BACK-END SYSTEMS

5.5.1 Data Attack

Data are processed in back-end systems. The data sent to the back-end system can pose several security threats, including data flooding and spurious data.

Data Flooding

A tag read by an RFID reader is a typical RFID event. These RFID events are collected by the RFID middleware and sent to back-end systems. These events can be collected from several locations within an enterprise or across enterprise boundaries, as depicted in the EPC Global network architecture. By accident, if a large number of tags which are not attached to products are placed in front of a reader, by default, the reader will collect tag data in its nominal read range, and then send the data to the back-end. Although it may be an accidentally mistake, the back-end will suffer a data flooding attack.

In another situation, a large number of counterfeited readers cooperating with distrusted middleware can cause data flooding intentionally. The readers collect a lot of data from tags and send it to the back-end at the same time. While distrusted middleware can handle all of the data, the back-end cannot afford the sudden flooding of data.

If the middleware buffers too many events, for instance if the back-end is halted for awhile, or if the middleware were controlled by an adversary, and then suddenly sends all of them to the back-end, it can cause a problem. Data flooding can make the system process slowly or halt the processor altogether. It can even break down the system.

Countermeasures

For a determined back-end system, the filtration of events, which are "of interest" and which are not, will prevent a lot of data flooding, whether accidentally or deliberate. In the first situation, placing the inventory of tag rolls in a radio-shielded environment can prevent the accidental flooding of the tag reader, and if the events (reading of the unused tag) can be filtered as not of interest, the back-end can refuse to process the events thereby avoiding data flooding.

A staging area where the events would be temporarily received from the middleware can be used as an "event buffer" for the back-end. Using Staging areas, the events in a queue can be processed, analyzed and sent to the right business process by the back-end. The back-end systems will be more robust by using staging areas and thus avoid flooding.

Spurious Data

There are two kinds of spurious data, one is spurious tag data, and the other is spurious events.

Spurious tag data: We know that counterfeited tags can be made to contain spurious tag data and passive attacks from wireless links can also add spurious data to the system. For back-end systems, these data may attack the back-end computer system, or disturb the normal database. In the category of data attack, these spurious data can be treated similarly to credit card fraud where a card is duplicated and used in multiple places at the same time.

Countermeasure: The key to this problem is putting extra effort into the check system in the back-end. For sensitive uses, such as ePassports and ePay RFID systems, anti-counterfeiting technology should be employed.

Spurious Events: A tag is read whenever it comes in the read range of a reader. This read is accepted by middleware and sent to the back-end system (e.g., a product was scanned by the cashier many times by mistake or deliberately). Middleware receives the RFID event; however, from a business standpoint, the read may be spurious and inventory that is already accounted for does not need to be accounted for again.

Countermeasure: Setting up certain patterns for the events can prevent this kind of attack. No single RFID event can be treated as genuine unless it follows a certain pattern. For back-end systems, it is essential to understand the context in which the event was generated and then correlate the events for the very same tag before making a business decision of what to do with the event. For example, in some retail corporations, the reading of products cannot exceed twice, if 2 or more of the same product need be scanned, the cashier must enter a number on the personal terminal machine to confirm the number of the same product being purchased.

5.5.2 Back-End Malware

Back-end malware can be launched by RFID viruses, worms and buffer overflow attacks. As we known, data stored on RFID tags are implicitly trusted. A typical tag contains a unique ID and may also contain some user-defined data. However, unauthorized modification of tag data or counterfeited tags with malicious data can launch attacks on RFID back-end system or RFID middleware from RFID tags. According to Rieback et al.'s research, it is not only possible to launch attacks but this can be done even from low-cost tags with memory capable of storing only 127 characters. Malicious code can take the form of either a worm or a virus, and can thus spread either through back-end network connections or through the RFID system itself. Other possible attacks that can be launched by RFID viruses include buffer overflow attacks. Thus a poorly designed back-end system and suspicious tag data could lead to harmful actions.

Rieback et al. maintain a website for RFID malware, giving some simple, practical examples of how an RFID virus and worms can be written to attack an RFID system through the use of an SQL injection attack. For more detailed information, please see http://www.rfidvirus.org/.

Viruses

In 2006, the paper "Is your cat infected with a computer virus? " attracted a lot of attention to RFID viruses, and recently, with the concept of the "internet of things," new research has been carried out with an RFID tag implanted in a man's body. Recent news reported that Dr. Mark Gasson, a senior research fellow at Reading University's Cybernetic Intelligence Research Group, has managed to attract further publicity with a variant of the same pointless experiment, featuring technology more commonly used to chip domestic pets and unspecified computer malware. Gasson surgically implanted an RFID chip infected with malware into his hand. He claimed this made him the first human to become "infected with a computer virus". Although he claimed that "that doesn't present the true nature of this, frankly, non-threat." RFID viruses have become an important issue in RFID security.

An RFID virus is RFID-based malware that autonomously replicates itself in other RFID entities such as RFID tags, middleware systems and back-end systems, without requiring a network connection, using only ordinary RFID system operations. As RFID tags are the carriers, RFID viruses can spread very quickly and cause dreadful loss. With or without a payload, RFID viruses modify or disrupt the workings of the back-end system. The payloads a virus can execute depend both on the self-replication mechanism and the database that is targeted. Once the newly-infected RFID tags are read by the reader and the data are sent to the middleware and back-end, they can infect other RFID systems (assuming use of the same software system). These RFID systems then infect other RFID tags, which infect other RFID software systems, and so on.

A virus performs two basic functions: it replicates itself and, optionally, it executes a payload. To replicate itself, the RFID virus uses the database. The details of replication depend on the database that is used, but, broadly, two classes of viruses can be distinguished: one uses self-referential queries, the other uses quines.

RFID viruses can attack the databases of back-end systems. The main objectives of these attacks are the following: to enumerate the database structure, to retrieve an authorized data, to make unauthorized modifications or deletions, etc. RFID tags with data written in a scripting language can perform an attack of this kind. Storage limitations have proven not to be a problem, as it is possible to do a lot of harm with a very small number of SQL commands. For an SQL database, RFID tags could contain data for a SQL injection attack. SQL injection is a type of code insertion attack, executing SQL code in the database that was not intended. In the previous section, it was shown that SQL code insertion can attack the back-end system by injecting the command ;shutdown-. Another example is that the SQL command drop table <tablename> will delete a specified database table. This kind of virus can be easily injected into a database with fewer than 16 bits of data. For a practical SQL insertion attack on a database, the attacker first inserts database operational instructions into the middleware system by manipulating an RFID tag with malicious code, and then through the XP_CmdShell of the database, realizes the SQL injection into the back-end system. Finally, the attacker can realize malicious data duplication by using the characteristic of annotation symbol and the UPDATE sentence.

Combined with several SQL insertions, the fragmentation attack proposed by Shankarapani et al. (2009) is possible to perform. Anti-virus schemes were proposed to detect the code insertion. However, with the increase in fragmentation, detection might prove to be even more difficult. With fragmentation, functionality and size can change depending on how many fragments are used and how the fragments are created and thus malware can escape detection by anti-virus schemes. Fragmentation attacks are also infiltrate attacks like SQL insertion. The attacker can insert benign fragments, which can pass through a secure system without much trouble. Vulnerabilities can then be exploited by many of these "benign" fragments. The malware can be almost anything that the attacker can imagine. Fragmenting is used to fit the malware into RFID tags. So the first step in the fragmentation attack, after creating the malware that will be executed on the target machine, is none other than to fragment it. The easiest method of fragmenting is to cut up the malware into pieces no bigger than the size of the tag minus the size of the header, and other information required to do the injection attack. SQL queries can be used to deliver the malware fragments.

Virus Spread

A truly self-replicating RFID virus is fully self-sufficient; only one infected RFID tag is required to spread the viral attack. In Rieback's dissertation, she gave a method of creating a fully self-sufficient RFID virus needing only one infected RFID tag. Some special computer programs and commands are used, such as Viral Self-Replication, Self-Referential Commands and Quines. Multiquines are used to create polymorphic RFID viruses which can change their binary signatures every time they replicate, hindering detection by anti-virus programs.

Viruses spread as follows: The infected RFID tag is read by the RFID system. While reading the tag "data" the malicious code is unintentionally executed by the middleware and back-end database. The injection code is thus appended to the content descriptions in the database. The data management system then proceeds to write these values into the data section of newly arrived (non-infected) RFID tags. The now-infected RFID tags are then sent on their way. The newly-infected tags then infect other establishments' RFID middleware and back-end databases, if those locations happen to be running the same RFID system. These RFID systems then infect other RFID tags, which infect other RFID systems, and so on.

Worms

A worm is a program that self-propagates across a network, exploiting security flaws in widely-used services. A worm is distinguishable from a virus in that a worm does not require any user activity to propagate. Worms usually have a "payload" which performs activities ranging from deleting files, to sending information via email, to installing software patches. One of the most common payloads for a worm is to install a "backdoor" in the infected computer, which grants hackers easy access to that computer system in the future. An RFID worm propagates by exploiting security flaws in online RFID services.

Worm infection: RFID worms do not necessarily require users to do anything (like scanning RFID tags) to propagate, although they will also happily spread via RFID tags, if given the opportunity. The RFID worm infection process begins when attackers or infected machines first discover RFID middleware servers to infect over the Internet. They use network-based exploits as a "carrier mechanism" to transmit themselves onto the target. One example is an attack against EPC-global's Object Naming Service (ONS) servers, which are susceptible to several common DNS attacks. These attacks can be automated, providing the propagation mechanism for an RFID worm.

RFID worms can also propagate via RFID tags. Worm-infected RFID middleware can infect RFID tags by overwriting their data with an on-tag exploit. This exploit causes new RFID middleware servers to download and execute a malicious file from a remote location. This file would then infect the RFID middleware server in the same manner as standard computer malware, thus launching a new instance of the RFID worm.

RFID-based buffer overflows can also exhibit worm-like behavior; they can leverage custom shell code to download and execute malware from a foreign location.

Buffer Overflow

Like middleware, back-end systems also suffer from buffer overflow. As many tags have severe storage limitations, resource rich tag simulating devices could be utilized to launch a buffer overflow attack.

Buffer overflows usually arise as a consequence of the improper use of languages such as C or C++ that are not 'memory-safe': functions without bounds checking (strcpy, strlen, strcat, sprintf, gets), functions with null termination problems (strncpy, snprintf, strncat), and user-created functions with pointer bugs are notorious buffer overflow enablers.

The life of a buffer overflow begins when an attacker inputs data either directly (i.e. via user input) or indirectly (i.e. via environment variables). This input data is deliberately larger than the buffer, so it overwrites whatever else happens to come after the buffer in memory. Program control data (e.g. function return addresses) is often located in the memory areas adjacent to data buffers. When a function's return address is overwritten, the program jumps to the wrong address upon returning. The attacker can then craft the input data such that the return address points to the data that caused the overflow in the first place, thus executing this code (either existing or customized shell code).

The suggested countermeasures to back-end malware are as followed:

Disable back-end scripting languages: RFID middleware that uses HTTP can mitigate script injection by eliminating scripting support from the HTTP client. This may include turning off both client-side (i.e. Javascript, Java, VBScript, ActiveX, and Flash) and server-side languages (i.e. Server-Side Includes).

Limit database permissions and segregate users: Access control of databases should be exerted. The database connection should use the most limited rights possible. Tables should be made read-only or inaccessible, because this limits the damage caused by successful SQL injection attacks. It is also critical to disable the execution of multiple SQL statements in a single query.

Use parameter binding: It is better to use stored procedures with parameter binding. Bound parameters (using the PREPARE statement) are not treated as a value, making SQL injection attacks more difficult.

Throughout the RFID interface, SQL injection and buffer overflow attacks, and attacks on the back-end in general can be successfully implemented. This shows that the RFID interface is a valid entry point for attackers. At the very least, the RFID interface can be used to insert information into the database, unless proper verification systems are in place to ensure that only legitimate tags are trusted. The interesting part of Rieback's research was the example of the code that infected the database, thus allowing it to write the replication code of any tag scanned after infection. In a large compatible system such as an airport, a single infected tag could wreak havoc worldwide. A lot of controversy was generated when this paper was released. RFID developers were quick to make this attack improbable, but they never said impossible.

Summarizing, an RFID tag is an unsecured and un-trusted data source. So the information obtained from such devices should be analyzed until there is sufficient evidence that the data is accurate. However, this is not a new concept. As in all information systems the input data should be examined to ensure that it will not cause problems. The back-end system should have sufficient checks and guards in place in its reader to read certain sizes and to validate the data using checksum techniques. There are also well known ways of preventing these kinds of attacks, such as bounds checking, checksum techniques for data size etc., just like countermeasures to middleware malware and buffer overflow as previously.

5.5.3 Attacks on Back-End Services

Application Layer Attack

An application layer attack targets application servers by deliberately causing a fault in a server's operating system or applications, which results in the attacker gaining the ability to bypass normal access controls. The attacker takes advantage of the situation, gaining control of the application, system, or network, and can do any of the following: read, add, delete, or modify data or the operating system; introduce a viral program that uses computers and software applications to copy viruses throughout the network; introduce a sniffer program to analyze the network and gain information that can eventually be used to crash or corrupt the systems and network; abnormally terminate data applications or operating systems; disable other security controls to enable future attacks.

Countermeasure: Access control is the main issue in application layer attacks. The best way to prevent application layer attacks is to use a secure gateway.

Attacks on ONS

ONS (Object Naming Service) is a service that, given an EPC, can return a list of network-accessible service endpoints pertaining to the EPC in question. ONS does not contain actual data regarding the EPC; it contains only the network address of services that contain the actual data. This information should not be stored on the tag itself; the distributed servers in the Internet should supply the information. ONS and EPC help locate the available data regarding the particular object.

Since ONS is a subset of a Domain Name Server (DNS), all threats to a DNS also apply to ONS. There are several distinct classes of threats to the DNS, such as Packet Interception, Query Prediction Name, Chaining or Cache Poisoning, Betrayal by Trusted Server, DOS attacks and Authenticated Denial of Domain Names.

Countermeasures: All the means to protect DNS infrastructure can be used to protect ONS like DNS Authentication, query rate limiting, packet capture and so on.

Since the back-end is often the furthest point away from the RFID tag, both in a data sense and in physical distance, it may seem far away from being a target for attacking an RFID system. However, it bears pointing out that they will continue to be targets of attacks because both the database and the services system are the heart of the RFID deployment, in other words, "where the money is." After all, back-end services security is vital to RFID deployment.

5.6 ATTACKS ON WIRED LINKS

Aside from the Wireless link threats discussed in Chapter 3, there are also wired link attacks on RFID systems. Attacks on the network between readers and middleware and between middleware and back-end are not specific to RFID systems, because they mostly employ standard network technology; the connections are mostly wired, and hence less easily accessed. On the other hand the networks may be large and span several continents, in the case of international companies for example. In this case, the threat analysis is the same as for any other large-scale, wired, Internet Protocol (IP)-based network.

5.6.1 Network Attacks

RFID systems are often connected to back-end databases and networking devices on the enterprise backbone. Therefore, these devices have the same vulnerabilities as general purpose networking devices. Flaws in the operating system and inattention to network protocol can be used by malicious attackers in order to launch attacks and compromise the back-end infrastructure. Attacks can cause data loss by power interruption and hijacking of the middleware.

For instance, the TCP Replay Attack is a typical network attack. The Replay attack is a type of attack that often happens in TCP/IP networks. To perform a replay attack, the attacker must first capture a portion of sensitive traffic, and then simply replay it back to the host in an attempt to replicate the transaction. This attack may also cause a target connection to drop. Random TCP sequence numbers and encryption like SSH and IPSec can help reduce the risk of this type of attack.

5.6.2 Cryptographic Attacks

In research on the security of RFID systems, it is always assumed that a secure wired communication link between the reader and the database can be established by the middleware. Common sense would suggest using an authentication protocol with a cryptographic algorithm. Thus attacks between the reader and the middleware, the middleware and the back-end are mainly attacks on the authentication protocols. Determined attackers are employing attacks such as MIM (Man-in-the-middle attack), replay attacks (by eavesdropping on the response message from either side, then retransmitting the message to the legitimate side), and forgery attacks (the simple copying of reader information by eavesdropping) to break the employed cryptographic algorithms and reveal or manipulate sensitive information.

Countermeasures: Network protocol attacks could be countered by hardening all components that support RFID communication, using a secure gateway and secure operating systems, disabling insecure and unused network protocols and configuring the protocols used with the least possible privileges. Cryptographic attacks can be hindered through the employment of strong cryptographic algorithms following open cryptographic standards and using a key of sufficient length.

5.7 DOS ATTACKS

Denial of Service (DoS) attacks mean attacks refers to disrupting the normal service of a system. DoS attacks are a major cause for concern in RFID environments. Not only are such attacks possible, they are also very easy, un-sophisticated, and cheap to carry out.

Attack 1 (Jamming the operating communication channel): This attack is carried out by sending a stream of packets to the middleware or the back-end system so that the network's bandwidth or processing capacity is swamped and subsequently denies access to regular clients. Another possibility is to disturb the communication between readers and tags by using jamming transmitters. In this case, the attacker becomes active by emitting signals. If the communications link is thus too noisy, normal operation of the RFID system is no longer possible. Note that the attacker needs not have any knowledge of the communication protocols used.

Attack 2 (Maliciously using the "blocker tag"): This attack is realized by replying to every request during the singulation process like a blocker tag, causing the readers to always detect a collision and be unable to singulate tags. Deliberately blocked access and subsequent denial of service for RFID tags may be caused by malicious use of "blocker tags" or the RFID Guardian. Both approaches were proposed to safeguard RFID communications against unauthorized access and privacy threats.

Attack 3 (Deactivating or destroying a tag): A system can be halted if the reader is no longer able to read tags. It is possible to 'fry' tags (rendering them useless) in adverse conditions (e.g., exposure to very high temperatures, very strong electromagnetic fields, etc.); these attacks can be curbed using Faraday cages. Tags can be destroyed by having malicious readers overload them with more data than they can handle; these attacks can be curbed by using the reader revocation technique. Destruction of tags can also be done in several ways, for example by physically removing the tag's antenna or by frying the tag in a microwave oven or utilizing an RFID-Zapper. There are variety of possibilities for rendering tags unusable: mechanical or chemical treatment or highly powered electromagnetic waves are effective means of causing permanent service disruption. Very high or very low temperature or other environmental forces can lead to permanent malfunction, too. In practice, vandalism could be the cause of mechanical demolition of readers. Massive mechanical bending of tags can also render them inoperable. Removal of the antenna of a reader or tag also makes the devices inoperable. There are vendors of RFID tags that use tag antennas with predetermined breaking points. By stripping the upper layer of the RFID label, the tags can be rendered inoperable by customers to protect their privacy.

Attack 4 (Unauthorized use of LOCK commands): LOCK commands are included in several RFID standards in order to prevent unauthorized writing on RFID tags' memories. Depending on the applied standard, the lock command is applied by a predefined password and can have permanent or temporary effects. Moreover, since RFID middleware includes networking devices, an adversary may take advantage of the system's limited resources and cause a denial of service in the RFID middleware. Malicious deactivation of tags is usually done by utilizing inherent weaknesses or by built-in deactivation commands. For example, Australian researchers have found a weakness in "first-generation RFID tags" that allows a DoS attack to be performed whereupon the tags grant the researchers access to their memory.

DoS attacks are generally very hard to defend against and there are therefore no good mechanisms to thwart DoS attacks all together. But some countermeasures can be used to lessen the attacks.

Countermeasure 1: Facilities and transportation with radio wave blocking abilities can be used to prevent any interruption from outside.

A DoS attack can be assumed to be in progress if the number of perceived RFID tags exceeds some reasonable threshold (for example, 1,000 tags at a checkout line). Such threshold detection is simple and robust, as it does not rely on the exact behavior of the malicious blocker tag. In other words, this approach would work for either universal or selective blocker tags of a malicious kind.

More sophisticated detection mechanisms might rely on the use of prescribed tag ID ranges. For example, the reader could be connected to a database listing every valid tag in the range of identifiers associated with a particular manufacturer (corresponding, for example, to the "EPC manager" in an EPC). A tag whose identifier lies within the range but is not on the list could be identified as fraudulent. If tag identifiers are at least partially random, it will be hard for an attacker to guess a valid product identifier. This defense is also not fool proof; for example, it does not protect against spoofing valid tag identifiers that have been recorded previously by the attacker. In practice, this approach would also rely on access to manufacturer databases, which may be impractical in retail settings.

Countermeasure 2: DoS attacks are often easy to detect and the attacks can therefore often be stopped before they do too much harm. For example, crude jamming can easily be detected by passively listening to the operating frequency. Blocking of the singulation algorithm can also be detected as it signals an extreme density of tags in the interrogation zone. Altogether, countering DoS attacks implies the use of automated detection mechanisms as well as manual countermeasures such as well established control routines, etc. As with many other detection systems, a trade-off between false positives and false negatives must be made.

Another important point to notice is that equipment that may be used to perform DoS attacks may not always be unwanted. For example, an RFID-Zapper may be a valuable tool for a user wishing to destroy tags inside lawfully bought clothing or other merchandize to protect himself from tracking and thus protect his right of privacy. The problem arises when such tools are used for malicious purposes.

Denial of Service attacks and traffic analysis are severe security threats in all types of networks including wired and wireless. While theoretically these types of attacks can be countered, the scarce resources of RFID tags make their defense problematic and they remain an open research question.

CONCLUSION

RFID systems are exposed to a variety of threats. In this chapter, we have looked at some possible computer system attacks that can be made against RFID systems. In general, there are many ways

Table 1.

Entity	Attacks		Countermeasures
Tag	Data Integrity		Employ digital signature scheme
			Read only
			Employ a READ/WRITE protection system
	Illegitimate Reading		Physical measures
			Cryptographic measures
	Spoofing, Counterfeiting & Mimicking		Track & trace
			Restrict access to the "correct" information
			Use authentication and authentication protocols
Reader	Impersonating and Counterfeiting		Reject abnormal tag information
			Random transmit frequency
			Read detectors
			Use authentication protocol
Middleware	Buffer Overflows		Bounds checking
	Malware		Sanitize the input
			Isolate the RFID middleware server
			Review source code
Back-End	Data attack	Data Flooding	Filtration of the tag data
			Use staging areas
		Spurious Data	Anti-counterfeiting technology
			Refuse event in an abnormal pattern
	Malware	viruses	Disable back-end scripting languages
		worms	Limit database permissions and segregate users
			Use parameter binding
			others: check the input and review source code, etc
		Buffer Overflows	Bounds checking
	Attacks on Back End Services	Application Layer Attacks	Use a secure gateway
		Attacks on ONS	Means to protect DNS
Wired Link	Network Protocol Attacks		Secure gateway, operating systems and others
	Cryptographic Attacks		Strong cryptographic algorithms and key of sufficient length
Denial of Services(DoS)			Use facilities and transportation with radio wave blocking
			abilities to prevent any interruption from outside
			More sophisticated detection mechanisms

of attacking RFID systems. There are also many ways of defending against the attacks. The various attacks against RFID systems have also been made against individual subsystems. However, the increased cleverness of those who attack RFID systems will probably lead to blended attacks. Thus, different countermeasures can also be used at the same time. Table 1 summarizes the attacks and the various countermeasures. Moreover, fine-tuning the setup of the infrastructure can aid in making the whole system more secure. So although there is no full solution that works in all cases, combining multiple solutions will essentially yield a safe overall system.

FUTURE RESEARCH DIRECTIONS

Since RFID systems are actually computer driven systems, all conventional attacks such as viruses, worms, back-door and Trojan attacks can be mounted to attack RFID systems. Future research should be to counter the traditional attacks on computer systems which are used in RFID infrastructure. New computer system attacks on RFID systems by mounting conventional attacks should be considered. How to frustrate these attacks on practical deployments will be a big challenge to the spread of RFID systems and the expanding technology of the internet of things.

REFERENCES

Chinnappa Gounder Periaswamy, S. (2007). *Fingerprinting RFID tags*. Unpublished master dissertation, University of Arkansas, Arkansas.

Juels, A., Rivest, R. L., & Szydlo, M. (2003). *The blocker tag: Selective blocking of RFID tags for consumer privacy*. Paper presented at the 10th ACM Conference on Computer and Communications Security, CCS 2003, October 27, 2003 - October 31, 2003, Washington, DC.

Juels, A., Syverson, P., & Bailey, D. (2005). *High-power proxies for enhancing RFID privacy and utility. Paper presented at the Privacy Enhancing Technologies*. 5th International Workshop, PET 2005, Revised Selected Papers, 30 May-1 June 2005, Berlin, Germany.

Rieback, M. R. (2008). *Security and privacy of radio frequency identification*. Unpublished doctoral dissertation, Vrije Universiteit, Amsterdam.

Rieback, M. R., Crispo, B., & Tanenbaum, A. S. (2005). *RFID guardian: A battery-powered mobile device for RFID privacy management*. Paper presented at the Information Security and Privacy 10th Australasian Conference, ACISP 2005, 4-6 July, Berlin, Germany.

Rieback, M. R., Crispo, B., & Tanenbaum, A. S. (2006). *Is your cat infected with a computer virus?* Paper presented at the 4th Annual IEEE International Conference on Pervasive Computing and Communications, PerCom 2006, March 13, 2006 - March 17, 2006, Pisa, Italy.

Shankarapani, M., Sulaiman, A., & Mukkamala, S. (2009). Fragmented malware through RFID and its defenses. *Journal in Computer Virology*, *5*(3), 187–198. doi:10.1007/s11416-008-0106-0

Traub, K., et al. (2010). *The EPCglobal architecture framework*. Retrieved from http://www.gs1.org/gsmp/kc/epcglobal/architecture/architecture_1_4-framework-20101215.pdf

ADDITIONAL READING

Avoine, G. (2009). *The future security challenges in RFID*. Paper presented at the 3rd International Workshop on RFID Technology - Concepts, Applications, Challenges - IWRT 2009, In conjunction with ICEIS 2009, 6-7 May 2009, Setubal, Portugal.

Blass, E.-O., & Molva, R. (2009). *New directions in RFID security*. Paper presented at the iNetSec 2009 - Open Research Problems in Network Security: IFIP WG 11.4 International Workshop, Zurich, Switzerland, April 23-24, 2009, Revised Selected Papers.

Chen, C.-Y., Kuo, C.-P., & Chien, F.-Y. (2009). *An exploration of RFID information security and privacy*. Paper presented at the 2009 Joint Conferences on Pervasive Computing, JCPC 2009, December 3, 2009 - December 5, 2009, Tamsui, Taipei, Taiwan.

Czeskis, A., Koscher, K., Smith, J. R., & Kohno, T. (2008). *RFIDs and secret handshakes: Defending against ghost-and-leech attacks and unauthorized reads with context-aware communications*. Paper presented at the 15th ACM conference on Computer and Communications Security, CCS'08, October 27, 2008 - October 31, 2008, Alexandria, VA, United States.

D'Arco, P., Scafuro, A., & Visconti, I. (2009). *Revisiting DoS attacks and privacy in RFID-enabled networks*. Paper presented at the 5th International Workshop on Algorithmic Aspects of Wireless Sensor Networks, ALGOSENSORS 2009, July 10, 2009 - July 11, 2009, Rhodes, Greece.

Dalton, G. C. Jr, Edge, K. S., Mills, R. F., & Raines, R. A. (2010). Analysing security risks in computer and radio frequency identification (RFID) networks using attack and protection trees. *International Journal of Security and Networks*, *5*, 87–95. doi:10.1504/IJSN.2010.032207

Dimitriou, T. (2009). RFID security and privacy. In Kitsos, P., & Zhang, Y. (Eds.), *RFID security* (pp. 57–79). Springer, US.

Duc, D. N., Lee, H., Konidala, D. M., & Kim, K. (2009). *Open issues in RFID security*. Paper presented at the International Conference for Internet Technology and Secured Transactions, ICITST 2009, November 9, 2009 - November 12, 2009, London, United Kingdom.

Eagles, K., Markantonakis, K., & Mayes, K. (2007). *A comparative analysis of common threats, vulnerabilities, attacks and countermeasures within smart card and wireless sensor network node technologies*. Paper presented at the Information Security Theory and Practices,Smart Cards, Mobile and Ubiquitous Computing Systems, First IFIP TC6/WG8.8/WG 11.2 International Workshop, WISTP 2007, 9-11 May, Berlin, Germany.

Garcia-Alfaro, J., Barbeau, M., & Kranakis, E. (2008). *Security threats on EPC based RFID systems*. Paper presented at the International Conference on Information Technology: New Generations, ITNG 2008, April 7, 2008 - April 9, 2008, Las Vegas, NV, United States.

Garfinkel, S., & Rosenberg, B. (2005). *RFID: Applications, security, and privacy*. Addison-Wesley Professional.

Gounder Periaswamy, S. C., Bharath, S., Chagarlamudi, M., Estes, S., & Thompson, D. R. (2007). *Attack graphs for EPCglobal RFID*. Paper presented at the 2007 IEEE Region 5 Technical Conference, TPS, April 20, 2007 - April 22, 2007, Fayetteville, AR, United stAtes.

Guo, J.-H., Yang, H.-D., & Deng, F.-Q. (2008). Intrusion detection model for RFID system based on immune network. *Journal of Computer Applications*, *28*, 2481–2484. doi:10.3724/SP.J.1087.2008.02481

Henrici, D. (2008). *RFID security and privacy* (*Vol. 17*). Berlin, Germany: Springer.

Knospe, H., & Pohl, H. (2004). RFID security. *Information Security Technical Report*, *9*(4), 39–50. doi:10.1016/S1363-4127(05)70039-X

Laurie, A. (2007). Practical attacks against RFID. *Network Security*, *2007*, 4–7. doi:10.1016/S1353-4858(07)70080-6

Lim, T.-L., & Li, T. (2008). *Exposing an effective denial of information attack from the misuse of EPCGlobal standards in an RFID authentication scheme*. Paper presented at the 2008 IEEE 19th International Symposium on Personal, Indoor and Mobile Radio Communications, PIMRC 2008, September 15, 2008 - September 18, 2008, Poznan, Poland.

Mitrokotsa, A., Rieback, M. R., & Tanenbaum, A. S. (2008). *Classification of RFID attacks*. Paper presented at the 2nd International Workshop on RFID Technology - Concepts, Applications, Challenges - IWRT 2008, 12-13 June 2008, Madiera, Portugal.

Periaswamy, S. C. g., Bharath, S., Chagarlamudi, M., Estes, S., & Thompson, D. R. (2007). *Attack graphs for EPCglobal RFID*. Paper presented at the 2007 IEEE Region 5 Technical Conference, 20-21 April 2007, Piscataway, NJ, USA.

Qiujian, Z., & Xiaomei, W. (2009). *SQL injections through back-end of RFID system*. Paper presented at the 2009 International Symposium on Computer Network and Multimedia Technology (CNMT 2009), 18-20 Jan. 2009, Piscataway, NJ, USA.

Rong, C., & Cayirci, E. (2009). RFID security. In John, R. V. (Ed.), *Computer and information security handbook* (pp. 205–221). Boston, MA: Morgan Kaufmann. doi:10.1016/B978-0-12-374354-1.00013-3

Rotter, P. (2008). A framework for assessing RFID system security and privacy risks. *IEEE Pervasive Computing/IEEE Computer Society [and] IEEE Communications Society*, *7*, 70–77. doi:10.1109/MPRV.2008.22

Schaberreiter, T., Wieser, C., Sanchez, I., Riekki, J., & Rong, J. (2008). *An enumeration of RFID related threats*. Paper presented at the 2nd International Conference on Mobile Ubiquitous Computing, Systems, Services and Technologies, UBICOMM 2008, September 29, 2008 - October 4, 2008, Valencia, Spain.

Soppera, A., Burbridge, T., & Molnar, D. (2006). RFID security and privacy - Issues, standards, and solutions. In Steventon, A., & Wright, S. (Eds.), *Intelligent spaces* (pp. 179–198). London, UK: Springer. doi:10.1007/978-1-84628-429-8_12

Staake, T., Michahelles, F., & Fleisch, E. (2008). *The application of RFID as anti-counterfeiting technique: issues and opportunities RFID Technology and Applications* (pp. 157–168). New York, NY: Cambridge University Press.

Sulaiman, A., Mukkamala, S., & Sung, A. (2008). SQL infections through RFID. *Journal in Computer Virology*, *4*, 347–356. doi:10.1007/s11416-007-0075-8

Thornton, F., Haines, B., Das, A. M., Bhargava, H., Campbell, A., & Kleinschmidt, J. (2006). *RFID security*. Rockland, MA: Syngress Publishing, Inc.

Thornton, F., & Sanghera, P. (2007). *How to cheat at deploying and securing RFID*. Burlinton, MA: Syngress Publishing Inc.

Xiao, S., Lang, W.-M., & Hu, D.-H. (2007). Research on the security solutions of RFID. *Microcomputer Information*, *32*, 210–211.

Xu, H. (2007). *Security analyses for the passive RFID systems deployed by HJM model tree.* Paper presented at the International MultiConference of Engineers and Computer Scientists, 21-23 March 2007, Kwun Tong, China.

Yan, Q., Li, Y., Li, T., & Deng, R. (2009). *A comprehensive study for RFID malwares on mobile devices.* Paper presented at the 5th Workshop on RFID Security (RFIDsec 2009 Asia), Taipei, Taiwan.

Zhao, W., Liu, X., Li, X., Liu, D., & Zhang, S. (2009). *Research on hierarchical P2P based RFID code resolution network and its security.* Paper presented at the 4th International Conference on Frontier of Computer Science and Technology, FCST 2009, December 17, 2009 - December 19, 2009, Shanghai, China.

Zulkharnain. (2006). Reverse engineering RFID. *Academic Open Internet Journal, 19*, 48-65.

KEY TERMS AND DEFINITIONS

Computer System Attack: An attack is realization of threat, the harmful action aiming to find and exploit the vulnerability of RFID computer system.

Cryptographic Attack: A cryptographic attack is a method for circumventing the security of a cryptographic system by finding a weakness in a code, cipher, cryptographic protocol or key management scheme. This process is also called "cryptanalysis". For RFID system, the cryptographic scheme is usually lightweight or even ultra-lightweight, which is vulnerable to cryptographic attack.

EPC Global Network: The typical RFID network created by EPC Global. It is a suite of network services that enable the seamless sharing of Radio Frequency Identification (RFID)-related data throughout the supply chain comprised by a set of standards of RFID system.

RFID Buffer Overflow: An RFID buffer overflow, is an anomaly where a program for RFID system, while writing data to a buffer, overruns the buffer's boundary and overwrites adjacent memory.

RFID Malware: Malware is short for malicious software, is programming designed to threat the security of the computer system. By extension, RFID malware is malware that is transmitted and executed in an RFID system.

RFID Virus: An RFID virus is an RFID-based computer virus that autonomously self-replicates its code in RFID system, without requiring additional network connection or payload.

RFID Worm: An RFID worm is an RFID-based exploit that abuses a network connection to achieve self-replication. RFID worms may propagate by exploiting online RFID services, can also spread via RFID tags. The worm infected RFID software can then "infect" new RFID tags by overwriting their data with a copy of the RFID worm code.

ENDNOTES

1. EPCglobal Network, http://www.epcglobalinc.org
2. EPCglobal Discovery Services Standard (in development), http://www.epcglobalinc.org/standards/discovery
3. ALR-9900 Enterprise RFID Reader Family, http://www.alientechnology.com/docs/products/DSALR9900+.pdf
4. Jamming device against RFID smart tag systems, United States Patent 7221900, http://www.freepatentsonline.com/7221900.html
5. Cloud computing, http://en.wikipedia.org/wiki/Cloud computing

Section 3
Existing Solutions

Chapter 6
An Overview of Cryptography

Ehsan Vahedi
University of British Columbia, Canada

Vincent W.S. Wong
University of British Columbia, Canada

Ian F. Blake
University of British Columbia, Canada

ABSTRACT

As Radio Frequency Identification (RFID) devices become ever more ubiquitous it is very likely that demands on them to provide certain types of security such as authentication, confidentiality, and privacy and encryption for security, depending on the application, will increase. This chapter gives a brief overview of cryptographic techniques and protocols. Given the often limited complexity and power of RFID devices, much effort has been devoted to devising so-called "lightweight" cryptographic techniques for such devices, and a few of these are considered in this chapter. Even public key techniques to provide services such as identification and digital signatures have been proposed for some scenarios involving RFID devices, although such devices will obviously require significant computing power. While such applications are seemingly beyond currently available technology, given the speed at which technology is able to yield computational increases at reasonable cost and device size, it seems prudent to consider such protocols at this point.

6.1 INTRODUCTION

Radio frequency identification (RFID) devices are finding ever increasing applications, a trend that is likely to grow in the years ahead. Such devices vary in complexity from a few gates for such applications as grocery item identification, costing a few pennies, to several thousand gates for more sophisticated applications such as passports, costing a few dollars. It is likely that as the number of applications grows, the need for some form of security for many of them will become important. If past experience is a guide, it is also likely that the cost of these devices will decrease and their complexity/capabilities will increase.

Thus the problem of devising suitable cryptographic algorithms for these devices has been widely considered. Such work often goes under the name of *lightweight cryptography*, meaning the investigation of techniques to lower the number of gates required on the RFID chips to implement variations of standard algorithms while not sacrificing too much in the way of algorithm security.

DOI: 10.4018/978-1-4666-3685-9.ch006

The purpose of this chapter is to provide a very brief overview of those cryptographic techniques that seem likely to find application in such efforts. A few of the more promising results on lightweight cryptography are then discussed.

The next section gives a brief outline of those results from number theory that are essential for an understanding of cryptographic techniques, especially those of public key cryptography. The following section gives a brief outline of the standard elements of a cryptographic toolkit i.e. the set of cryptographic primitives that are available to a cryptographer to implement security in a given scenario. Virtually all of these algorithms have been incorporated into standards, a subject that will be mentioned again later. These elements include in particular such systems as stream and block ciphers as well as hash functions. A discussion of one-way functions then follows. These are the standard functions that are easy to compute but computationally infeasible to invert and include the standard functions of discrete logarithm and integer factorization. The notion of public key systems that use the one-way functions are introduced, followed by a description of certain standard cryptographic protocols that are likely to be useful in situations involving RFID devices. Of course the more sophisticated protocols will require a higher gate count and hence be applicable only for the high end applications. These last two sections are included as much for providing a standard notation as providing a framework for the next sections on lightweight cryptography. The chapter concludes with comments on the future research possibilities for these interesting devices.

Throughout the chapter extensive use has been made of two important sources. The first is the website of the National Institute of Standards and Technology (NIST), the arm of the US government that is concerned with providing security services and specifying standards for those communicating with any of its agencies. NIST has been providing an invaluable service in establishing national cryptographic standards, through their Federal Information Processing Standards (FIPS) publications, that have invariably been adopted by other countries around the world. The standards go through a rigorous process of evaluation by other government services and open discussion. As noted, virtually all of the important primitives and protocols have been included in these standards and these will be noted when discussed. In addition they are eminently readable documents providing essential reading for cryptographers.

The other important source for this work is the book (Menezes et al., 1996), a remarkable work that has provided researchers and practitioners alike an invaluable source for cryptographic theory and implementations. It will be referred to frequently as needed.

6.2 NUMBER THEORETIC PRELIMINARIES

The notion of public key cryptography relies very heavily on certain number theoretic ideas. It is beyond the scope of this chapter to present these in detail. A brief overview of the necessary facts is given, without any proofs, in the hope the reader will be able to appreciate the basic ideas behind the concepts and how they relate to the cryptographic systems described later, as well as their potential value to RFID systems. This is of course a tall task since it amounts to a review of public key cryptography. More complete treatments of the material are given in many books (e.g. Menezes et al., 1996; Hoffstein et al., 2008; Katz et al., 2008; Smart, 2003; Stinson, 1995). Much material is omitted.

Denote by $\mathbb{Z}$ the set of all integers $\{\cdots,-2,-1,0,1,\cdots\}$ and by $\mathbb{Z}_n$ the set of integers modulo the positive integer *n*,

$$\mathbb{Z}_n = \{0,1,2,\cdots,n-1\}$$

where multiplication and addition is modulo n. In the case n is composite this is a ring rather than a field since some elements lack multiplicative inverses. For two integers a, b, the *greatest common divisor (gcd)*, denoted (a, b), is the largest positive integer dividing both a and b. If for two integers a and b, $(a, b) = 1$ they are called *relatively prime*. Similarly the *least common divisor (lcd)* is the smallest positive integer that both a and b divide, often denoted by $[a, b]$. It follows easily that $[a, b] = ab / (a, b)$. The *fundamental theorem of arithmetic* states that every positive integer has a unique factorization into powers of prime numbers, up to ordering of the primes.

Denote by $\varphi(n)$ the number of integers less than n that are relatively prime to n, i.e.

$$\varphi(n) = \left|\{j \mid 1 \le j \le n,\ (j, n) = 1\}\right|.$$

This is referred to as the *Euler Totient function*. It is straightforward to establish that

$$n = \prod_{i=1}^{k} p_i^{e_i} \Leftrightarrow \varphi(n) = \prod_{i=1}^{k} p_i^{e_i - 1}(p_i - 1).$$

In the case that n is a prime p, $\varphi(p) = p - 1$, i.e. all the nonzero elements of $\mathbb{Z}_p$ are relatively prime to p. Similarly if $n = p{\cdot}q$ for two primes p and q, then $\varphi(n) = (p-1)(q-1)$.

The inverse of an element $a \in \mathbb{Z}_n$, if it exists, is an element denoted by $a^{-1} \in \mathbb{Z}_n$ such that

$$a{\cdot}a^{-1} \equiv 1 \pmod{n}.$$

It is not difficult to show that an element $a \in \mathbb{Z}_n$ has an inverse in $\mathbb{Z}_n$ if and only iff $(a, n) = 1$ i.e. iff a and n are relatively prime. Denote by $\mathbb{Z}_n^*$ the set of elements in $\mathbb{Z}_n$ that have inverses i.e.

$$\mathbb{Z}_n^* = \{j \mid 1 \le j \le n - 1,\ (j, n) = 1\}.$$

Clearly $\left|\mathbb{Z}_n^*\right| = \varphi(n)$. The set $\mathbb{Z}_n^*$ is a multiplicative group modulo n. A classic result of number theory is that this group is *cyclic*, that is, it has a single generator, say g, and is written $\mathbb{Z}_n^* = \langle g \rangle$, iff $n = 2, 4, p^k$ or $2p^k$ for some odd prime p. This implies, for instance, that $\mathbb{Z}_p^*$ is cyclic. The set of elements $\mathbb{Z}_p$ forms a *finite field* with each nonzero element having a multiplicative inverse. Finite fields are finite sets of elements with two operations, usually denoted addition and multiplication, that satisfy the axioms of a field, in particular the existence of multiplicative inverses of nonzero elements. It can be shown that such finite fields can only have a number of elements of the form p^k for some prime p and positive integer k. Furthermore it can be shown that for any prime p and integer k, there will exist a finite field with p^k elements. Such a finite field will be denoted $\mathbb{F}_q$, $q = p^k$. In the case of integers modulo a prime, perhaps the simplest type of finite field, we will write interchangeably $\mathbb{Z}_p$ or $\mathbb{F}_p$. A finite field of size p^k for a prime p and positive integer k can be generated by considering the set of polynomials of degree $< k$ where addition is coordinate-wise and multiplication is modulo a fixed irreducible polynomial of degree $< k$ over $\mathbb{F}_p$. It is not difficult to show that such polynomials always exist. Any two finite fields of the same order can be shown to be isomorphic. Hence up to isomorphism there is only one finite field of a given order.

In the case when $\mathbb{Z}_n^*$ is cyclic, if $a \in \mathbb{Z}_n^*$ is a generator of $\mathbb{Z}_n^*$ then a^j is also a generator of $\mathbb{Z}_n^*$ if $(j, \varphi(n)) = 1$. Hence the number of generators of $\mathbb{Z}_n^*$ in this case is $\varphi(\varphi(n))$. It can also be shown that when $\mathbb{Z}_n^*$ is cyclic then a is a generator if and only if $a^{\varphi(n)/p} \not\equiv 1 \pmod{n}$ for each prime p dividing $\varphi(n)$.

In algebraic objects that support a Euclidean metric one has the notions of division of elements and remainder. In $\mathbb{Z}$ for two elements $a,b \in \mathbb{Z}, b \leq a$ we can write

$$a = q \cdot b + r \ , \ 0 \leq r \leq b$$

and we refer to q as the *quotient* and r as the remainder. This is referred to as the *Euclidean algorithm (EA)*. An *extended* version of this algorithm, the *extended Euclidean algorithm (EEA)*, finds (by repeated application of the EA) the gcd of two elements $a,b \in \mathbb{Z}$ and $s,r \in \mathbb{Z}$ such that

$$d = (a,b) = r \cdot a + s \cdot b.$$

The EEA immediately provides a means of computing inverses of elements $a \in \mathbb{Z}_n^*$ for arbitrary n. Since by definition $a \in \mathbb{Z}_n^* \Leftrightarrow (a,n) = 1$, the EEA can be used to find integers r, s such that $1 = r \cdot a + s \cdot n$ and hence $a \cdot r \equiv 1 \pmod{n}$ i.e. the inverse of $a \pmod{n}$ is $a^{-1} \equiv r \pmod{n}$.

Consider a set of congruences $x \equiv a_i \pmod{n_i}$, $i = 1, 2, \cdots, k$ where the moduli n_i are relatively prime i.e. $(n_i, n_j) = 1, \ i \neq j$. Then there is a unique solution to these equations modulo $n = \prod_j n_j$ given by

$$x = a_1 N_1 M_1 + \cdots + a_k N_k M_k \ ,$$
where $N_i = n / n_i \ , \ M_i \equiv N_i^{-1} \pmod{n_i}$.

This is referred to as the *Chinese remainder theorem (CRT)*. This is a fundamental result of number theory used widely in cryptography for many purposes, including efficient implementation of the RSA public key system where RSA are the initials of the three authors of the system, Rivest, Shamir and Adleman.

For an arbitrary n and element a in $\mathbb{Z}_n^*$, the *order* of a is the least positive integer t such that $a^t \equiv 1 \pmod{n}$. A classical result of number theory is that

$$a^{\varphi(n)} \equiv 1 \pmod{n}$$

for any positive integer n and $a \in \mathbb{Z}_n^*$. The result is referred to as *Euler's theorem*. It says that the order of any element in $\mathbb{Z}_n^*$ divides $\varphi(n)$. In the particular case that n is a prime p, it reduces to

$$a^{p-1} \equiv 1 \pmod{p}$$

a result referred to as *Fermat's little theorem*.

The solutions to quadratic equations in $\mathbb{Z}_n$ over finite fields are of interest in cryptography. An element $a \in \mathbb{Z}_n$ is said to be a *quadratic residue mod* n if there exists an $x \in \mathbb{Z}_n$ such that $x^2 \equiv a \pmod{n}$. Otherwise it is called a *quadratic nonresidue*. The set of quadratic residues mod n are denoted Q_n and nonresidues $\bar{Q}_n$. In the case of $\mathbb{Z}_p^*$, if g is a generator then g^i is also a generator for i even and hence it can be shown that $|Q_p| = |\bar{Q}_p| = (p-1)/2$.

In the case of $n = pq$, the product of two primes, a situation of interest to cryptography, if $x^2 \equiv a \pmod{n}$ has a solution, then it has 4 solutions. More generally, if $n = \prod_{i=1}^{k} p_i^{e_i}$ then if $a \in Q_n$ there are 2^k solutions to $x^2 \equiv a \pmod{n}$. One can use the CRT for this problem to solve the quadratic equation modulo the primes (an easy problem) and combine the solutions with the CRT to give solutions modulo n.

A technique to work with quadratic residues is to define the *Legendre symbol* for integer a and prime p, $\left(\frac{a}{p}\right)$ as

$$\left(\frac{a}{p}\right) = \begin{cases} 0, & if \ p \mid a \\ 1, & if\, a \in Q_p \\ -1, & if\, a \in \bar{Q}_p \end{cases}$$

The properties of these symbols have been widely investigated. Particularly useful for their calculation is that $\left(\frac{ab}{p}\right) = \left(\frac{a}{p}\right)\left(\frac{b}{p}\right)$ and the fact that for $a \equiv b \pmod p$, $\left(\frac{a}{p}\right) = \left(\frac{b}{p}\right)$. Of particular interest for computing Legendre symbols is the *Law of Quadratic Reciprocity* (for which there are an inordinately large number of proofs) which says that for p and q distinct odd primes we have

$$\left(\frac{p}{q}\right) = \left(\frac{q}{p}\right)(-1)^{(p-1)(q-1)/4}.$$

Thus the computation can be done very simply by alternately computing the Legendre symbol using a sequence of steps with smaller primes.

A generalization of the Legendre symbol for the case where the modulus n is nonprime is the Jacobi symbol $\left(\frac{a}{n}\right)$ which is defined in terms of the prime factorization of n. Namely if $n = \prod_{i=1}^{k} p_i^{e_i}$ then

$$\left(\frac{a}{n}\right) = \left(\frac{a}{p_1}\right)^{e_1} \cdots \left(\frac{a}{p_k}\right)^{e_k},$$

where the symbols on the right hand side are Legendre symbols. Again, these symbols enjoy many useful properties which make their computation relatively straightforward. In particular, we have:

$$\left(\frac{ab}{n}\right) = \left(\frac{a}{n}\right)\left(\frac{b}{n}\right)$$

$$\left(\frac{a}{n}\right) = \left(\frac{b}{n}\right) \quad if \quad a \equiv b \pmod n$$

$$\left(\frac{m}{n}\right) = \left(\frac{n}{m}\right)(-1)^{(m-1)(n-1)/4}$$

Thus even when the factorization of n is unknown it is possible (and easy) to compute Jacobi symbols. Notice the fact that $\left(\frac{a}{n}\right) = 1$ does not imply that the equation $x^2 \equiv a \pmod n$ has a solution. Thus if $n = pq$, this equation will have a solution if and only if $\left(\frac{a}{p}\right) = 1$ and $\left(\frac{a}{q}\right) = 1$. The solution is then achieved via the CRT. However $\left(\frac{a}{n}\right)$ can also equal 1 if $\left(\frac{a}{p}\right) = -1$ and $\left(\frac{a}{q}\right) = -1$ but in this case the equation will have no solutions.

Following the notation in (Menezes et al., 1996), denote by J_n the set of $\{a \in \mathbb{Z}_n^* | \left(\frac{a}{n}\right) = 1\}$ for n odd. For example, for $n = pq$, recall that

$$\begin{aligned} Q_n &= \{a \in \mathbb{Z}_n^* \ | x^2 \\ &\equiv a \pmod n \ has \ a \ solution\} \\ &= \{a \in Q_p, \ a \in Q_q\} \end{aligned}$$

Thus $J_n \setminus Q_n$ is the set of integers $\{a\}$ for which the equation $x^2 \equiv a \pmod n$ has no solutions even though $\left(\frac{a}{n}\right) = 1$.

It will be clear that there is a need in cryptography to be able to generate very large prime numbers. Typically one chooses a large integer n and tries to determine if it happens to be a prime. There are a variety of deterministic tests which always give the correct answer to this question. One could use such a test and repeatedly choose large integers of the required size until meeting with success. However such tests tend to be complex and inefficient. More efficient techniques to test whether an integer is prime or not are probabilistic in nature i.e. they give an answer that has a probability of at least p of being correct - hence at most $1-p$ of being incorrect.

The Miller-Rabin primality test is one such algorithm which is in very common use (and is actually recommended in several standards as the preferred method for choosing primes). A brief description of the test is given. Suppose the integer n chosen is in fact prime and suppose $n-1=2^s r$, r odd. It can then be shown that for any integer $a \in \mathbb{Z}_n^*$ either $a^r \equiv 1 \pmod{n}$ or there is at least one value of j, $0 \leq j \leq s-1$ for which $a^{2^j r} \equiv -1 \pmod{n}$. Thus if these conditions are tested for the given integer n and neither condition is true, then n must be composite. However if the integer passes the test (i.e. $a^r \equiv 1 \pmod{n}$ and there is a value of j such that $a^{2^j r} \equiv -1 \pmod{n}$) then there is a chance that n is a prime. In fact if n is composite then at most 1/4 of the integers a, $1 \leq a \leq n-1$ will pass the given test. Thus if the test is repeated k times, by successively choosing values of $a \in \mathbb{Z}_n^*$ at random, the probability the compositeness of n will not be detected by this test is at most $(1/4)^k$. For example, if the test is repeated 30 times (as recommended in some standards) the probability the chosen integer passes all tests yet is not a prime will be at most $(1/4)^{30} \sim 10^{-18}$.

Many of the algorithms for integers, such as the EA and EEA are easily seen to extend to polynomials over a finite field where the degree of the polynomial serves in place of the magnitude of the integer. This aspect is not pursued here.

The finite fields of interest to cryptography seem limited to $\mathbb{F}_p$ and $\mathbb{F}_{2^p}$ i.e. the finite field with 2^p elements for some prime p. The reasons for this will emerge in later discussions. For an arbitrary finite field with q elements, the multiplicative subgroup $\mathbb{F}_q^*$ is in fact cyclic, i.e. there is an element $\alpha \in \mathbb{F}_q^*$ such that $\mathbb{F}_q^* = <\alpha>$. Since this element necessarily has order $q-1$, there are in fact exactly $\varphi(q-1)$ generators for $\mathbb{F}_q^*$.

6.3 A CRYPTOGRAPHIC TOOLKIT

This section introduces the various component systems that might be required to implement cryptographic protocols. These include the notions of stream and block ciphers and hash functions.

6.3.1 Stream Ciphers

A *Verman cipher*, also referred to as a *one-time pad*, is a sequence containing purely random bits. The bits are XORed to the message bits and transmitted. If the recipient has the same one-time pad, they are able to XOR them to the cipher bits to restore the message. The problem, of course, is that the one time pad must be available at the receiver in order for decryption to take place and hence must be transmitted securely to the recipient at some prior time.

A stream cipher is an attempt to emulate this system using a device such as a shift register, that generates a stream of output symbols, typically bits, that is XORed to the data. While many different types of such systems have been considered, only those that are shift register based are considered here. Their low-complexity implementations tend to make them popular for constrained devices and wireless applications. There are two distinct such systems (Menezes et al (1996)), *synchronous* and *self-synchronous*.

In a synchronous cipher, the stream generator uses only an internal initial state of a system to generate the bit stream which is then XORed to the message bits to provide the encrypted bits for transmission. This situation is shown in Figure 1. The key stream here is often taken to be an initial state on a register used in the stream generator, as discussed below. The decryptor, knowing the key k and the initial state is able to generate the same stream of bits and XOR's them to the cipher stream to restore the message bits.

This system can be implemented in a very efficient manner and has the feature that the encryptor and decryptor have to be synchronized. If the channel changes a few bits, only the corresponding message bits will be affected, and be incorrect. If an encrypted bit is dropped or an extraneous bit is inserted into the encrypted stream, the system will lose synchronism and the output bits will be incorrect from then on. It should be noted that the system can be easily generalized. For example, rather than XOR'ing the bit stream to the message stream, a more complex Boolean function could be used. The stream generator could also be implemented in a non-binary manner and the symbol stream could be converted to binary at the output.

In a self-synchronizing system the bit stream generator is produced depending only on the key k and a fixed number of previously generated encrypted bits. The general form of such a system is shown in Figure 2.

As with the synchronous system, both receiver and transmitter have to be in possession of the key k. If a single bit is changed in transmission, the stream generator at the receiver will produce erroneous bits for the length of the memory of the system. After this number of bits, the generator will again produce correct bits. Hence the reception of an incorrect bit will have limited propagation and then the system will self-synchronize. The same is true if the bits are deleted or inserted - correct reception will resume after a sufficient number of correct bits are received.

In the above systems, the stream generator is often implemented with linear feedback shift registers (LFSRs) which are now briefly discussed. Only the binary case is considered and a figure of such an LFSR is given in Figure 3. There are many ways of representing such a shift register. In the one shown, the register consists of n cells, each cell containing either a 0 or 1. The contents of the $i-$th cell are either fed into the lower XOR (a modulo 2 adder) $(a_i = 1)$ or not $(a_i = 0)$. The lower XOR adder computes the sum of the contents of those cells fed back. At the next clock cycle, the contents of the cells are shifted one cell to the right and the bit computed by the XOR adder is shifted into the left most cell. The bit that was in the right most cell is a bit of the stream generator. The system is *linear* as the XOR operation is a linear one in the field of two elements.

A few simple comments on such LFSRs are noted. The binary $n-$tuple of the contents of the

Figure 1. General form a synchronous cipher system: (a) encryption, (b) decryption

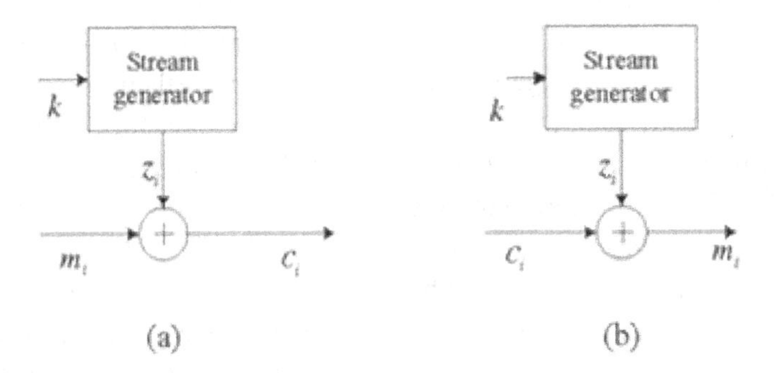

Figure 2. General form a self-synchronous cipher system: (a) encryption, (b) decryption

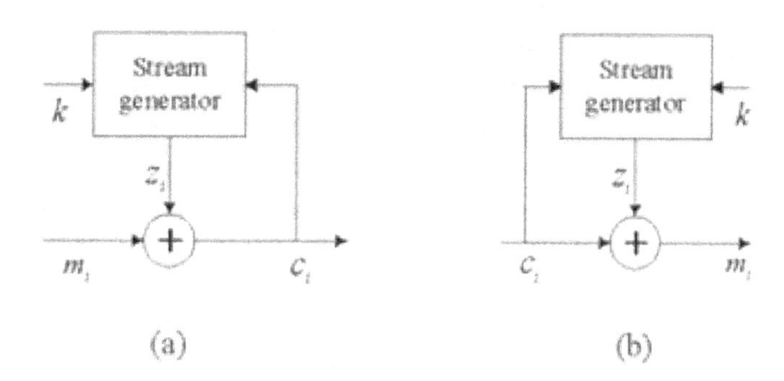

register at a given time is referred to as the *state* of the system. If the register is ever in the all-zero state, it remains in that state. The system produces a periodic output sequence since once in a given state it produces exactly the same output as the last time it was in that state. Clearly the period can be at most $2^n - 1$ since the all-zero state must be avoided. The properties of the sequence generated by this system are crucially dependent on the feedback coefficients a_i. Let the feedback polynomial be

$$f(x) = a_0 + a_1 x + a_2 x^2 + \cdots + a^{n-1} x^{n-1} + x^n.$$

Working over the field of two elements $\{0,1\}$, often referred to as the finite field $GF(2)$ or $\mathbb{F}_2$. The polynomial is said to be *irreducible* if it cannot be factored (over $\mathbb{F}_2$) into factors of degree at least one. It is said to be *primitive* if

$$f(x) \,|\, (x^{2^n - 1} - 1) \quad \text{but } f(x) \nmid (x^k - 1), \quad \text{for any } k < 2^n - 1$$

where division takes place in $\mathbb{F}_2[x]$, the set of polynomials over $\mathbb{F}_2$. We say that $a(x)$ divides $b(x)$ if there is a polynomial $g(x)$ such that $b(x) = a(x)g(x)$. It is a simple matter with available software to generate both irreducible and primitive polynomials of very high degree (far higher than would be used in practice).

Such LFSR's form the basis of many stream ciphers. It is easily shown that such a sequence has 'random-like' or *pseudo-random* properties

Figure 3. Linear feedback shift register

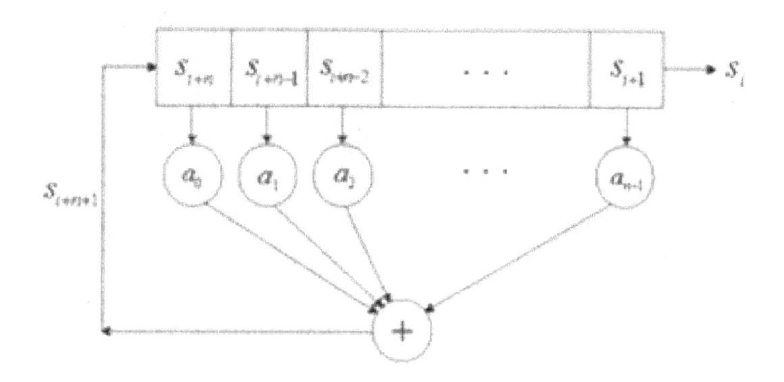

in terms of the number of subsequences of 0's and 1's. From a cryptographic point of view however, they are not very secure: knowing on the order of *2n* bits of the output of a LFSR is sufficient to determine the feedback connections and hence all future outputs. Thus some modifications are needed for their use in cipher systems. One such possibility is to replace the XOR adder of Figure 3 with a carefully chosen nonlinear Boolean function of the n inputs. This can be effective (although it can change the period of the output sequence to at most 2^n).

To discuss general binary sequences, one can define the *linear complexity* of a given sequence $\{s_i, i = 1, \cdots, n\}$ as the length of the shortest LFSR that generates the sequence. There is a computational algorithm to determine the complexity, the *Berlekamp-Massey* algorithm, (Menezes et al., 1996) which is of fundamental importance to coding theory as well.

To derive a useful stream cipher from such LFSRs one would like to generate a sequence with large period, large linear complexity and good statistical properties, in the sense that it probabilistically resembles a random sequence (and hence the generating mechanism is not easily deduced byobserving the output sequence). One method to achieve this is to use ℓ LFSR's of lengths $n_1, n_2, \cdots, n_\ell$. The overall period of such a system is then the least common multiple of all the periods $2^{n_i} - 1$ (under appropriate conditions). The final output is generated from the ℓ LFSR outputs by a suitably chosen Boolean function on ℓ variables.

There are many other techniques to use LFSR's to produce effective stream ciphers. There have also arisen many non-LFSR techniques. For example, one could use a *linear congruential generator* by choosing integers a, b and m and for a given initial "state" z_0 form the sequence

$$z_i \equiv a z_{i-1} + b \pmod{m}.$$

Such systems are commonly used as the core of random number generation for computer systems. They have been well studied in a variety of contexts. For a binary output, convert the symbol stream to its binary equivalent.

Another very popular and well regarded stream cipher system is the RC4 system (due to Rivest of MIT). While the details of the system are proprietary, a description of it is given in (Smart, 2003).

6.3.2 Block Ciphers

We define an $n-$bit block cipher as a function that maps $n-$bit blocks of plaintext, p to $n-$bit blocks of ciphertext c, with a $k-$bit key K as an input parameter such that the mapping is invertible. Encryption is denoted $e_K(p) = c$ and decryption as $d_K(c) = p$. Since the same key is used for decryption (although the decryption will differ from the encryption in certain obvious ways) such a system is often referred to as *symmetric key encryption* (as opposed to *asymmetric key encryption* to be discussed below).

A great many symmetric key encryption systems have been proposed over the years and many adopted into standards for special purposes. They are analyzed under various types of assumptions. It is always assumed the adversary knows the details of the encryption system, lacking only knowledge of the key. The minimum assumption is that the adversary is able to observe ciphertext blocks as they pass on the channel. It might further be assumed that the adversary has some number of plaintext-ciphertext pairs available. Further it might be assumed the adversary is able to obtain ciphertexts of given plaintexts of his choice. This might even be done in some kind of adaptive manner, the adversary choosing the next plaintext depending on the response from previous pairs. A variety of other assumptions can also be made for security analysis of block ciphers, as appropriate for the situation. Good discussions of these are in (Menezes et al (1996), Smart (2003)), and (Stinson (1995)).

Our treatment here will be necessarily brief. The only algorithms that will be discussed will be the *Data Encryption Standard (DES)* algorithm (now largely of historical significance) and the *Advanced Encryption System (AES)* algorithm which replaced DES. The DES algorithm remains of importance for the principles it laid down. The AES algorithm is currently the one that finds most use, being the official cipher to be used in communicating with the US government. Before that discussion, we note the four *modes of operation* in which a block cipher may be used. While originally developed in connection with DES, they are valid for any block cipher and standardized in the FIPS 81 document (FIPS 81, 1980).

The four modes of operation for block ciphers are labeled

1. Electronic codebook (ECB)
2. Cipher block chaining (CBC)
3. Cipher feedback (CFB)
4. Output feedback (OFB)

The first two modes are shown in Figure 4 (E for encryption and D for decryption). The other two modes involve more intricate block shifting and partial block operations and the reader is referred to (FIPS 81 (1980),Smart (2003), and Menezes et al (1996)) for their complete description. Each mode has certain characteristics making one more appropriate than another in certain situations. In ECB, the encrypted blocks are independent of one another and hence will not hide repeated patterns in the original data if synchronized with the block boundaries. CBC is self synchronizing in the sense that an error in a cipher block will affect the current block and the next block but not subsequent blocks. Of course, encrypting two identical data streams with the same initial vector (IV), results in the same ciphertext. The properties of the other two modes can be characterized in a similar manner.

DES was basically developed at the IBM Research Center at Yorktown Heights, NY in response to a call by the US government which was interested in developing standards for the emerging field of cryptography (an unusual initiative for the time) and called for submissions of algorithms. The original algorithm submitted by IBM was referred to as Lucifer and was chosen from among other submissions. The National Security Agency (NSA) modified the design (to strengthen it, although many in the academic community had suspicions at the time) and produced DES (FIPS 46 (1977)). Originally designed for use for five years, it proved an important and useful development and survived until the new millennium, being eventually superseded by the AES in 2001, to be discussed below.

DES operates on 64 bit data blocks with a 56 bit key (expanded through parity checks to 64 bits). Many criticized the choice of such a low key size but it was nonetheless a very effective system. It operates with 16 rounds and its general system characteristics are shown in Figure 5. The components shown in the figure are briefly discussed. The plaintext block of 64 bits first undergoes an initial permutation (IP) (all components of the system are completely specified in the standard) and the resulting block divided into left and right blocks, L_0, R_0, of 32 bits each. The following steps are repeated 16 times . At the $i-$th step, the right block R_i becomes the left block, L_{i+1}, at the next stage. For each of the 16 cycles, 48 bits of the key are chosen according to a specified schedule in the standard. These 48 key bits are fed into a function (f in Figure 5) along with the block R_i. The first step of the function f expands the block R_i to 48 bits. The resulting 48 bits are XORed with the 48 bits of the key, K_i, (a different 48 bits at each round) and fed into 8 *S-boxes*, specified in the standard. These boxes were carefully chosen to satisfy certain properties, in particular to give resistance to certain cryptanalytic techniques. These boxes have

Figure 4. Two modes of operation for a block cipher (a) electronic codebook (ECB), (b) cipher block chaining (CBC)

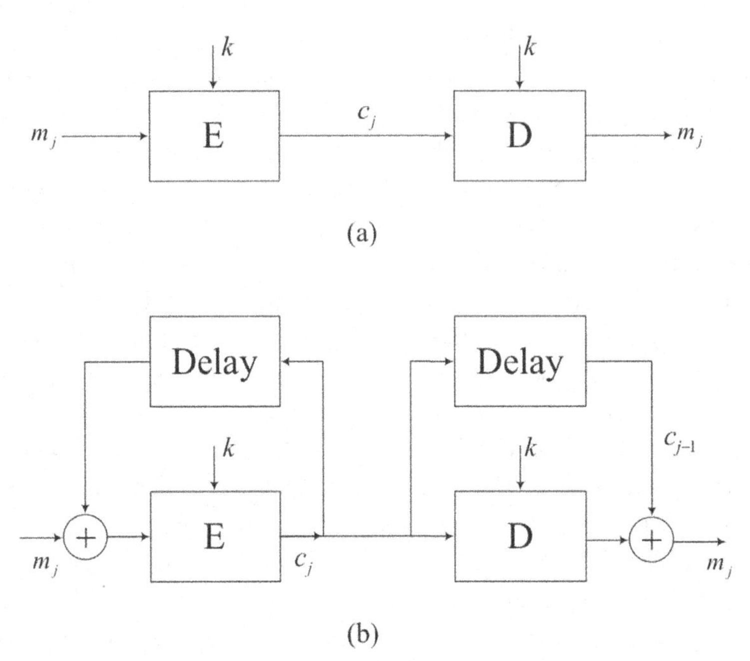

six input bits each and 4 output bits to restore the output to 32 bits. After permutation these bits are XOR'ed with the 32 bit block L_i to form R_{i+1}. After 16 rounds the output block undergoes the inverse of the initial permutation to produce the cipher block. The omitted details of this description can be found in many places such as (FIPS 46, 1977; Smart, 2003; Menezes et al., 1996).

It has been observed that the round structure of DES and the function f, through the S-boxes, reflect in a direct manner the principles of diffusion and confusion that Shannon espoused in his classic paper (Shannon, 1948), suggesting that a strong encryption scheme ought to include interleaved layers of *diffusing* the data throughout the block with layers of complex functions (confusion). It should also be noted that each of the functions mentioned in the description has an inverse, used in the decryption process, since the overall cipher has to be invertible.

As noted already, DES was in use until the early 2000's. Techniques for multiple DES encryptions, referred to as double and triple DES, expanded the key to essentially 112 and 168 bits, respectively. However, these systems proved too slow for many applications and in the fall of 1997 NIST made a public call for submissions for an algorithm to replace DES. It was to support data block sizes of length 128, 192 or 256 bits with key sizes of 128, 192 and 256 bits. Numerous submissions were made by research groups around the world and, after extensive evaluations, taking into account speed and implementation ease as well as security analysis (by research groups at NSA) the eventual winner was a cipher system referred to as Rijndael by its inventors J. Daemen and V. Rijmen of Belgium. While quite a different system than DES, it does share some features with it. It proceeds in 10, 12 or 14 rounds depending on the key size and has a kind of S-box to provide nonlinear complexity (or confusion). It has a strong mathematical flavor to it, using arithmetic in the finite field $\mathbb{F}_{2^8}$ in places. A brief overview of the algorithm is given here. The reader is referred to

Figure 5. Basic structure of DES

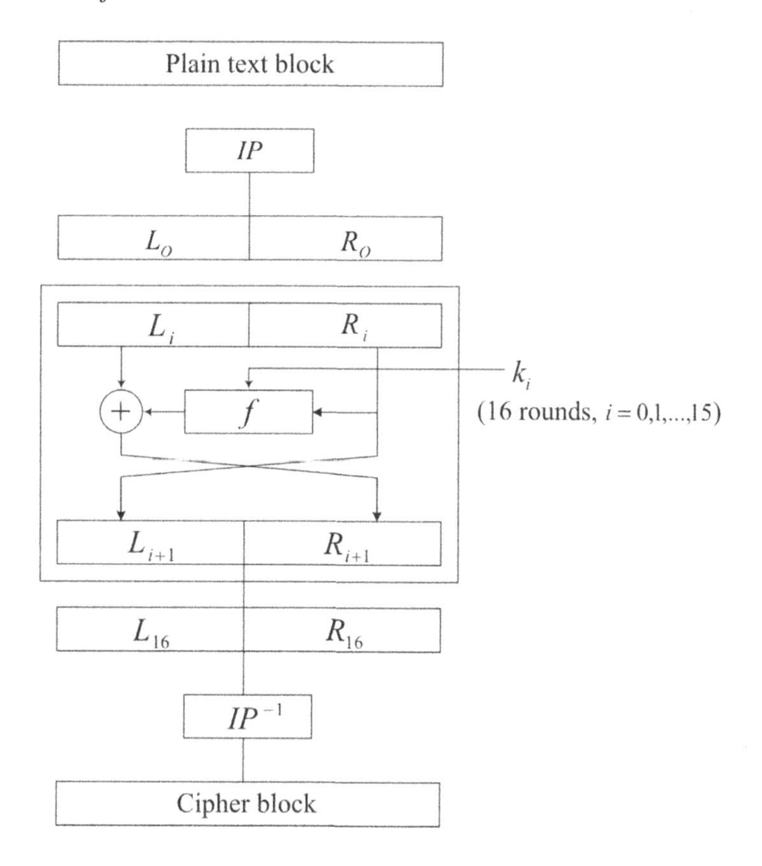

the NIST standard document for details (FIPS 197, 2001).

The cipher maintains a *state matrix*, a 4×4 array of bytes set initially to the input (assuming a 128 input block - or 16 bytes). Similarly each key round, which uses 128 bits obtained from the key in a specified manner, is represented as a 4×4 matrix of bytes. The four basic operations in AES are as follows:

1. **AddRoundKey:** As noted, in each round 128 bits is obtained from the encryption key, interpreted as a 4×4 matrix of bytes and XOR'ed to the state matrix to update that matrix.
2. **SubBytes:** There is a single S-box (different ones for encryption and decryption – but in each case only one) which is used as a type of substitution box for each byte of the state matrix. This mapping of bytes is a mathematical operation over $\mathbb{F}_{2^8}$.
3. **ShiftRows:** The rows of the state matrix are shifted, the i – th row being shifted i places to the left, $i = 0, 1, 2, 3$.
4. **MixColumns:** Each column is operated on by an invertible operation in $\mathbb{F}_{2^8}$.

The principles of confusion and diffusion are evident in this description, similar to, but quite different in form, from DES.

It should be emphasized that while DES was an important cipher (and still remains in numerous legacy systems, especially its stronger version 3DES) and AES is almost certainly the most important system for symmetric key ciphers in use today, there are many other ciphers that have found use in a variety of systems. These include such systems as RC5, RC6 (block ciphers from

RSA Inc. - only RC4 is a stream cipher), FEAL, IDEA, SAFER etc.

6.3.3 Hash Functions

Hash functions are functions with certain properties that map a binary string of arbitrary length to a binary string of fixed length i.e.

$$\begin{aligned} h : \{0,1\}^* &\rightarrow \{0,1\}^n \\ x &\mapsto h(x) \quad . \end{aligned}$$

It is usually the case that the input string $x \in \{0,1\}^*$ has a length, denoted by $|x|$, that is very much greater than the output fixed length, n. Such functions play a critical role in cryptography although certain properties are required of them as discussed below. Typically the set of such functions can be either *keyed* or *unkeyed*, depending on the intended use. A keyed hash function will be denoted $h_k(\cdot)$ for a given key $k \in K$ for a key space K.

A hash function (unkeyed) can be used to form a *manipulation detection (or digest) code (MDC)*. In this case, the hash value $h(m)$ for a given message $m \in \{0,1\}^*$, is an $n-$bit 'representative' or 'digest' of the message. From the properties of hash functions noted below it is unlikely (*or computationally infeasible*) for an adversary to produce another $m' \neq m$ such that $h(m) = h(m')$ and thus one has some confidence the message has not been tampered with.

A keyed hash function is one parameterized by a key space K and can provide a *message authentication code (MAC)*. In this case, a recipient of a pair $(m, h_k(m))$ who knows the key k has confidence that the creator of the pair knows the key and the message m is thus authentic and has not been tampered with in transmission. These notions will be elaborated on later.

The subject of hash functions is a large one and we content ourselves here with a very brief discussion of their properties and construction. Consider the hash function $h(\cdot)$.

1. h is said to be *preimage resistant* if it is computationally infeasible to find any x such that $y = h(x)$ given essentially any output y.
2. h is said to have *second preimage resistance* if it is computationally infeasible to find a value x', for a given x, such that $h(x) = h(x')$.
3. h is said to have *collision resistance* if it is computationally infeasible to find any two values of inputs x, x' such that $h(x) = h(x')$.

The hash function is often referred to as a *one-way hash function (OWHF)* if it has properties 1 and 2 above. It is referred to as a *collision resistant hash function (CRHF)* if, in addition, it has property 3.

The three properties listed have relationships among them, some subtle. For example it is not difficult to see that if a hash function is collision resistant it is also 2nd preimage resistant.

Notice that if, in a given application, the domain of the hash function is of size $2^\ell, \ell \gg n$, the hash function will, on average, have $2^{(\ell-n)}$ inputs, a very large number typically, which map to the same output $n-$tuple. However, the pre-image resistance property of the hash function requires it be computationally infeasible to find any of these inputs.

The construction of efficient hash functions is a difficult problem and the subject of much research. Numerous functions proposed that appeared secure were subsequently found to be insecure.

Many of the hash functions in current use are *iterated hash functions* where a compression function is applied in an iterative manner operating on fixed size inputs. Suppose the compression function f operates iteratively on input blocks of size $n+m$ to produce an output of size n which in most applications will be at least 160 and, more recently 256, 384, or 512. It is noted that there is a so-called *birthday attack* or *square root* attack on hash functions which has a complexity of $O(n^{1/2})$ which will yield a preimage in this number of computations. The term "birthday attack" derives from the birthday paradox where it can be computed that it is more likely that two people among n have the same birthday than seems evident at first glance. With current technology the number of computations required to break a system should be on the order of at least 2^{80}, although this number should be used with caution. Accepting this number implies the hash function should have an output size of at least 160 bits.

To describe a typical iterated hash function, divide the input x into blocks of size m, padding the last block as necessary. Assume s blocks result. Often a final block is added using the binary expansion of the number of blocks. The hash function is then computed (Preneel, 2010) as:

$$\begin{aligned} H_0 &= IV\ (fixed\ initial\ vector) \\ H_i &= f(x_i \mid\mid H_{i-1}), \quad i = 1,2,\cdots,s \\ h(x) &= g(H_s) \end{aligned}$$

The compression function $f(\cdot)$ compresses the $m+n$ input bits to n output bits at each iteration. The final output function $g(\cdot)$ is a safeguard against an attacker being able to compute $h(\ x \mid\mid y\ )$ from $h(x)$ and y without knowing x (although this function is sometimes missing from many current hash functions) (Preneel, 2010).

An influential hash function construction of the above type, on which many later functions are based, is the Merkle (1990) hash function whose description is omitted here.

There are a few hash functions whose security is based on the assumed difficulty of a mathematical problem (e.g. Charles., et al 2006). However so far these have been too slow for use in practice.

As noted earlier, the US government, through its agencies, in particular NIST, has led the way in developing standards which are to be used for certain levels of communications with all its branches. In particular, the hash functions commonly referred to as SHA-1 (FIPS 180-1, 1995), which produces a 160 bit output, was announced in 1995. It uses 4 rounds of 29 steps each. It was superseded in 2003 by a family of hash functions, collectively referred to as SHA-2, which produces outputs of size 256, 384 and 512 bits. SHA-256 uses 64 rounds of single steps while SHA-512 uses 80 rounds of single steps. NIST is currently in the final stages of determining a new standard SHA-3, chosen from open submissions from around the world through a process of workshops. The European Union also has an active interest in developing standards for a variety of cryptographic primitives.

6.4 ONE-WAY PROBLEMS FOR CRYPTOGRAPHY

The first notion of an *asymmetric* cryptographic system arose in the context of key distribution in the classic paper of Diffie and Hellman (1976) as follows. They sought a function that had an asymmetry in terms of computing the function and inverting it i.e. easy to compute but difficult to invert. The candidate function they proposed was the exponential function. Specifically, consider a prime p and consider the set of nonzero elements $\mathbb{F}_p^* = \{1,2,\cdots,p-1\}$. This set forms a cyclic group under multiplication and suppose α is a generator of the group i.e.

$$\mathbb{F}_p^* = \{1,2,\cdots,p-1\} = \{1,\alpha,\alpha^2,\cdots,\alpha^{p-2}\}$$

For a given integer j, $1 \le j \le p-2$, it is easy to compute α^j. By considering the binary expansion of j the complexity of computing α^j is at most $O(\log_2(j))$ squarings and multiplications. On the other hand, given $g = \alpha^j \in \mathbb{F}_p^*$ it appeared a fairly difficult problem to determine j. Diffie and Hellman argued that if users A and B each chose a random integer, a and b respectively, keeping them secret, A could compute $g_A = \alpha^a \bmod p$ (and B $g_B = \alpha^b \bmod p$), assuming the element α and prime p are public knowledge. If A and B exchange these quantities (g_A and g_B respectively), they are each able to compute $K = g^{ab} \bmod p$ (e.g. A knows a and receives g^b from B and can compute $(g^b)^a$), which they can use as a common key for a symmetric encryption system (such as AES). The key exchange protocol (called the Diffie-Hellman protocol) is shown in Box 1.

An adversary, knowing the public information p and g and seeing g^a and g^b transmitted on the channel, must determine g^{ab}. This is referred to as the *Diffie-Hellman problem (DHP)*. The determination of a knowing g^a, g, p, is referred to as the *discrete logarithm problem (DLP)*. If one can solve the DLP then one can certainly solve the DHP. It is referred to as the DLP in an obvious analogy to the ordinary logarithm functions to base g. It is not known in general if it might be possible to compute g^{ab} from g^a and g^b without taking *discrete logarithms* (DLOGs).

Rather than work in the field $\mathbb{F}_p^*$, it is preferable to work in a general cyclic group of prime order, $G =< g >$, $|G| = p$, and much of asymmetric cryptography takes place in such a setting. The prime order assures the group is cyclic, with no nontrivial subgroups etc. that might be exploited by an adversary. One attack on the DLP is the *baby-step-giant-step* (BSGS) algorithm which proceeds (in an arbitrary cyclic group) as follows. Let $m = \sqrt{p}$ and suppose the discrete logarithm of the element $\gamma = g^x$ is required, for a given generating element g of the cyclic group G. One makes a table of the pairs $(i, g^i), i = 1, 2, \cdots, m-1$ and note that γ can be written as $\gamma = g^{jm+k}$ for some positive integers j and k. To solve the DLP, do the following steps: compute $\gamma \cdot g^{-m}$ and check to see if the result is in the table of pairs computed. If not, repeat the process. If the result is in the table after ℓ steps and $g^{(x-\ell m)} = g^s$ in the table, then it is easily computed that $x = \ell m + s$. The algorithm has space and time complexity $O(\sqrt{p})$ and applies to any cyclic group. A more efficient algorithm in terms of space for such a problem is the *Pollard rho algorithm* for which we refer to (Menezes et al., 1996).

For certain groups however, there is a more efficient algorithm to solve the DLP. Consider the previous group of $\mathbb{F}_p^*$ i.e. the integers modulo p. In such a setting, one can define the notion of smoothness of elements. An integer x is said to be smooth with respect to the integer $y < x$ if all the prime divisors of x are less than y. It turns out that groups that have such a notion of smoothness of elements, admit a technique referred to as *index calculus* for solving the DLP. The technique is more involved than would be reasonable to explain in this chapter. It is sufficient to note that

Box 1.

	A		B	
A chooses a compute :	g^a	$\rightarrow$	g^a	
	g^b	$\leftarrow$	B chooses b, compute :	g^b
Common key:	g^{ab}		g^{ab}	

the DLP in groups which admit a notion of smoothness of elements will have a computational complexity (again we follow the notation of (Menezes et al (1996))) $L_{2^m}[1/3, c]$ for DLOG's in $\mathbb{F}_{2^m}^*$ for $c < 1.587$ while for logs in $\mathbb{F}_p^*$ it is $L_p[1/3, c]$ for $c = (64/9)^{1/3} = 1.923$ where

$$L_n[\alpha, c] = O(\exp((c + o(1)) (\ln n)^{\alpha}(\ln \ln n)^{1-\alpha})),\ 0 < \alpha < 1.$$

The behavior of this function is called *subexponential* (in $\ln n$), being less than exponential but more than polynomial (i.e. polynomial for $\alpha = 0$ and exponential for $\alpha = 1$).

Notice that the baby-step-giant-step algorithm has an exponential complexity in $\ln(p)$ since $\sqrt{p} = \exp(1/2\ln(p))$ and so the DLP problem in groups which admit a smoothness of elements have a substantially lower complexity.

In practice, for the DLP in $\mathbb{F}_p^*$ one would choose the prime p to be of the form $p = 2q + 1$ for some prime q. In this case, $\mathbb{F}_p^*$ has order $2q$ and hence has subgroups of order 2 or q, the largest possible.

Apart from the DLP, the other number theoretic problem most used in public key cryptography is that of integer factorization. Typically the numbers of interest are those of the form the product of two primes i.e. $n = pq$ where p and q are primes. Such numbers will be referred to as *RSA modulii* for reasons that will become apparent. Interestingly, algorithms used for integer factorization have invariably been modified to provide an algorithm for the DLP in $\mathbb{F}_p^*$, although the authors know of no proof that this must be so. The best general algorithm for factorizing is the *general number field sieve (GNFS)*. We refer to (Cranfield et al (2001)) for a description of this algorithm. The complexity of this algorithm is $L_n[1/3, (64/9)^{1/3}]$ as for the DLP for primes. While other number theoretic problems are of interest for certain problems of cryptography, much of public key cryptography is concerned only with RSA modulii.

Another problem that is sometimes useful is the square root problem which has some surprising features. The equation

$$x^2 \equiv a \pmod{p}$$

for an odd prime p will have solutions if and only iff $\left(\frac{a}{p}\right) = 1$. In this case it is relatively easy to find solutions. For example (Menezes et al., 1996), if $p \equiv 3 \pmod{4}$ then

$$s \equiv a^{(p+1)/4} \pmod{p}$$

is a solution. For any prime there is an efficient random algorithm to solve the equation, when it has solutions. Thus taking square roots modulo a prime is an easy problem.

On the other hand, consider n an RSA modulus i.e. $n = pq$ for two primes p, q. Then it is easy to see that finding solutions for $x^2 \equiv a \pmod{n}$ is equivalent to factoring n, which is assumed to be a hard problem. Suppose an algorithm is available to give a square root of a, $\pmod{n}$, say y. Then

$$x^2 \equiv y^2 \pmod{n} \quad \text{or}$$
$$(x + y)(x - y) \equiv 0 \pmod{n}$$

and hence one of $(x + y, n)$ and $(x - y, n)$ will be nontrivial and give a factor of n. Thus factoring an RSA modulus is probabilistically equivalent to factoring.

Finally, additive subgroups of elliptic curves are mentioned. It will turn out that no concept of smoothness in such groups has been found and hence the best known algorithm to solve a DLOG problem in them will be equivalent to the BSGS - and hence of square root complexity in the size of the group. The cases for finite fields of odd characteristic are slightly different than for characteristic 2 and we discuss only the characteristic 2 case here and (Blake et al., 1999) is a convenient reference for this material. Only the briefest of treatments is given which we hope will yield an understanding of why such curves are of interest in cryptography.

Consider the equation (over $\mathbb{F}_{2^n}$) :

$$y^2 + xy = x^3 + ax^2 + b,\ a \in \{0, \gamma\},\ b \in \mathbb{F}_{2^n}^*,$$

where γ is an element of $\mathbb{F}_{2^n}$ of trace 1 over $\mathbb{F}_2$. It can be shown that as a and b vary over their domains, a total of $2(2^n - 1)$ nonisomorphic curves are obtained, accounting for all such curves. The set of solutions of this equation, for fixed a and b,

$$E_{a,b}(\mathbb{F}_{2^n}) = \{(x, y) \,|\, y^2 + xy = x^3 + ax^2 + b\}$$

is an elliptic curve. There is a natural addition of points on this curve: suppose $P_1 = (x_1, y_1)$ and $P_2 = (x_2, y_2)$ are two points on the curve. Then their sum $P_3 = P_1 + P_2 = (x_3, y_3)$ is given by

$$\begin{aligned} x_3 &= \lambda^2 + \lambda + a + x_1 + x_2 \\ y_3 &= (x_1 + x_3)\lambda + x_3 + y_1 \end{aligned}$$

where

$$\lambda = \begin{cases} \dfrac{y_1 + y_2}{x_1 + x_2} & x_1 \neq x_2 \\ \dfrac{x_1^2 + y_1}{x_1} & x_1 = x_2 \end{cases}$$

These equations can be described as follows: to add two points P_1 and P_2 on the curve, one draws a straight line through them. The line, from the form of the curve equation, must intersect the curve in a unique third point. This is equivalent to the statement that a cubic (the right hand side of the defining equation) with two given roots has a unique third root. Reflecting this intersection point about the x-axis yields the point P_3. It should be mentioned that the additive group identity element is the *point at infinity*, $\mathcal{O}$, a point regarded as being at infinity either far up or far down the y – axis - not obtainable from the form of the equation given earlier.

Thus the points on the elliptic curve form an Abelian (commutative) group. Determining the order of this group usually used for cryptographic purposes, say of order 2^{160} to 2^{400}, was considered somewhat of a challenge and the Schoof algorithm (Blake et al., 1999) was often used. Improvements to point counting algorithms have been steady over the past decade to the point where it is regarded as a problem of modest complexity for curves of order well beyond those used for cryptography.

As noted, while the previous versions of the DLOG problem have been in multiplicative groups, the elliptic curve DLOG problem is in an additive group and can be stated as follows: Given a point $P \in E_{a,b}[\mathbb{F}_{2^n}]$ and a point $aP = P + P + \cdots + P$ a times determine a.

For such a problem, elliptic curves with large order (for a given finite field - assumed of characteristic 2 although the theory for odd characteristic is very similar) are needed. It can be shown that elliptic curves over $\mathbb{F}_{2^n}$ have an order of the form:

$$|E_{a,b}[\mathbb{F}_{2^n}]| q + 1 - t,\ |t| < 2\sqrt{q}\ ,\ q = 2^n.$$

It is also possible to show that the order of the curve (characteristic 2) is either of the form 0 or $2 \pmod 4$ depending on whether a has trace 0 or 1 (over $\mathbb{F}_2$). Thus it is not possible to obtain prime order curves in this case. However, somewhat surprisingly, it was always possible to obtain curves of order twice or four times a prime respectively.

Why are elliptic curves of interest? It turns out that a notion of smoothness for the points of an elliptic curve has not to this date been formulated. Hence there is no known index calculus approach to solving the DLP on elliptic curves. Thus the best algorithms to date to solve this DLOG problem is essentially equivalent to the BSGS algorithm of complexity of the square root of the curve order, $O(2^{n/2})$. To see the implications of this, we compare the complexity of factoring an integer (RSA modulus) with N bits to finding a DLOG on an elliptic curve over a finite field of order 2^n, or n bits, which we also take to be the order of the curve. Ignoring constants in the order notation, which turn out to change things very little, the comparison is between (Blake et al., 1999)

$$C_1 = 2^{n/2} \quad \text{and}$$
$$C_2 = \exp(cN^{1/3}(\log(N \log 2))^{2/3})$$

Equating these two expressions (equating the complexity of the two problems) an expression is obtained that gives the relationship between the number of bits needed for the order of the elliptic curve group compared to the number of bits in the RSA modulus. The result is (Blake et al (1999))

$$n = \beta N^{1/3}(\log(N \log(2))^{2/3}, \ \beta \approx 4.91$$

The results of the previous section on one-way functions such as integer factorization and DLOG problems, both in finite fields and elliptic curves, have set the stage for their application to the creation of *asymmetric* or public key encryption systems, which are introduced in this section.

6.4.1 RSA Encryption

Recall Euler's theorem which states that for a positive integer n we have:

$$a^{\varphi(n)} \equiv 1 \pmod{n}, \quad a \in \mathbb{Z}_n^*.$$

RSA encryption uses modulii n of the form pq for two odd primes p and q. Suppose an integer e is chosen in $\mathbb{Z}_n^*$ so that $(e, \varphi(n)) = 1$. Using the EEA an integer d can be found so that $e{\cdot}d \equiv 1 (\bmod\, \varphi(n))$. It follows that for any integer $a \in \mathbb{Z}_n^*$ we have

$$a^{ed} \equiv a^{k\varphi(n)+1} \equiv a \pmod{n}.$$

While the equation is true for any integer n our interest will only be for RSA modulii, $n = pq$. In all this work we have equated bit strings with integers without mention i.e. the message bit string m will be equated with the integer m via a base two expansion.

With the above remarks it is clear that if a user A declares the integer pair n, e as their *public* parameters and holds the corresponding e as their *private* or *secret* parameter, as well as the facorization of n, another user B wishing to send a message m can encrypt the message as

$$c \equiv m^e \pmod{n}.$$

Knowing d, user A is then able to compute

$$c^d \equiv (m^e)^d \equiv m^{ed} \equiv m \pmod{n}$$

and recover the message.

Several comments should be made on this *public key system.*

If an adversary is able to compute d from the pair (n, e) the system could be broken as follows.

Since

$$ed \equiv 1 \pmod{\varphi(n)}$$

or

$$ed-1 \equiv k\varphi(n) \pmod{n},$$

suppose

$$ed-1 = 2^s t$$

and recall that

$$a^{ed-1} \equiv a^{2^s t} \equiv 1 \pmod{n}.$$

It can be shown that $a^{2^{s-1}t} \not\equiv \pm 1 \pmod{n}$ for at least half of the elements of $\mathbb{Z}_n^*$ (note that squaring $a^{2^{s-1}t} \pmod{n}$ gives $1 \pmod{n}$). Thus for any such element $a \in \mathbb{Z}_n^*$, the gcd $(a^{2^{s-1}t}-1, n)$ gives a nontrival factor of n. Thus if one is able to find d one is able to probabilistically factor n. The converse is of course trivial: if one is able to factor $n = pq$ one is able to determine $\varphi(n) = (p-1)(q-1)$ and hence find d.

It has been noted in the literature that one should choose the RSA modulus with some care and the following criteria have been suggested: the prime p should be such that (a) $p-1$ has a large prime factor r (for example one might choose $p = 2r+1$), (b) $p+1$ should have a large prime factor and (c) $r-1$ should also have a large prime factor. Such a prime is called (e.g., Menezes et al., 1996) a *strong prime*. The reasons for the conditions is to lessen the likelihood of certain attacks being successful. In general there seems no proof that such primes make the RSA system more secure. However, intuitively they seem like reasonable precautions and cost little in terms of computational cost.

There are a variety of other considerations regarding the use of the RSA system and possible attacks for certain types of parameter choices (see references such as Menezes et al., 1996; Stinson, 1995; Smart, 2003; Katz et al., 2008). These are generally easy to avoid and the system has been very successful in practice and continues to be in wide use.

It was previously noted that taking square roots modulo n was computationally equivalent to factoring n. Rabin used this observation to formulate the public key system whose security is formally equivalent to factoring n, as follows. The public key is an RSA modulus $n = pq$ and the secret key is the factors p and q. For user B to send message m to user A with public key $n_A = p_A q_A$, $x \equiv m^2 \pmod{n}$ is sent. To decrypt, user A takes square roots modulo p_A and q_A respectively and combines them using the CRT. This yields 4 square roots, one of which is the correct message m. The security of this system is computationally equivalent to factoring.

6.4.2 El Gamal Encryption

A technique to use the DLP as the basis for a public key encryption system was first realized by El Gamal (1985). In its general form let G be a multiplicative group of order n (usually taken to be a prime), with generator g i.e. $G = \ < g >$. The public key is the pair g, g^a for some randomly chosen integer a. The private key is the integer a. A user sending message m, represented as an element in G, to the receiver with this public key, chooses a random integer k and computes $u = g^k$ in G as well as $v = m \cdot (g^a)^k$ and sends the encryption $c = (u, v)$. The recipient, knowing the private key a, computes $w = u^a = g^{ak}$ in G and $v \cdot w^{-1} = m$.

This version of the encryption takes place in any finite cyclic group and hence the elliptic curve version is actually covered by this description, although in additive notation. For completeness it is repeated in the next subsection.

6.4.3 Elliptic Curve Encryption

For the elliptic curve $E(\mathbb{F}_q)$ over the finite field $\mathbb{F}_q$, define the set of points of order ℓ as

$$E_\ell[\mathbb{F}_q] = \{P \in E(\mathbb{F}_q) \mid \ell P = \mathcal{O}\},$$

where $\mathcal{O}$ is the point at infinity. As usual, ℓ is normally taken to be a large prime. The set $E_\ell(\mathbb{F}_q)$ is called the set of *torsion points of order* ℓ.

El Gamal encryption with elliptic curves follows precisely the same recipe as above. User Alice (A) chooses $P_A \in E_\ell(\mathbb{F}_q)$ and a randomly chosen integer a and forms $Q_A = a \cdot P_A$. With the elliptic curve $E(\mathbb{F}_q)$ understood, Alice's public key is the pair of points (P_A, Q_A) and secret key is the integer a. The security of the system depends on the intractability of the DLP problem, finding a from P_A and Q_A. For user B, Bob, to encrypt a message M (assumed point on the curve), he chooses a random integer k and forms the points $S = k \cdot P_A$ and $R = M + k \cdot Q_A$ and transmits the pair of points (R, S) to Alice. From S Alice is able to form $a \cdot S = a \cdot k \cdot P_A = k \cdot Q_A$ (without knowing k) and hence retrieve the message by forming $R - k \cdot Q_A = M$.

6.5 CRYPTOGRAPHIC PROTOCOLS

The notions of one way functions and public key systems discussed in the previous sections are used here as tools to devise useful cryptographic protocols i.e. sequences of steps to implement certain functions to achieve security in a public environment. For most of the protocols discussed, both the standard and elliptic curve versions will be given.

In practice many of the protocols mentioned here use the notion of a *trusted third party (TTP)* or *central authority (CA)* in a networked environment to allow participants in the network to have confidence that the keys that are being used are in fact valid. One incarnation of such a concept is that of a *public key infrastructure (PKI)* for which the CA has its own publicly known RSA modulus (for example) n_{CA} and exponent e_{CA} It then maintains a register with *certificates* for each user which binds that users modulus and encryption key to the user ID by signing the combination with its own private key. Any user in the system may then verify the validity of a particular user's public parameters with their identity. Such considerations are ignored here since they are unlikely to be of value to the RFID scenario.

6.5.1 Key Exchange

The Diffie-Hellman (DH) key exchange prototol was discussed earlier. A major problem with that protocol is the *man-in-the-middle* attack. Suppose users A and B wish to communicate and initiate the DH protocol. An eavesdropper E inserts herself between them and intercepts all messages, pretending to be B to A and A to B. All messages go through E who establishes (different) keys with A and B, subsequently encrypting and decrypting all messages to each party. A technique to avoid this attack was devised by Menezes, Qu and Vanstone, referred to as the MQV protocol (Smart, 2003). The protocol uses a CA and both long term public/private keys as well as shorter term *ephemeral* public/private keys to provide an efficient way to establish an authenticated common key. The details are omitted.

Many sophisticated key distribution algorithms have been developed for large network environments. Some of these require the use of certificates and PKIs discussed earlier, although some do not. The reader is referred to the references for a discussion since they are unlikely to be of value in the RFID environment of interest here.

6.6.2 Digital Signatures

The notion of a digital signature is central to many applications of cryptography, particlulary in the financial sector. There are many different signature systems. Consider first the RSA digital signature and suppose each user in the system has an RSA modulus (for user A, say n_A) and encryption and decryption exponents e_A and d_A, respectively. For A to sign a message m, she computes $s_A(m) \equiv m^{d_A} \pmod{n_A}$. Any recipient of $s_A(m)$ can then retrieve A's public key information n_A, e_A and verify that

$$(s_A(m)^{e_A}) \equiv (m^{d_A})^{e_A} \equiv m^{k\varphi(n_A)+1} \equiv m \pmod{n_A}.$$

In order to create such a signature, the signer would need knowledge of the decryption key, d_A, presumably a computationally infeasible task to obtain from the public information available. Thus only user A could have created the signature.

Such signatures have legal status in many countries in the sense that one can take a message and signature (from a legally recognized PKI) to court to verify that it was signed by a member of the PKI network - and not by anyone else.

A few comments on the signature scheme are in order. If $m < n_A$ (taken as integers) then the message m can be recovered from the signature and such a scheme is referred to as *signature scheme with message recovery*. If $m > n_A$ the message is recovered only modulo n_A. In many applications the signature $s_A(m)$ is appended to the message and the pair $(m, s_A(m))$ transmitted. Such a scheme is sometimes referred to as a *digital signature scheme with appendix*.

More common perhaps is the use of a hash function with signatures. In this case one uses a hash function, such as one of the SHA family, of the message to produce $h(m)$ of some fixed length. Interpreting this function as an integer one can then form a signature

$$s_A(m) \equiv (h(m))^{d_a} \pmod{n_A}.$$

In some cases one might want to pad the hash value $h(m)$ with some predetermined binary sequence as the bit length of the hash might be considerably shorter than the RSA modulus.

There are many other digital signature schemes and note is made of two of them, the El Gamal and the NIST standard Digital Signature Algorithm (DSA) signature schemes. Just as El Gamal realized how to use the DLP for a public key encryption scheme, he also used similar ideas to create a digital signature scheme. In this scheme user A generates a large prime p_A and let g be a generator of the cyclic group $\mathbb{F}_{p_A} = <g>$. A random integer a is chosen and $y = g^a \in \mathbb{F}_{p_A}^*$. The public key is then the triple (p_A, g, y) and the secret key is the integer a. To sign a message m, a random integer $k \in [1, p-2]$, $(k, p-1) = 1$ is chosen (a different integer for each signature) and let k^{-1} be its inverse modulo $p-1$ (assumed to exist). Let $r = g^k$ and $s \equiv (h(m) - ar)k^{-1} \pmod{p-1}$ (where $h(m)$ is interpreted as an integer). The signature is the pair (r, s) which is appended to the message. To verify the signature on receipt of $((r, s), m)$, the public information (p, g, y) is retrieved and the hash $h(m)$ computed. From the pair (r, s) compute

$$u \equiv y^r r^s \equiv g^{ar} g^{ks}$$
$$\equiv g^{ar+k(h(m)-ar)k^{-1}} \equiv g^{h(m)} \pmod{p}.$$

The quantity $g^{h(m)}$ can be computed independently and if equal to the above computed value, the signature is accepted.

Schnorr (Schnorr (1990)) modified this scheme in the following important way. Suppose, after choosing the prime p one chooses a second prime q, such that $q | p-1$ and operates in the cyclic subgroup of $G = <g> = \mathbb{F}_p^*$, say

$G_1 = \ < g_1 >$ where $|G_1| = q$. The point of this modification is that the security of the scheme is believed to be equivalent to taking discrete logarithms in $\mathbb{F}_p^*$ that is, although the order of the cyclic group is $q, p \gg q$, in order to use index calculus one would have to use integers in the full group $\mathbb{F}_p^*$ rather than $\mathbb{F}_q^*$. It has the advantage of more efficient computations in the smaller group while maintaining the security characteristics of the larger group. In other words, as far as is known, the most efficient method for finding discrete logarithms in the smaller group is by a square root algorithm, such as BSGS in the smaller group, as opposed to using index calculus for the larger group. While the index calculus in the large group is subexponential it is more than compensated for by the larger size of the group.

The Schnorr technique was adapted when NIST formed the DSA standard (FIPS 186, 2000). In this standard one chooses a prime number p with 512+64 ℓ bits, where ℓ is an integer between 1 and 8. A second prime q is formed of 160 bits (a security level of 2^{80} is assumed), such that $q | p-1$ and let g be a generator of the the group of order q. A secret key a is chosen and $y = g^a \pmod{p}$. The public information is then (p, q, g, y). To sign a message a random integer k is chosen in $[1, q-1]$ and $r \equiv (g^k \pmod{p}) \pmod{q}$ is computed. The integer $s \equiv k^{-1}(h(m) + ar) \pmod{q}$ is also computed. The signature, as with El Gamal, is then (r, s).

The elliptic curve version of this system, (referred to as ECDSA) is also included in (FIPS 186, 2000) and although essentially an additive version of DSA, has a few details which should be noted. Let $E(\mathbb{F}_p)$ be an elliptic curve over $\mathbb{F}_p$ and $P = (x, y)$ a point of order q, P and q public information. An integer $a \in [1, q-1]$ is chosen as the secret key and $Q = aP$ part of the public key. To sign a message m, the user chooses a random integer $k \in [1, q-1]$ and forms $kP = (x_1, y_1)$ and sets $x \equiv x_1 \pmod{q}$ (since $x_1 \in \mathbb{F}_p$). The signature for message m is then (r, s) where

$$s \equiv k^{-1}(h(m) + xr) \pmod{q}.$$

The verification of the signature is straightforward. Given the public key Q of the signer, message m, (r, s) and $h(m)$, one computes

$$u \equiv h(m)s^{-1} \pmod{q}, \quad v \equiv rs^{-1} \pmod{q}$$
$$uP + vQ = s^{-1}(h(m) + rx)P = kP = (x_2, y_2)$$

If $x_2 \equiv r \pmod{q}$ the signature is accepted as valid.

There are numerous other signature schemes in the literature, many of which do much more than simply sign a document. The above discussion should be sufficient for the present purposes.

6.5.3 Message/Entity Authentication

There are a large variety of message and entity authentication techniques available. At the simple end of the spectrum one might be interested in verifying that a message has not been altered, either maliciously or by the channel. Applications such as the Internet include a *cyclic redundancy check (CRC) code* that appends a short digest of the packet. If the packet has not been altered in transmission, the digest computed by the receiver will match the received digest. Such systems designed only to detect errors or tampering of the message are oftern referred to as *message detection codes (MDC)*.

A *message authentication code (MAC)* is similar but goes further in tying the message to a particular user. Many if not most such constructions use hash functions (to produce HMAC's) and great care is needed to give the constructions protection from misuse. For example (Menezes et al., 1996), suppose a MAC is given by

$M = h(k||m)$ where h is a hash fucntion (with suitable collision properties), k a user's key, and m a message. If one extended the message by one block, y (assuming a block based hash function) it would be possible to produce $M' = h(k \,||\, m \,||\, y)$ from M and y without knowing the key k. Similarly, appending the key at the end of a message allows another type of attack (e.g., see Menezes et al., 1996; Katz et al., 2008). One conclusion is that the key must be involved with both the beginning and end of the construction. Thus a construction (Menezes et al., 1996)

$$h_k(x) = h(k \quad p \quad x \quad k)$$

where p is a padding of the key to one block length, is of interest. A more detailed analysis of the system suggests that a more secure system (Katz et al (2008),Menezes et al (1996)) would be of the form:

$$k_k(x) = h(k \,||\, p_1 \,||\, h(k \,||\, p_2 \,||\, x))$$

where p_1 and p_2 are paddings to respect the block length of the hash function. While it suffers from two applications of the hash function it has better security properties.

Many other constructions for such keyed MACs are available.

6.5.4 Identification Protocols

In an identification protocol typically one is asked to prove they are in possession of a secret without divulging the actual secret. For example, in the Fiat-Shamir identification scheme below, one is asked to prove they know the factorization of a particular RSA modulus without leaking any information as to the factors themselves.

The idea behind this protocol is briefly discussed. It uses the notions of *zero knowledge proofs* and *interactive proofs*. Suppose user A is to prove her identity to user B. Before the process begins it is assumed the CA has chosen primes p and q and publishes the modulus $n = pq$ in some central directory, keeping the factorization secret. Each user chooses a secret s relatively prime to n (the chances of choosing one that is not, is very small indeed). User A forms the integer $v_A \equiv s_A^2 \pmod{n}$ and gives it to the CA. Note that the CA is able to take the square root of v_A since it knows the factorization of n. For any adversary however, the task of finding s_A is equivalent to factoring n. The process of A identifying herself to B then is equivalent to A being able to convince B she knows the square root of her public key, v_A, modulo n, without leaking any information as to her square root in the process. To accomplish this, the following steps are repeated a number of times, say t times (see Box 2).

User B then verifies by squaring the received value in the third step and verifying that the value is correct (knowing both $v_A \equiv s_A^2 \pmod{n}$ and r^2). The point of requiring a choice in the second step is that if another user, say C, wanted to impersonate A and knew that B would always choose $e = 1$ then in the first step they could generate a random r and send r^2 / v_A to B. If

Box 2.

$A \rightarrow B : x \equiv r^2 \pmod{n}$	A choosesarandom integer r and sends it to B
$A \leftarrow B : e \in \{0,1\}$	B chooses a random exponent , 0 or 1, sends it to A
$A \rightarrow B : y \equiv rs^e \pmod{n}$	A responds with either r or $rs \pmod{n}$

user B chooses $e = 1$ then in step 3 user C responds with $y \equiv r \pmod{n}$. User B would compute $xv_A \equiv r^2 / v_A \cdot v_A \equiv r^2 \pmod{n}$ and compare this to $y^2 \equiv r^2 \pmod{n}$ and hence achieve agreement. Thus there is a probability of 1/2 of discovering an impostor and, over t trials, a probability of at least $1 - 1 / 2^t$.

Ohta and Okamoto (Ohta et al (1988)) gave an interesting *extended* version of this protocol that is also useful.

Another interesting identification protocol that finds application in constrained environments is the GPS protocol (Girault, 1991), an RSA based system where the tag (or prover) has an RSA modulus $n = pq$ and produces a public key $v = g^{-s} \pmod{n}$ for the secret key $s \in \{0, 1, \cdots, 2^{\sigma} - 1\}$ with a public base g. To initiate the protocol the prover chooses a random integer r and computes $x \equiv H(g^r \pmod{n})$ which is transmitted to the verifier who then chooses a random integer c which is transmitted back to the prover. The prover computes $y = r + (s \cdot c) \pmod{n}$ which is transmitted back to the verifier. The verifier accepts the identity if

$$H(g^y \cdot v^c \pmod{n}) = x$$

This is a simplified version of the protocol which omits several details on the ranges of the various parameters used.

There are numerous identification protocols. One more is discussed here, the Schnorr identification protocol (Menezes et al., 1996; Schnorr, 1990), perhaps a more conventional protocol than the Fiat-Shamir protocol discussed above (and was originally designed for the more lightweight environment of smart cards). The protocol is discrete logarithm based and it supposes that p is a large prime such that the prime $q | p - 1$ and let g generate the group of order q in $\mathbb{F}_p^*$. A CA is assumed with a capability to produce signatures that bind identities to public keys of users. A user A in the network chooses a random integer a and computes $v_A = g^{-a}$. The CA creates the certificate for A, $\{I_A, v_A, s_{CA}(I_A, v_A)\}$. For user B to verify the identity of A, A first chooses a random integer k and computes g^k which is sent to B along with A's certificate. B chooses a random integer r say in some interval such as $[1, 2^t]$, for t sufficiently large. User A computes $y = k + ar \pmod{q}$ and sends it to B. User B then verifies that the received value g^k equals the computed value $g^y v_A^r$.

6.5.5 Other Protocols

There are many other interesting cryptographic protocols that are likely too complex to implement to be of interest to low complexity environments like RFID, at least in the near future. For example, the area of *secure multiparty computation* considers problems where each participant of a network has a variable to be used in a particular computation involving all variables in such a way that a result can be announced without any participant divulging any information on their variable. The "prototype" of such a computation is the so-called *greater-than* or *millionaires problem* where two participants each have a value and the object is for them to decide who has the larger variable without divulging any information as to their value (beyond that implied by the actual result). The application of such a protocol to auctions is immediate. Here, a network of participants might wish to determine the winner, without divulging the value of the winning bid or the losing bids. Similar techniques can be applied to electronic voting.

Many other interesting scenarios are also possible to consider for implementating cryptographic protocols but at this point they have little value for the RFID environment. These include such applications as electronic voting and digital cash.

6.6 LIGHTWEIGHT CRYPTOGRAPHY

Given the limited computing power of current RFID devices, it is not surprising that the number of researchers interested in devising 'lightweight' cryptographic schemes suitable for them has grown exponentially. The number of papers on this topic has grown so quickly over the past decade and is now so vast that any attempt to survey them in a comprehensive manner is futile. Rather an attempt is made here to select certain representative approaches to the problem that seem promising in the various scenarios envisioned and discuss their approaches. Thus the material is far from comprehensive, intended only to give a flavor of the field.

The following subsection discusses approaches to implementing lightweight versions of cryptographic primitives such as DES, AES and hash functions. Approaches to authenticating RFID devices are not discussed here. The number of references in the literature on this topic is particularly large, and they are addressed elsewhere in the volume. The next subsection considers how far progress has been made in reducing the complexity requirements for public key systems which would allow such functionality as signatures and authenticated key exchange to be considered. It seems somewhat ambitious to consider these for devices currently on the market but the level of attention they have received in the literature suggests they be included. Of course such systems refer exclusively to elliptic curve systems as RSA or discrete logarithm systems seem well beyond even future projections for the capabilities of RFID.

As mentioned, the computational effectiveness of RFID and their range of applications, is likely to increase significantly in the future and so a rather optimistic view for them has been adopted here in terms of the topics covered.

6.6.1 Lightweight Techniques for Block Ciphers and Hash Functions

A fundamental requirement of the application of cryptography to secure protocols for RFID is a lightweight block cipher and hash function.

Lightweight Block Ciphers

Three approaches to implementing a lightweight block cipher for RFID applications are: (i) implementing standard ciphers such as DES or AES in particularly efficient ways; (ii) modifying such ciphers to a less than complete version, for example, reducing the number of rounds or (for DES) modifying the S-boxes; (iii) devising completely new algorithms specifically for their lightweight properties. Although very efficient implementations of DES and AES have been devised, at this point they seem only suitable for high end RFID devices. Modifying these algorithms is an interesting approach but suffers somewhat from a lack of thorough security analysis. Nonetheless the approach does benefit from having a respectable pedigree and will no doubt be a useful approach to pursue. The third approach of devising completely new algorithms specifically for a lightweight application suffers from a similar comment. Examples for all three approaches will be briefly discussed here.

Of the approach of implementing a full DES or AES algorithm, one of the more impressive results is the work of (Feldhofer et al., 2004). They implemented the full AES (128 bit key and 128 bit data blocks) algorithm with an 8 bit architecture to conserve power. The result is an implementation with 3595 gate equivalents (GEs), requiring 1016 clock cycles per data block clocked at 100kHz. While still an expensive proposition in terms of utilization on an RFID device, it is showing promise for potential applications for authentication and identification protocols. In comparison, an implementation of DES (56 bit key and 64 bit data block) quoted on the NIST website gives approximately 3000 GE's and 28 clock cycles (although a 2,300 GE implementation is noted in Eisenbarth et al., 2007).

For the next approach, one might modify an existing block cipher such as DES or AES. One possibility is to use a *key prewhitening filter* at the input and output (Eisenbarth et al., 2007) which adds very little to the implementation complexity. The resulting system is referred to as DESX. While this extends the key length (from 56 to 118 bits) it does little to decrease complexity.

Another approach is to modify DES or AES in more significant ways. This is the approach that yields DESL (DES light). Here the idea is to replace the 8 S-boxes of DES with a single carefully designed S-box. The single box is designed to resist linear and differential cryptanalysis, among other attacks. It is reported in (Eisenbarth et al (2007)) that such an approach can result in an implementation of 1850 GEs.

The third approach is to design a block cipher from scratch with a view to implementation simplicity. This is the approach of the system PRESENT, described in the thesis of Poschmann (2006) and the works of (Bogdanov et al., 2007 and Rolfes et al., 2008). Specifically designed for lightweight applications, it is a substitution-permutation network with an input block length of 64 bits that supports key lengths of 80 and 128 bits. It is a rounds-based architecture (31 rounds plus a final round key round for post-whitening). To give the flavor of the algorithm without detail or description of the operations, it is described as:

```
generateRoundKeys
for i=1 to 31 do
    addRoundKey(STATE, K_i)
    sBoxLayer(STATE)
    pLayer(STATE)
end for
addRoundKey(STATE, K_32)
```

Three implementations of the algorithm are given in (Rolfes et al., 2008). The smallest of the three, a bit serialized version, requires only 1000 GE's. The security of the system is discussed in (Bogdanov et al., 2007) and in particular its resistance to linear and differential cryptanalysis and to key schedule attacks is noted.

Lightweight Hash Functions

Hash functions are crucial components for cryptographic applications, particularly for digital signatures and authentication algorithms. Of course a block cipher is easily modified to yield an iterative hash function by appropriately recursing on feedback from the output and including partial blocks at the input. As noted in the previous chapter, care must be taken in this technique.

This tends to be a computationally expensive approach for hash functions. In applications requiring both a block cipher and a hash function it might be inefficient to include both in the system. In applications requiring only a hash function more efficient approaches exist. Typically for RFID applications a hash function will not be called upon to hash large amounts of data - indeed, typically only one block of input will be required. Also, the property of collision resistance is not required since the number of devices will typically be known by the reader. One-wayness may be all that is required. In addition, the security level desired may well be less than typically assumed for signature applications, say 64 bits of security rather than 256 bits. Many hash schemes assume some type of random padding. This would introduce problems of ensuring the bits generated are truly random - not an easy task in hardware and expensive in terms of GEs.

There have been a very large number of hash functions proposed, both for the conventional and lightweight environments. The best implementations to date for conventional hash functions such as SHA-1 and SHA-256, as well as such older ones such as MD5, have (Bogdanov et al., 2008) on the order of 8,000 to 11,000 GEs, a considerable number that renders them unsuitable for RFID applications. An interesting family of hash functions, the HB family (Hopper et al., 2000), has been more successful although some attacks have recently been suggested for this family.

Shamir (Shamir (2008)) has recently proposed an intriguing hash function specifically designed for lightweight environments. Only the briefest of descriptions of this system is given. It will likely play an important role in future discussions. It is based on the Rabin encryption system where to encrypt the message m (an integer) one computes $c \equiv m^2 \left(\bmod n\right)$ where n is often taken to be an RSA modulus. It was noted previously that breaking this system is computationally equivalent to factoring the modulus n. For an authentication protocol for RFID devices, one might assume the ID S of the tag is on the order of 64 to 128 bits long. The reader, in attempting to authenticate the tag, sends a challenge R of perhaps 32 to 64 bits long. For security in the Rabin system a modulus of over 1000 bits is required. The use of a *mixing function* $M(S,R)$ is proposed (Shamir (2008)) to thoroughly mix and expand the two arguments S,R to the size of the modulus to prevent small integer attacks on the squaring system. One possibility suggested in (Shamir, 2008) is to use nonlinear shift registers in a certain manner for this mixing function and to compute $m = M(S,R)$. It is also noted there that for the purposes of modest security requirements of RFID devices only on the order of 32 to 64 bits of the resulting $m^2 \left(\bmod n\right)$ (where n is on the order of 1000 bits) need be transmitted for authentication since it is highly unlikely that these are correct if the computation is invalid. Thus only this number of bits need be computed, further reducing system complexity. Clever techniques to accomplish this are given (Shamir, 2008).

Finally the choice of modulus is discussed. It is suggested that the modulus be of the form $2^n - 1$ and particular attention is focused on the modulus $2^{1277} - 1$. It turns out that the factorization of this integer is completely unknown i.e. no factors of this integer are known - although it is not a prime. It is argued that if a reasonable mixing function is used then it is unlikely the overall system can be broken from knowledge of the transmitted digits *even if the modulus was completely factored at some point in the future!* There are some subtleties in this description and the original paper should be consulted for details. Clearly the GE requirements of this system should be very modest indeed. Doubtless this system will be developed further.

6.6.2 Lightweight Techniques for Public Key Systems

Of the three approaches to public key systems, RSA, discrete logarithms in finite fields and elliptic curve systems, only the latter has been seriously considered for most lightweight applications. Indeed, from its inception, elliptic curve systems have been promoted particularly for constrained environments such as smart cards and mobile applications. To this point they are the only type of public key systems the authors are aware of that have been considered for RFID devices.

In applying elliptic curves to RFID the approach has really been no different than their application to any other environment, namely that great effort has been made to make all aspects of the implementation of the system efficient. The algorithms developed fall into two basic categories: very efficient algorithms for finite field arithmetic, and efficient algorithms for basic elliptic curve algorithms such as point multiplication and point doubling, as discussed in the last chapter. An enormous amount of effort has been expended on both problems and it would serve little purpose to attempt to summarize the work here. The results obtained, of course, are of value to any implementation of elliptic curve systems, not just those for constrained environments.

There are now numerous papers on this topic (e.g., Batina et al., 2007; 2006; 2009; 2008; Lee et al., 2008), among many others. Typically standards call for the restriction to either prime fields or fields on the order of 2^p for p a prime. Fields of characteristic 2 tend to have advantages over prime fields due to carry free arithmetic. Restriction to the binary characteristic and fields of the order of 2^n for n composite have not been favored by the security establishment since such "composite" fields invite the application of the Weil descent attack on such curves. However it is noted in (Batina et al., 2009) that such an argument does not apply to fields on the order of $2^{2 \cdot p}$ for p prime and these are investigated there.

It is usually taken (Blake et al., 1999) that an elliptic curve system with field size 2^{163} has approximately the same security level as an RSA system with RSA modulus 1024 bits, providing approximately 80 bits security, typically the minimum one might consider for non-constrained environments. The work (Batina et al (2009)) considers also the "composite" fields $2^{2 \cdot 67}$ and $2^{2 \cdot 71}$ as interesting fields where certain efficiencies in the field operations may be exploited. A table listing the types of attacks against certain curves is also given in that work. As a result of these and other works on efficient implementations of elliptic curve cryptosystems it appears that such systems typically require on the order of 10,000 to 20,000 GEs. This is well beyond the range normally assumed of an RFID device, even a high end one, but as noted previously such work is regarded as valuable for future evolution of such devices. The overview of this subsection is very far from comprehensive and has only skimmed the surface of the very large literature on the subject, hopefully giving an indication of the approaches that are available.

CONCLUSION

This chapter has provided a brief overview of cryptography and its potential application to RFID systems through the research on lightweight cryptogaphy. While such devices may range in complexity from a few tens of gates to several thousand, it seems unsafe to predict that the number of gates in future versions will increase significantly. Thus this overview has included protocols that may be beyond the range of interest for current RFID technology but might well become of interest in the future. It is difficult not to be optimistic about the future of such devices given their ever increasing power and hence their enormous potential in an ever increasing array of applications. Thus part of the rationale here has been to include approaches that may at this point seem to have high computational requirements in the hope that future devices will rise to the occasion.

REFERENCES

Ahmed, E. G., Shaaban, E., & Hashem, M.. (2010). Lightweight mutual authentication protocol for low cost RFID tags. *International Journal of Network Security and its Applications, 1,* 27-37.

Batina, L., Guajardo, J., Kerins, T., Mentens, N., Tuyls, P., & Verbauwhede, I. (2006). *An elliptic curve processor suitable for RFID tags.* 1st Benelux Workshop on Information and System Security (WISSec 2006), Belguim.

Batina, L., Guajardo, J., Kerins, T., Mentens, N., Tuyls, P., & Verbauwhede, I. (2007). *Public-key cryptography for RFID tags*. Fifth Annual International Conference on Pervasive Computing and Communications Workshop. New York.

Batina, L., Guajardo, J., Preneel, B., Tuyls, P., & Verbauwhede, I. (2009). Public key cryptography for RFID tags and applications. In Kitsos, P., & Zhang, Y. (Eds.), *RFID security: Techniques, protocols and system-on-chip design*. Springer-Verlag.

Batina, L., Mentens, N., Sakiyama, K., Preneel, B., & Verbauwhede, I. (2007). *Public key cryptography on the top of a needle*. IEEE International Symposium on Circuits and Systems (ISCAS 2007), Special Session: Novel Cryptographic Architectures for Low-Cost RFID. New Orleans.

Batina, L., Seys, S., Preneel, B., & Verbauwhede, I. (2008). Public key primitves. In *Wireless sensor network security* (pp. 77–108). IOS Press.

Blake, I. F., Seroussi, G., & Smart, N. (1999). *Elliptic curves in cryptography*. Cambridge University Press.

Bogdanov, A., Knudsen, L. R., Leander, G., Paar, C., Poschmann, A., & Robshaw, M. J. B. … Vikkelsoe, C. (2007). PRESENT: An ultra-lighweight block cipher. In P. Paillier & I. Verbauwhede (Eds.), CHES 2007: LNCS (pp. 450-466). Santa Barbara.

Bogdanov, A., Leander, G., Paar, C., Poschmann, A., Robshaw, M., & Seurin, Y. (2008). Hash functions and RFID tags: Mind the gap. In Oswald, E., & Rohatgi, P. (Eds.), *CHES 2007, LNCS 5154* (pp. 283–299). doi:10.1007/978-3-540-85053-3_18.

Calmels, B., Canard, S., Girault, M., & Sibert, H. (2006). Low-cost cryptography for privacy in RFID systems. *Smart Card Research and Advanced Applications, LNCS, 3928*, 237–251. doi:10.1007/11733447_17.

Charles, D. X., Goren, E. Z., & Lauter, K. E. (2006). *Cryptographic hash functions from expander graphs*. 2nd NIST Hash Functions Workshop.

Cranfield, R., & Pomerance, C. (2001). *Prime numbers: A computational perspective*. Springer-Verlag.

Diffie, W., & Hellman, M. (1976). New directions in cryptography. *IEEE Transactions on Information Theory*, *22*, 644–654. doi:10.1109/TIT.1976.1055638.

Eisenbarth, T., Paar, C., Poschmann, A., Kumar, S., & Uhsadel, L. (2007). A survey of lightweight-cryptography implementations. *IEEE Design & Test of Computers*, *24*, 522–533. doi:10.1109/MDT.2007.178.

ElGamal, T. (1985). A public key cryptosystem and a signature scheme based on discrete logarithms. *IEEE Transactions on Information Theory*, *31*, 469–472. doi:10.1109/TIT.1985.1057074.

Feldhofer, M., Dominikus, S., & Wolkerstorfer, J. (2004). Strong authentication for RFID systems using the AES algorithm. In Joye, M., & Quisquater, J.-J. (Eds.), *CHES 2004, LNCS 3156* (pp. 357–370). Boston. doi:10.1007/978-3-540-28632-5_26.

FIPS. 46. (1977). Data encryption standard. Federal Information Processing Standard. NIST, Department of Commerce, Washington, D.C.

FIPS. 81. (1980). DES modes of operation. Federal Information Processing Standard, NIST, Department of Commerce, Washington, D.C.

FIPS. (1995). *Secure hash standard. Federal Information Processing Standard, NIST* (pp. 180–181). Washington, D.C.: Department of Commerce.

FIPS. 186. (2000). Digital signature algorithm. Federal Information Processing Standard, NIST, Department of Commerce, Washington, D.C.

FIPS. 197. (2001). Advanced encryption standard. Federal Information Processing Standard, NIST, Department of Commerce, Washington, D.C.

FIPS. (2003). *Secure hash standard. Federal Information Processing Standard, NIST* (pp. 180–182). Washington, D.C.: Department of Commerce.

Girault, M. (1991). Self certified public keys. In Davies, D. (Ed.), *EUROCRYPT '91, LNCS 547* (pp. 490–497). UK.

Hankerson, D., Menezes, A. J., & Vanstone, S. (2004). *Guide to elliptic curve cryptography*. Springer Verlag.

Hoffstein, J., Pipher, J., & Silverman, J. (2008). *An introduction to mathematical cryptography*. Springer Press Verlag.

Hopper, N. J., & Blum, M. (2000). *A secure human-computer authentication scheme* (Report CMU-CS-00-139).

Katz, J., & Lindell, Y. (2008). *Introduction to modern cryptography. Chapman and Hall*. CRC Press.

Lee, Y. K., Sakiyama, K., Batina, L., & Verbauwhede, I. (2008). Elliptic-curve-based security processor for RFID security. *IEEE Transactions on Computers*, *57*, 1514–1527. doi:10.1109/TC.2008.148.

McLoone, M., & Robshaw, M. J. B. (2007). Public key cryptography and RFID tags. In Abe, M. (Ed.), *CT-RSA 2007 LNCS 4377* (pp. 372–384). San Francisco.

Menezes, A. J., Van Oorschot, P. C., & Vanstone, S. A. (1996). *Handbook of applied cryptography*. CRC Press. Retrieved from http://www.cacr.math.uwaterloo.ca/hac/

Merkle, R. C. (1990). A fast software one-way hash function. *Journal of Cryptology*, *3*, 43–58. doi:10.1007/BF00203968.

Ohta, K., & Okamoto, T. (1988). Practical extension of Fiat-Shamir scheme. *Electronics Letters*, *24*, 955–956. doi:10.1049/el:19880650.

Peris-Lopez, P., Hernandez-Castro, J., Tapiador, J. M. E., & Ribagorda, A. (2006). *LMAP: A real lightweight mutual authentication protocol for low-cost RFID tags*. Workshop on RFID Security (RFIDSec'06), Austria.

Poschmann, A. (2009). *Lightweight cryptography: Cryptographic engineering for a pervasive world*. Dissertation, Doktor-Ingenieur, Ruhr-University Bochum.

Poschmann, A., Leander, G., Schramm, K., & Paar, C. (2006). *New lightweight crypto algorithms for RFID*. Workshop on RFID Security (RFIDSec'06), Austria.

Preneel, B. (2010). The first 30 years of cryptographic hash functions and the NIST SHA-3 competition. In Pieprzyk, J. (Ed.), *CT-RSA 2010, LNCS 5985* (pp. 1–14). doi:10.1007/978-3-642-11925-5_1.

Rolfes, C., Poschmann, A., Leander, G., & Paar, C. (2008). Ultra-lightweight implementations for smart devices - Security for 1000 gate equivalents. In Grimaud, G., & Standaert, F. X. (Eds.), *CARDIS 2008 LNCS 5189* (pp. 89–103). doi:10.1007/978-3-540-85893-5_7.

Schnorr, C. (1990). Efficient identification and signatures for smart cards. *Advances in Cryptology, Crypto '89. LNCS*, *435*, 239–252.

Shamir, A. (2008). SQUASH - A new MAC with provable security properties for highly constrained devices such as RFID tags. In Nyberg, K. (Ed.), *FSE 2008 LNCS 5086* (pp. 144–157). doi:10.1007/978-3-540-71039-4_9.

Shannon, C. E. (1948). Communication theory of secrecy. *The Bell System Technical Journal*, *28*, 656–715.

Smart, N. (2003). *Cryptography: An introduction*. McGraw-Hill.

Stinson, D. (1995). *Cryptography: Theory and practice*. CRC Press.

Tsudik, G. (2007). A family of dunces: Trivial RFID identification and authentication protocols. In N. Borisov & P. Golle (Eds.), PET 2007, LNCS 4776 (pp. 45-61).

Vajda, I., & Buttyan, L. (2003). *Lightweight authentication protocols for low-cost RFID tags*. 2nd Workshop on Security in Ubiquitous Computing.

Washington, L. C. (2003). *Elliptic curves: Number theory and cryptography*. Chapman and Hall, CRC Press.

Chapter 7
Identification and Authentication for RFID Systems

Behzad Malek
Ryerson University, Canada

Ali Mir
Ryerson University, Canada

ABSTRACT

In this chapter, the author briefly reviews the various attacks on existing identification and authentication schemes and describes the challenges in their design for RFID systems. The chapter categorizes the RFID identification and authentication schemes into two general categories: cryptographic and non-cryptographic solutions. Cryptographic solutions are based on symmetric or asymmetric cryptography systems. Depending on the resources available on the RFID tags, algorithms based on standard cryptography cannot be utilized in an RFID system and new cryptographic algorithms must be designed. However, there remain security challenges in protecting the RFID systems that cannot be solved solely by relying on cryptographic solutions. The chapter also reviews these challenges and looks at the countermeasures based on non-cryptographic solutions that would further protect RFID systems.

BACKGROUND

Advancements in technology have enabled mass production of cheap, miniaturized RFID transponders (tags) that have become rampant in every application ranging from animal/cargo tracking to labeling items in stores and to payment systems. All types of RFID transponders have non-volatile memory storing the identification data and some additional data. The identification data might be equivalent to a Universal Product Code (UPC) code that uniquely identifies the RFID transponder. Additional data can be stored in the tag to carry more information about a product, such as its description, category, manufacturer, expiry date, price and other useful data. The main task of RFID transponders is to securely transmit data from the memory and to confidently identify a tag. In other words, an RFID scanner (reader) should be able to find the RFID transponder in its reading range and recognize its identity, based on the data transmitted from the tag.

RFID transponders are wirelessly activated and usually scanned without being noticed. Every

DOI: 10.4018/978-1-4666-3685-9.ch007

time an RFID transponder is scanned, it (almost always) responds immediately with the same identification number. The tag receives a specific service, depending on its identity. The service varies greatly from one application to another. It can range from simply matching the identity of the transponder to a price in a retail store to granting entrance access through a secured door in a building.

There are many technical challenges facing the designers and researchers in making a robust RFID system. These challenges are mainly due to the physical constraints of the RFID devices and their limitations on sophisticated measures. In this chapter, we review some of the challenges in choosing an RFID technology and designing a suitable identification mechanism.

Organization: The challenges in designing an identification scheme in an RFID system are described in Section 6.2. Later, in Section 7.2, various attacks on authentication schemes are introduced. In Section 7.3, symmetric authentication schemes based on standard cryptography are partially reviewed. Authentication schemes based on asymmetric, standard cryptography are summarized in Section 7.4. RFID systems have unique requirements that demand security solutions that cannot be provided by standard cryptography, and new designs are needed. Custom, specialized cryptographic and non-cryptographic solutions for RFID systems are given in Section 7.5 and Section 7.6, respectively. Finally, this chapter is concluded in the last section with recommendations for the design of an RFID authentication system.

7.1 IDENTIFICATION IN RFID SYSTEMS

RFID systems communicate via electromagnetic waves and are categorized as radio systems. All radio systems operate in a narrow band to avoid signal interference with other radio systems. Therefore, available frequencies and transmitted power in every radio system, including RFID systems, are heavily regulated. These regulations and restrictions directly affect an RFID system in reading range, memory and the applicable standards. In this section, we briefly review each characteristic.

7.1.1 Reading Range

In general, there are three types of transponders: *passive*, *semi-passive* and *active*. Passive tags have no battery in their circuitry, and they rely solely on the reader to provide the power for the tag to operate. Active and semi-passive tags use internal batteries to power their circuitry. An active tag also uses its battery to broadcast radio waves to a reader, whereas the power to broadcast in semi-passive tags is supplied by the reader. Usually, active tags operate in higher frequencies and have a longer reading range than passive and semi-passive tags.

A specific range of frequencies is reserved in every country or region for industrial, scientific or medical applications that are classified as Industrial-Scientific-Medical (ISM) bands. RFID systems operate in the ISM bands as well. Three major frequency ranges are usually defined within ISM bands for RFIDs: Low Frequency (LF), High Frequency (HF) and Ultra High Frequency (UHF).

LF: Frequencies below 135 kHz are in the LF band. They are low frequency, long wavelength signals. The propagation conditions in these frequencies make LF systems preferable for long-distance, low cost transponders for applications such as livestock tracking. Low frequencies have low absorption rate or high penetration depth in non-metallic materials, which is useful for transponders to be implanted in animals (Finkenzeller (2003)). Such transponders have a low power consumption rate due to the low operating frequency.

HF: The 13.553-13.567 MHz range is located in the middle of the short wavelength range and is referred to as the HF band. The propagation conditions in this frequency ranges are suitable for designing low cost and medium speed transponders (Finkenzeller (2003)). Common RFID systems in the HF band operate with a 13.56 MHz frequency, which is suitable for fast data transmission (typically 106 kbits/s) and high speed operations used in implementing cryptographic functions (Finkenzeller (2003)).

UHF: Frequencies in the range between 888-889 MHz and 902-928 MHz (in the USA and Australia) or 2.400-2.4835 GHz (in Europe) are categorized as the UHF band (Finkenzeller (2003)). This range is suitable for applications that demand very fast transmission rates, long reading ranges and high clock frequencies.

The operating frequency directly impacts the transmission rate and power consumption of the RFID system. Higher frequencies yield greater transmission rates, but they require higher transmission power as well. Table 1 lists the maximum permissible transmission power of the electric field (E) and the available reading range for RFID systems.

Various frequencies impose different restrictions in terms of speed and amount of data that can be transmitted in an RFID system. The designer of an authentication scheme for RFID systems has to be aware of the operating frequencies, in order to utilize the available resources within the limits of the chosen technology.

Table 1. Permissible field strengths for RFID systems (Finkenzeller 2003)

Frequency (MHz)	Max. E field	Distance (m)
1.705-10.000	100 μ V/m	30
13.553-13.567	10 mV/m	30
26.960-27.280	10 mV/m	30
40.660-40.700	1 mV/m	3
49.820-49.900	10 mV/m	3
902.0-928.0	50 mV/m	3
2435-2465	50 mV/m	3

7.1.2 Memory

All types of RFID transponders – passive, semi-passive and active – have some non-volatile memory to store their identification code and information data. The size of the available memory ultimately determines the price of the RFID tag and the design of the authentication protocol. There are two types of memory on RFID tags in general: read only memory (write once) and read/write memory. The former is utilized to store an unchangeable identification number and maybe some other descriptive data on a tag. The size of the read only memory varies from 128 bits that only fit a UPC number to a few kilobytes in microwave transponders. Memory cells can store written data for an extended period (typically 10 years) without a power supply (Finkenzeller, 2003).

The second type of memory is the read/write memory that is used for general purpose applications, such as storing a price, an item's description, an expiry date, encryption keys, random numbers and any other dynamic data that changes frequently. Read/write memories are also available in different capacities from just one byte, in the so-called 'pigeon transponder', to 64 kilobytes in microwave transponders used in complex applications. Although the size of memory in RFID tags differs greatly, the amount of memory available on the low-cost RFID tags, such as EPC Gen 2 tags (Global, 2010), is very limited (usually less than one kilobyte).

7.1.3 Timing

Timing plays a very important role, as many attacks can be prevented if the timing is factored into the design of an authentication scheme. For example, if the response from an RFID transponder is suspiciously delayed for too long, it is possible that someone is trying to spoof a remote transponder by maliciously relaying the actual response from the real transponder in the relay attack. We will see later, in Section 7.6.2, that precise measuring of time is one of the main protection mechanisms against the relay attack.

The timing factors in an RFID technology are governed by the underlying standard. As an example, the ISO 14443A standard (ISO 14443 (2008) and ISO 15693 (2006)) specifies that the reader must periodically search for new transponders by sending out a Type A Request (REQA) command signal. The minimum time allowed between the start bits of two consecutive REQA commands depends on the frequency of the carrier signal (f_c) and is defined as $7000 / f_c$. With a carrier frequency set to 13.56MHz as in ISO 14443, the minimum delay is about $500\mu s$. The tag within the vicinity of the reader must respond to the REQA command within 5 ms after receiving an unmodulated carrier signal (ISO 14443 (2008)). To ensure bit synchronization, a response would need to adhere to the Frame Delay Time (FDT) being used as set out in the standard. FDT is specified by the standards (ISO 14443 (2008) and ISO 15693 (2006)), and the minimum FDT should be around $91\mu s$ or $86\mu s$ depending on the last data bit sent by the reader. Therefore, the reader and transponder must ensure that the start bit of the response is aligned to a valid FDT value. The reader will expect the transponder's response to start after FDT. In ISO 14443, the Frame Waiting Time (FWT) specifies the time within which a transponder shall start its response after the end of the reader's data. The FWT ranges from $300\mu s$ to $5s$ with a default value of 4.8ms. Clearly, the timing standards are necessary for compliance and inter-operability of various RFID systems. However, these standards might prevent precise measuring of time, which is necessary for (almost) all security mechanisms, including distance-bounding protocols (Hancke & Kuhn, 2008 and Singelée & Preneel, 2007). Therefore careful attention must be paid to the timing issues in RFID systems.

7.2 AUTHENTICATION IN RFID SYSTEMS

As mentioned earlier, remote identification is the main task of RFID tags. Nevertheless, identification should be performed with some level of confidence and trust. Authentication is the first line of defense against wireless attacks in an RFID system.

7.2.1 Wireless Attacks in RFID Systems

It is generally perceived that the security of a system is only as strong as its weakest link. Any authentication scheme has to guarantee a level of security in RFID systems. The RFID system comprises multiple components, including the wireless link, tag and the reader. The authentication scheme can be attacked at various locations. The data being exchanged between the tag (bearer of secret data) and the reader (verifier of secret data) are transmitted over a wireless link. Securing transmission of data over a wireless link is a classical security problem. The wireless link has to be secured against eavesdropping, replay attacks, relay attacks and tampering.

Eavesdropping: This is the simplest type of attacks, where the attacker is passive and only listens to the communication data being exchanged between the tag and the reader. Since the communication takes place over a wireless channel, it is trivial for the attacker to capture messages being transmitted over the air. Moreover, it is possible to covertly trace an RFID tag across multiple readers violating the privacy of an RFID tag's bearer. Also, confidentiality of sensitive data communicated over a public channel has to be protected against eavesdropping attacks.

Replay attack: The attacker in the replay attack launches an eavesdropping attack first, attempting to listen to the data being exchanged between a legitimate tag and the reader. Then it tries to repeat the data exchange, in order to impersonate the legitimate tag or the reader. In other words, the attacker tries to use the messages exchanged in previous rounds to compromise the new authentication round. To secure RFID tags against the replay attacks, the authentication mechanism has to be randomized. That is, every new authentication session has to be different from the previous sessions.

Relay attack: This is a slightly different attack from the replay attack. The attacker in the relay attack only intercepts the communication data for a current authentication round and redirects them to legitimate tags or readers. In other words, the tag and the reader remotely authenticate one another, without knowing that the attacker is communicating with the reader via the signals received from a remote tag. If the attacker has relayed the signals promptly, it can covertly surpass the authentication process. It is very challenging to detect a relay attack and to deter it, as the RFID tags immediately respond to any query signal and their response signals can be easily captured in the air. There are a few existing non-cryptographic solutions that safeguard authentication schemes against the relay attacks in RFID systems.

Tampering: In the tampering attack, the attacker is active. Not only does the attacker eavesdrop on the data exchanged between the tag and the reader, but it also tries to change the data in a meaningful way. A proper change in the communicated data may yield access to the attacker and change the intended result of the authentication. For example, the attacker may succeed in changing the price of an item from $200 to $20, which is recorded in an RFID tag mounted on to an item. Use of cryptographic algorithms can help alleviate tampering with confidential data in RFID tags. However, RFID tags are very limited in size, memory, and computation power. Therefore, most classical cryptographic algorithms will not fit in low cost RFID tags, and new light-weight algorithms must be designed.

7.2.2 Authentication Types

Authentication is about adding trust to the identification procedure. In any authentication scheme, the entity being authenticated is referred to as the prover and entity checking the identity of the prover is referred to as the verifier. Authentication mechanisms provide a proof of validity of an identity to a verifier. Authentication schemes for RFID systems can be categorized into two main types, depending on the number of entities being authenticated in a scheme: *unilateral* and *mutual authentication*.

Unilateral or one way authentication proves the identity of only one entity, i.e. the prover, to the verifier. In RFID systems, a unilateral authentication scheme usually verifies the identity of the RFID tag to the reader. Authenticating the RFID tags to the reader is referred to as Forward Link Authentication (FLA). In FLA, the prover is the RFID tag and the verifier is the RFID reader. Similarly, if the unilateral authentication scheme verifies the identity of the reader (prover) to the RFID tag (verifier), it is referred to as Backward Link Authentication (BLA).

Unilateral authentication schemes are mostly realized by a challenge-response protocol as follows: there are two main entities involved as shown in Algorithm 1. The prover A and the verifier B share the same private key K. B sends a random number r_B as a challenge to A. Then, A encrypts the random number with the shared key K, denoted by $\mathcal{E}_K(r_B)$ in Algorithm 1, and sends it back as the response to B's challenge. B verifies the result by checking if the encryption of r_B with K in its possession yields the same result as it received from A. It should be noted that A can use any one-way function (i.e. hash function), not necessarily an encryption function to calculate $\mathcal{E}_K(r_B)$.

In mutual authentication, both the prover and the verifier are authenticated to each other. Mutual authentication works similarly but requires an additional round. The mutual authentication processes are shown in Algorithm 2 and it works as follows: B sends a random number r_B to A. Then, A generates another random number r_A. It encrypts both r_B and r_A with the shared key K, denoted by $\mathcal{E}_K(r_B, r_A)$ in Algorithm 2, and sends the result to B. On the other side, B decrypts the message and verifies if r_B is returned. Note that to recover r_A from A's response, $\mathcal{E}_K$ has to be an invertible function and cannot be replaced with any one way function. So far, A is authenticated to B if the verification by B passes. In order to authenticate B to A, B has to recover r_A from $\mathcal{E}_K(r_B, r_A)$ first. Then, it makes some changes to the random numbers and encrypts them with K. This is shown by $\mathcal{E}_K(r_A, r_B)$ in Algorithm 2. On the other side, A verifies the result and authenticates B if a correctly (modified) r_A is recovered from B's response.

Mutual authentication protocols are more costly than unilateral authentication protocols, as they require an additional random number generator at A, an extra round of communication between A and B and an invertible function for encryption and decryption instead of one-way functions. The performance of unilateral and mutual authentication schemes varies greatly depending on the cryptographic functions being used in the procedure. Various cryptographic functions have different performances in terms of the size of secret keys, the mechanism to share keys, the space of random numbers, communication bit rates and power consumption.

Algorithm 1. Unilateral position

1. $B : r_B \rightarrow A$
2. $A : \mathcal{E}_K(r_B) \rightarrow B$
3. B : check $\mathcal{D}_K(\mathcal{E}_K(r_B)) = r_B$

Algorithm 2. Unilateral position

1. $B : r_B \rightarrow A$

$A : \mathcal{E}_K(r_B, r_A) \rightarrow B$

B : check $\mathcal{D}_K(\mathcal{E}_K(r_B, r_A)) = r_B$

B : recover r_A

5. $B : \mathcal{E}_K(r_A, r_B) \rightarrow A$

A : check $\mathcal{D}_K(\mathcal{E}_K(r_A, r_B)) = r_A$

7.3 SYMMETRIC AUTHENTICATION SCHEMES

In symmetric cryptosystems, the sender and the receiver in the system, such as the prover and the verifier in an authentication scheme, share the same cryptographic key for their operations. Trust is established between the parties based on the common cryptographic key that is being shared. Many unilateral/mutual authentication schemes have been proposed for RFID systems that are based on a symmetric encryption algorithm or a hash function. Challenge-response authentication protocols provide a high level of security and are standardized in ISO/IEC 9798-2: Information Technology Security techniques Entity Authentication Mechanisms (1993). Many authentication schemes proposed for RFID systems are based on a classical symmetric algorithm or cryptographic hash function and implement a challenge-response protocol (Feldhofer & Rechberger, 2006; Feldhofer, Dominikus, & Wolkerstorfer, 2004;

Molnar & Wagner, 2004; Pramstaller, Mangard, Dominikus, & Wolkerstorfer, 2005; ISO/IEC 9798-2, 1993).

7.3.1 MAC/Hash-Based Authentication

A Message Authentication Code (MAC) is a keyed hash function in which a cryptographic key is needed to obtain the same (hash) output for a given input. MAC functions use symmetric keys, and they have been widely used in authentication schemes. Almost any identification scheme can be turned into an authentication method by adding a MAC and a random number generator. A general MAC-based authentication scheme works as follows: the reader challenges transponders by sending a random number. Each transponder first combines its identity with the challenge random number. Then it responds by calculating the MAC of the combination. If the reader and transponder share the same key in calculating the MAC, the transponder is authenticated successfully. In this setup, the transponder has to send its actual identity to the reader. As we will see later, this creates a privacy problem, since an adversary can track the transponder at different locations.

An authentication scheme based on MAC functions is given by Molnar and Wagner (2004). They propose a randomized MAC based scheme to authenticate RFID transponders. Under this scheme, the reader first challenges the transponder with r_1. Then the transponder generates another random number r_2 and responds to the reader's query with $ID \oplus f_k(r_1, r_2)$, where f_k is a MAC function with a shared key k. The protocol works without a central database and does not reveal the transponder's identity to eavesdroppers. Nevertheless, the reader and transponder need to have exchanged two random numbers (r_1, r_2) before sending $ID \oplus f_k(r_1, r_2)$ to the reader. Remember that r_1 is sent to the transponder in the reader's challenge, but r_2 is communicated in plaintext with the reader. Therefore, this scheme is a unilateral, FLA scheme.

Most RFID transponders are capable of executing cryptographic hash/MAC functions, as they require very simple binary operations (Stinson, 2002). However, current commercial RFID transponders do not usually implement MAC functions due to higher production cost, as a cryptographic hash function requires additional gates on the transponder. The implementation analysis of various hash functions is given in a research paper by Feldhofer and Rechberger (2006). Common cryptographic hash functions like MD4, SHA-1 and SHA-256 require between 7,350 and 10,868 gates on an RFID transponder. The implementation results of common cryptographic hash functions are summarized in Table 2, showing the feasibility of embedding a hash functions on RFID transponders. Common hash/MAC functions have small footprints in terms of power consumption and number of gates to be implemented in an RFID tag. The resources required to implement a hash/MAC function are well within the constraints of EPC Gen 2 transponders (Global, 2010).

7.3.2 AES-Based Authentication

Encryption algorithms provide message integrity protection against eavesdropping attacks, and they are necessary to implement a mutual authentication in RFID systems. One of the most popular encryption functions in symmetric cryptosystems is the Advanced Encryption Standard (AES) (FIPS 197, 2001). It is considered highly secure, well suited for hardware implementations and standardized for security applications.

In (Feldhofer, Dominikus & Wolkerstorfer, 2004), a strong authentication method based on AES is proposed. The proposed scheme is a unilateral authentication method to authenticate RFID tags. The challenge here is to modify AES not only to fit into RFID transponders, but also to comply with timing standards of RFID systems (FIPS 197, 2001; ISO/IEC 9798-2, 1993). According to the standards, an RFID transponder must respond to the challenge message within $320\mu s$ after a query signal, otherwise it has to remain silent and its response is not received. This time is not enough for encrypting a challenge using the AES algorithm. In order to preserve power in the RFID circuit, the frequency of the internal clock of the transponder must be reduced from 13.56 MHz to 100 kHz to (about $10\mu s$) (Feldhofer et al., 2004). This means that only 32 clock cycles are available during the waiting time for the response. Using the standard data unit, the minimal number of clock cycles to perform an AES-128 encryption or decryption is 65. According to (Pramstaller, Mangard, Dominikus & Wolkerstorfer (2005)), four clock cycles are required for transferring the data units, 54 clock cycles are needed to perform the nine normal AES rounds and seven are required for the final round on a normal size chip. If a smaller size chip is used, the number of clock cycles will be more (Pramstaller, Mangard, Dominikus & Wolkerstorfer (2005)). In Table 3, different implementations of the AES algorithm for RFID tags are compared with each other.

If we take the implementation of AES with the smallest footprint, the time needed for encrypting data (92 clock cycles for the minimum version) is almost three times longer than the time to respond for the tag. Feldhofer et al. (2004) propose an interleaving challenge-response protocol to quickly authenticate multiple transponders by multitasking. This is useful when many RFID transponders are authenticated at the same time. In their scheme, the reader first retrieves all unique IDs of the transponders communicating with the reader. Then it organizes the transponders in a list and challenges the first transponder (Tag_1) in the list with C_1. After receiving the challenge, Tag_1 immediately starts the encryption of C_1 without sending any response. In the meantime, the reader sends more challenges to other transponders it had discovered in the beginning. They also start encrypting their challenge messages right after they are received. In the time that Tag_1 is busy encrypting C_1, the reader continues challenging other transponders. When the encryption of C_1 is finished, Tag_1 waits for the reader's request to send $R_1 = \mathcal{E}_K(C_1)$ as its response. After the reader has sent the challenge messages to all the tags in its list, it sends the requests for response in the same order.

Only RFID transponders are authenticated to the reader in this scheme. Every reader, legitimate or illegitimate, could query the tags to receive the sensitive data. A mutual authentication scheme based on AES encryption is proposed in (Feldhofer, Dominikus, & Wolkerstorfer, 2004) to ensure that only authenticated readers can received the sensitive data. In the proposed scheme, the tag picks a random number as the challenge C_t and includes that in the response message

Table 2. Summary of implementation results of hash functions

Version	Clock Cycles	Area [GE]	$I_{mean} (\mu A)$ 100 KHz
SHA-256	1,128	10,868	15.87
SHA-1	1,274	8,120	10.68
MD5	612	8,400	-
MD4	456	7,350	-

Table 3. Summary of the performance of different AES-128 modules

Version	Clock Cycles	Area [GE]	$I_{mean}(\mu A)$ 100 KHz
AES by Feldhofer, Wolkerstorfer & Rijmen (2005)	1,032	3,400	8.15
Minimum	92	8,541	-
Standard	65	11,205	-
High performance	35	15,850	-

$(\mathcal{E}_K(C_t, C_r))$, which is encrypted by the AES encryption function. Using the symmetric key K, the reader has to decrypt $\mathcal{E}_K(C_t, C_r)$ to retrieve C_t. The reader authenticates itself by sending $\mathcal{E}_K(C_r, C_t)$ back to the tag. Note that changing the position of C_t and C_r in $\mathcal{E}_K(\cdot,\cdot)$ implies a modification on C_t and C_r combined together. This is usually a simple modification, such as a concatenation or addition of C_t and C_r by a fixed number. The tag uses K to decrypt $\mathcal{E}_K(C_r, C_t)$ and to check the modification of C_t and C_r performed by the reader. If the modification matches with the tag's modification of C_t and C_r, it authenticates the reader otherwise rejects it. This scheme is considered as highly secure and well suited for hardware implementations.

Authentication schemes based on symmetric cryptosystems are well suited for hardware implementations specially in resource constrained devices, as they only require simple binary operations. On the other hand, there are other issues inherent in symmetric cryptosystems. In symmetric cryptosystems, the total number of shared keys increases quadratically with the number of parties in the system. Every two parties (the prover and the verifier) has to share a unique key, otherwise, parties cannot be separately identified/ authenticated. If multiple parties share the same key, compromise of one party exposes the keys of all other parties.

Cryptographic keys should be updated frequently to remove compromised keys and maintain a high level of security in the system. This requires a symmetric cryptosystem to change the keys of two parties in every update, since the same secret key has to be shared between two parties. Keys have to be communicated to all parties through a secure channel. However, building secure communication channels increases the communication overhead in the system and sometimes it is not possible to build such a channel. It is therefore important to consider the strengths and weaknesses of symmetric cryptosystems and use them wisely in applications.

RFID authentication schemes based on symmetric cryptography are useful for closed systems, where there are a few parties and updating the shared keys is feasible through a secure channel. In a closed system, all parties are usually controlled by one central authority, which is responsible for distributing cryptographic keys and their updates in the system. For example, a car immobilizer is a closed system, where the same cryptographic key is imbedded in the immobilizer and the car at the manufacturer. Therefore, the manufacturer has the role of the central authority and is responsible for distributing the keys.

Key management in symmetric cryptosystems can quickly become unmanageable in large scale applications. For example in a luggage tracking system in an airport application, tracking the RFID tags is expanded over various geographic locations, which are managed under different authorities. Consequently, readers and transponders from different authority domains have to securely share keys and communicate with one another. In this scenario, symmetric cryptosystems are no longer sufficient and we have to use asymmetric cryptography.

7.4 ASYMMETRIC AUTHENTICATION SCHEMES

There are various asymmetric cryptographic algorithms in the literature. They vary greatly in terms of size, security, power consumption and communication overheads. Among many asymmetric cryptosystems, one can refer to *RSA* (Menezes, van Oorschot, & Vanstone, 1996), *Rabin* (Lenstra & Verheul, 2001), *Elliptic Curve Cryptography (ECC)* (Boneh & Franklin, 2003; Blake, Seroussi, & Smart, 2005; Hankerson, Menezes, & Vanstone, 2004), and *NTRU* (Hoffstein, Pipher, & Silverman, 1998; Hoffstein & Silverman, 2003). RSA (Menezes, van Oorschot, & Vanstone, 1996) and Rabin cryptosystems (Lenstra & Verheul, 2001) are based on the difficulty of factoring a large composite number of two large prime numbers. RSA has been widely implemented and used in software and over the internet. However, ECC (Hankerson, Menezes, & Vanstone, 2004; Blake, Seroussi, & Smart, 2005) is based on the difficulty of solving the Discrete Logarithm Problem (DLP) over large elliptic curve groups. ECC is well suited for small devices, as it offers shorter key lengths for the same level of security as RSA and it yields great savings in memory (Gura, Patel, Wander, Eberle, & Shantz, 2004). The NTRU cryptosystem (Hoffstein, Pipher, & Silverman, 1998; Hoffstein, Pipher, & Silverman, 2001) is based on finding very short vectors in lattices of very high dimension (lattice reduction problem) and it is very efficient in hardware implementations. It has been shown that it is possible to implement some of the asymmetric cryptosystems (e.g. ECC and NTRU) in low-cost RFID tags (Hoffstein & Silverman, 2003; Hoffstein, Howgrave-Graham, Pipher, Silverman, & Whyte, 2003; Kaps, 2006; Kim, Kim, & Park, (2007; Keller & Marnane, 2007; Feldhofer & Wolkerstorfer, 2007; Batina, Mentens, Sakiyama, Preneel, & Verbauwhede, 2006). Nevertheless, the price of tags will increase to accommodate more sophisticated circuitry and larger memory.

Gura et al. (2004) implemented elliptic curve point multiplication for 160-bit, 192-bit, and 224-bit NIST/SECG curves and RSA-1024 and RSA-2048 on an 8-bit micro-controller and they have analyzed the results. In Table 4, we have selectively included the comparison results between ECC and RSA. We refer the reader to (Gura, Patel, Wander, Eberle & Shantz (2004)) for further details. In Table 4, secp160r1, secp192r1 and secp224r1 refer to elliptic curve parameters recommended by the Standards for Efficient Cryptography Group (SECG) (Research (2000)).

There are other factors involved that affect the performance and acceptability of an asymmetric cryptosystem in RFID authentication. Besides the

Table 4. Average ECC and RSA execution times on the ATmega128

Algorithm	Time 8MHz	Data (bytes)	Code (bytes)	Security Bits
ECC secp160r1	0.81s	282	3,682	80
ECC secp192r1	1.24s	336	3,979	96
ECC secp224r1	2.19s	422	4,812	112
RSA-1024	10.99s	930	6,292	80
RSA-2048	83.26s	1,853	7,736	112

authentication type (unilateral or mutual), one of the key parameters is the complexity of encryption and decryption functions. We have seen in Table 4 that ECC outperforms RSA for small devices. However, the Rabin cryptosystem, which has the same security basis as RSA (factorization problem), is more advantageous than ECC as it has a very simple encryption function (Kaps, 2006). For a fair comparison of the performance of these cryptosystems, we pick parameters in each cryptosystem that yield the same level of security. A 512-bit Rabin's, according to Lenstra and Verheul (2001), provides a security level of around 60 bits. ECC secp112r1/secp112r2 with a 112-bit prime field provides a security level of 56 bits. The NTRU system with parameters as $(N,p,q)=(167,3,128)$ offers a security level of around 57 bits (Hoffstein & Silverman, 2003).

Table 5 summarizes a comparison between the resources consumed by the encryption algorithm in Rabin's scheme, NTRU and ECC. In the Rabin cryptosystem, the encryption can be done very efficiently, as it mostly takes a simple squaring operation or squaring and an addition. That is why the performance of Rabin for RFID tags is greatly superior over ECC. However, the decryption algorithm in Rabin's scheme is more complex and might not be fit into low-cost RFID transponders. Further analysis of Table 5 shows that for extremely resource constrained RFID tags, the NTRU cryptosystem appears to be the most appropriate choice. Nevertheless, the NTRU decryption function demands more resources than the encryption process.

Asymmetric cryptosystems, as well as symmetric cryptosystems, can be used in unilateral and mutual authentication schemes. As mentioned before, asymmetric-key (public key) cryptosystems are superior to symmetric-key cryptosystems in key management. Therefore they are highly recommended for implementation in large-scale, distributed systems. In the rest of this section, we describe authentication protocols that are based on the difficulty of the Elliptic Curve Discrete Logarithm Problem (ECDLP).

7.4.1 ECC-Based Authentication

The first ECC-based protocol is Schnorr's authentication scheme (Schnorr (1989)). An ECC variant of Schnorr's original scheme is shown in Algorithm 3. In this scheme, the tag (T) starts the authentication protocol by sending its certificate along with $X=rP$ to the reader R, where P is a public parameter point on an elliptic curve and r is a random number. The reader verifies the tag's certificate and checks if it is authorized by a trusted source (certificate authority). From the certificate, the reader retrieves the tag's public key $Q=sP$ where s is the tag's secret number that uniquely identifies the tag. The reader sends a random challenge number c to the tag. Then the tag calculates its response $y=r+(s\times c) \bmod n$ and returns y to the

Table 5. Comparison of encryption with Rabin's scheme, NtruEncrypt, and ECC

Factor	Rabin	NTRU (k=1)	ECC
Area (g.e.)	16,726	2,850	18,720
Clock Cycles	1,440	29,225	408,850
Avg. power $[\mu W]$	148.18	19.13	394.4
Throughput [kbits/s]	177.78	4.52	0.24

reader. The reader verifies if $yP - cQ = X$ before accepting the tag's identity.

By running Schnorr's protocol, the tag proves to the reader that it has s and therefore can be authenticated. s corresponds to the tag's public key by $Q = sP$. Note that Q had to be issued by a trusted authority in the first place, otherwise, the tag could have issued a pair of public/private key for itself and participated in the authentication protocol. Hence the reader must be able to verify the certificate authority's digital signatures on the tags' certificates.

A variation of Schnorr's scheme is proposed by Girault et al. (2006), where the modular reduction during the response calculation is eliminated and the operations are performed over integers (Z). In other words, the tag will no longer reduce $y = r + (s \times c)$ modulo n and calculating the response message is simplified for the RFID tag. Also, Okamoto (1992) proposed a more secure variant of Schnorr's protocol, which is shown in Algorithm 4. In Okamoto's scheme, the tag starts the protocol by sending $X = r_1P + r_2P$ to the reader along with a certificate containing $Q = (s_1 + s_2)P$. The tag proves its identity by responding with $y_1 = r_1 + (s_1 \times c)$ and $y_2 = r_2 + (s_2 \times c)$ to the reader's challenge number c. The reader then authenticates the tag by

Algorithm 3. Authentication based on Schnorr's scheme

1. T : pick r randomly
   T : compute $X = rP$
   $T : X, cert_T \rightarrow R$
   R : pick c randomly
5. $R : c \rightarrow T$
   T : compute $y = r + (s \times c) \bmod n$
   $T : y \rightarrow R$
   R : check if $yP - cQ = X$, o.w. reject

comparing X and $y_1P + y_2P - cQ$ together. The reader accepts the tag's identity if they are equal otherwise rejects it.

Hutter (2009) analyzed and compared Schnorr's, Okamoto's, and Schnorr's variant protocols with respect to the computational complexity, storage requirements and communication overhead imposed on the tags. The results are listed in Table 6. The results show that the number of clock cycles is comparable to the standard ECC implementation given in Table 5. Note that the number of clock cycles is greater than the standard ECC given in Table 5. This is because the ECC in Table 5 is optimized for a hardware implementation, whereas the implementation results of Table 6 are speculated based on a software implementation. Nevertheless, the results show that energy consumption, which is directly tied to the number of clock cycles, would be the limiting factor for implementations over RFID transponders.

The main focus in RFID authentication has been on sending encrypted messages from the RFID transponder to the reader. Nevertheless, in some applications, the tag may need to digitally approve of a message. In other words, the tag might need to apply a digital signature to a message.

Digital signatures can be deemed as the reverse operation of an encryption function. Driessen et al. (2008) have investigated the efficiency and suitability of digital signature algorithms to which we refer to evaluate the performance of digital signature schemes in small RFID devices. The study is based on the most popular digital signature schemes: DSA, a defacto standard for lightweight signature applications; NTRU Sign, a signature scheme based on NTRU Public Key Cryptography (PKC); and ECDSA, the elliptic curve equivalent of DSA. Driessen *et al.* used the Efficient and Compact Subgroup Trace Representation (XTR) of DSA (Lenstra & Verheul, 2000) for implementation which resulted in a threefold improvement in

Algorithm 4. Authentication based on Okamoto's scheme

1. T : pick r_1 and r_2 randomly
   T : compute $X = r_1P + r_2P$
   $T : X, cert_T \to R$
   R : pick c randomly
5. $R : c \to T$
   T : compute $y_1 = r_1 + (s_1 \times c) \bmod n$ and $y_2 = r_2 + (s_2 \times c) \bmod n$
   $T : y_1, y_2 \to R$
   R : check if $(y_1 + y_2)P - cQ = X$, o.w. reject

performance, as compared to standard DSA. The implementation results are listed in Table 7 listing the most power efficient scheme first. As indicated in Table 7, new cryptographic algorithms, e.g. NTRUSign, outperform standard cryptosystems and would be better candidates for small devices.

We have seen in this section that authentication in RFID systems can be achieved by means of asymmetric cryptosystems that do not have the problem of key management in large-scale applications. Nevertheless, they demand more resources and often require complex mathematical operations with large numbers inside small RFID tags.

7.5 SPECIALLY-DESIGNED AUTHENTICATION SCHEMES

Adding a standard encryption or signature function to the RFID tag's circuitry will significantly increase the price of the tag. Hash functions, encryption algorithms and signature schemes can be used for various purposes in the RFID tag, such as data integrity, authentication, and confidentiality and non-repudiation of data. In applications, where only an RFID authentication is needed, new schemes have been devised to minimize resources required and to reduce costs.

Standard cryptography, while guaranteed to provide a certain level of security, takes a great amount of power and a large size in RFID transponders. For very sensitive applications, expensive RFID transponders equipped with standard cryptographic functions might be justifiable. However, in other routine applications, where a minimal level of security is needed, expensive transponders are not suitable due to the budget restrictions. Remember that RFID systems are widespread mainly because they can cheaply replace bar-codes, while they provide additional functionalities.

Code-based cryptography is another very efficient and very high speed tool for designing light-weight authentication schemes, as compared to algebraic cryptography. Therefore, code-based cryptographic systems are a desirable choice for small devices. The first popular code-based system was HB+ by Juels and Weis (Juels & Weis (2005)). HB+ is a modification of the authentication protocol by Hopper and Blum (HB) that was originally designed to detect human users. HB+ is implementable on very small, low-cost RFID tags (Global (2010)); it only requires simplistic binary operations, such as XOR ($\oplus$) and dot-product ($\cdot$) and requires little memory as much as merely 800 bits (Juels & Weis (2005)). Although HB+ is implementable in EPC-Gen 2 tags (Global (2010)), it has a high false rejection rate – for 80 bits of security, the false rejection rate is

Table 6. Performance comparison of authentication protocols based on identification schemes

Parameter	Schnorr	Okamoto	Girault
Memory	100B	148B	100B
Computation	5,775	11,550	9,615
Clock Cycles	993,432	1,986,864	1,630,872

Table 7. Implementation results of NTRUSign, ECDSA, and XTR-DSA in the micro-controller

Scheme	ROM	RAM	Signature	Private Key	Public Key	E(Sign)	E(Verify)
NTRU Sign	11.3kB	542B	127B	383B	127B	22.3mJ	2.81mJ
ECDSA	43.2kB	3.2kB	40B	21B	40B	33.0mJ	33.8mJ
XTR-DSA	24.3kB	1.6kB	40B	20B	176B	34.7mJ	72.3mJ

estimated at 44% (Gilbert, Robshaw & Seurin (2008),Gilbert, Robshaw & Sibert (2005)). In HB+, the communication overheads increase linearly with the size of the security parameter. In other words, lowering the communication overheads (transmission rate) reduces the security of the scheme.

Gilbert et al. (2008) improved the HB+ protocol and proposed HB# by expanding the size of secret keys by increasing the transmission rate. Both HB+ and HB# (with some modifications) are specifically proposed for implementation in EPC-Gen 2 tags (Global, 2010). Thus, these two schemes have been widely used as a reference (Gilbert, Robshaw, & Seurin, 2008; Juels & Weis, 2005).

7.5.1 HB# Authentication Scheme

The HB# protocol is shown in Algorithm 5. In the HB# protocol, two $k_X \times m$ and $k_Y \times m$ matrices, denoted by X and Y respectively, are used. These matrices serve as the symmetric key shared between the tag and the reader. The tag uses a random binary source of noise. An $m-$bit vector v is added as a noise in which the probability of a bit being 1 is η.

The authentication process is performed in one round as follows: The tag sends a random $m-$bit vector b to the reader. Then the reader challenges the tag by sending a random $m-$bit vector a back to the tag. The tag responds to the challenge by calculating a noisy vector z as follows:

$$z = aX \oplus bY \oplus v,$$

where $\oplus$ denotes XOR of two binary vectors. For verification, the reader only has to count the number of positions, denoted by e, between the received z and a self-generated vector $aX \oplus bY$ that is in error. In other words, e equals the weight of $z \oplus aX \oplus bY$. If $e \leq t$ for some threshold t, the tag is authenticated. It is recommended in (Gilbert, Robshaw, & Seurin, 2008) to set $t = um$, for some $u \in (\eta, \frac{1}{2})$. Note that HB+ and HB# proved to be insecure against the man-in-the-middle attack (Gilbert, Robshaw, & Sibert, 2005), where the attacker alters the bits communicated between the tag and the reader.

There are other proposals in literature (Chien, 2007; Peris-Lopez, Castro, Estévez-Tapiador, & Ribagorda, 2009; Peris-Lopez, Hernandez-Castro, Estevez-Tapiador, & Ribagorda, 2006; Peris-Lopez, Castro, Estévez-Tapiador, & Ribagorda, 2006*b*; Peris-Lopez, Castro, Estévez-Tapiador, & Ribagorda, 2006*a*) promising efficient security for RFID systems. However, all of them have shown some security weaknesses against passive and active attacks (Castro, Estevez-Tapiador, Peris-Lopez, & Quisquater, 2008; Li & Wang, 2007; Li & Deng, 2007; Barasz, Boros, Ligeti, Loja, & Nagy, 2007*a*; Bárász, Boros, Ligeti, Lója, & Nagy, 2007*b*). The main challenge in designing new cryptosystems for RFIDs is to provide a proof of security and to build confidence in the proposed scheme.

The performance of HB# protocol (Gilbert, Robshaw, & Seurin, 2008) based on different parameter selection is given in Table 8. In order to reduce the required size of memory in the RFID tag, Gilbert et al. recommend using Toeplitz ma-

Algorithm 5. HB# authentication protocol for RFIDs

1. T : pick v randomly s.t.
$v \in \{0,1 \mid \text{prob}('1') = \eta\}^m$
T : pick randomly $b \in \mathbb{F}_{2^m}$
$T : b \rightarrow R$
R : pick randomly $a \in \mathbb{F}_{2^m}$
5. $R : a \rightarrow T$
T : compute $z = aX \oplus bY \oplus v$
$T : z \rightarrow R$
R : check $w(z \oplus aX \oplus bY) \leq t$

trices in place of random matrices. However, it is not clear by how much using Toeplitz matrices would reduce the security of the authentication protocol. Recently, it has been shown that the size of parameters in code-based cryptography can be drastically reduced, while preserving a high level of security (Berger, Cayrel, Gaborit, & Otmani, 2009). Further reduction in the size of parameters opens up the possibility of applying asymmetric code-based cryptography in RFID systems. There are other proposals in the literature by (Chien, 2007; Peris-Lopez, Castro, Estévez-Tapiador, & Ribagorda, 2009; Peris-Lopez, Hernandez-Castro, Estevez-Tapiador, & Ribagorda, 2006; Peris-Lopez, Castro, Estévez-Tapiador, & Ribagorda, 2006*b*; Peris-Lopez, Castro, Estévez-Tapiador, & Ribagorda, 2006*a*) promising great security and performance for RFID systems, but mostly are vulnerable to passive and active attacks (Castro, Estevez-Tapiador, Peris-Lopez, & Quisquater, 2008; Li & Wang, 2007; Li & Deng, 2007; Barasz, Boros, Ligeti, Loja, & Nagy, 2007*a*; Bárász, Boros, Ligeti, Lója, & Nagy, 2007*b*).

We have seen applications of hash/MAC functions, symmetric and asymmetric cryptography and a few new cryptosystems to protect the identification process in RFID systems. Securing RFID systems is not limited to cryptographic authentication schemes. Nevertheless, the relay attack can circumvent any cryptographic RFID authentication, and it must be prevented using other non-cryptographic solutions.

7.6 NON-CRYPTOGRAPHIC SOLUTIONS

Non-cryptographic solutions heavily rely on the physical characteristics of the RFID system and the communication channel. This is a fairly new topic of studies in RFID systems, and only two main categories of systems have been explored. The first category uses the intrinsic characteristics of the physical RFID tag using a Physical Unclonable Function (PUF). The second category uses the limitations in the communication channel (mostly time of signal propagation) to verify authentic RFID tags. This approach is usually referred to as the *distance bounding* technique. We briefly review both of these approaches in this section.

Table 8. Practical parameters of HB# for 80-bit security

k_X	k_Y	m	η	t	P_{FR}	P_{FA}	**Communication (bits)**
80	512	1,164	0.25	405	2^{-45}	2^{-83}	1,756
80	512	441	0.125	113	2^{-45}	2^{-83}	1,033

7.6.1 PUFs

The first category of identification mechanisms distinguishes RFID tags based on their physical characteristics of RFID tags. The uniqueness of every tag can be modeled as a PUF inside the tag. PUFs are tied to physical characteristics of an RFID tag's circuitry and are difficult to model precisely even if full physical scrutiny of the circuit is permitted. Reliable silicon PUFs have been proposed in (Lim, Lee, Gassend, Suh, Dijk, & Devadas, 2005 and Majzoobi, Koushanfar, & Potkonjak, 2009). A PUF computes its outputs by exploiting the inherent variability of wire delays and gate delays in its circuitry (Bolotnyy & Robins, 2007). These delays highly depend on some unpredictable factors, including manufacturing variations, quantum mechanical fluctuations, thermal gradients, electromagnetic effects, parasitics, and noise (Bolotnyy & Robins, 2007). Therefore, it is very difficult to model a PUF precisely even by using identical hardware. Moreover, PUF-based functions appear to be more resistant to hardware-tampering attacks and even some side-channel attacks, due to the difficulty in duplicating PUFs.

Bolotnyy and Robins (2007) propose a PUF-based MAC protocol to be implemented in RFID transponders. Although an important characteristic of a PUF is that it cannot be precisely modeled, there are three probabilistic parameters of a PUF that must be known. The first parameter, denoted by τ, represents the probability that the PUF's output for a random input is the same as the PUF's output of another chip for the same input. τ is usually obtained through empirical testing. The second parameter, denoted by μ, is the probability that the output of a PUF for a given input is different from the PUF's expected output in a controlled environment. Different variables, such as temperature, voltage, and manufacturing tolerances, can affect μ. The third parameter, denoted by λ, represents the difficulty for an adversary to model the PUF. Let us denote PUF by function p. For a message m, the PUF-based MAC computes $p(m)$ as the signature of the message.

Bolotnyy and Robins (2007) model the behavior of the PUF by function $p()$ and combine a unique token c as the secret key. The notation $p_c()$ represents a general PUF parameterized with c. Note that c can be generated by using a random number generator or even a counter in the RFID tag. This will make the prediction of PUF outputs more difficult, as $p_c()$ highly depends on the secret key c. In a signature scheme using a PUF, a random number r and the challenge message m are input to the PUF, and the PUF outputs $p_c(r,m)$. Bolotnyy and Robins (2007) use multiple PUF computations in the generation of a digital signature, in order to reduce the probability of forgery. In other words, the signature for a message m is comprised of $\{c, r_1, \cdots, r_n, p_c(r_1, m), \cdots, p_c(r_n, m)\}$, where $r_1, \cdots, r_n$ are different random numbers generated by the tag.

The number of PUF computations (n) depends on τ, μ and the application requirements. Bolotnyy and Robins (2007) quantify the reliability of the signature and the difficulty for an adversary to forge a signature using an equivalent PUF. Let us denote Pr_v as the probability that a valid signature is verified as authentic and Pr_f as the probability that a forged signature is incorrectly flagged as valid by the verifier. In other words, Pr_v represents the probability of a valid detection and Pr_f represents the probability of not recognizing a forged signature. The probability that at most t out of n PUF responses differ from the valid responses is calculated as follows:

$$Pr_v(n,t,\mu) = 1 - \sum_{i=t+1}^{n} \binom{n}{i} \mu^i (1-\mu)^{n-i}$$

On the other hand, the probability that at most t out of n responses differ from the responses of another tag is obtained by:

$$Pr_f(n,t,\tau) = 1 - \sum_{i=t+1}^{n} \binom{n}{i} \tau^i (1-\tau)^{n-i}$$

In (Lim, Lee, Gassend, Suh, Dijk, & Devadas, 2005), the empirical values for a PUF are given as $p:\{0,1\}^{64} \rightarrow \{0,1\}^{8}$ with $\tau = 0.4$ and $\mu = 0.02$. Setting $n = 30$ and $t = 3$, Bolotnyy and Robins (Bolotnyy & Robins (2007)) propose a signature-based authentication scheme with a valid detection probability of $Pr_v = 0.997107$ and a forgery non-recognition probability of $Pr_f = 0.000313$. These probabilities can be improved by picking sufficiently large t and n. The PUF-based scheme is very suitable for hardware implementation, since it requires the least number of gates to implement. This can be seen in Table 9, where an implementation of PUF is compared with other cryptographic primitives.

Although PUF functions are highly desirable in small, resource constrained RFID systems, they rely on the device physical characteristics, which are often very difficult to quantify in practice. The outputs of a PUF are only consistent probabilistically with the expected theoretical values. There PUF's outputs can be distorted due to noise and inconsistencies in the circuitry (Bolotnyy & Robins, 2007). This is still an ongoing research topic in the literature.

7.6.2 Distance Bounding Protocols

Physical distance can serve as a measure of trust and privacy in RFID systems. In some applications, users can be granted access based on their perceived location. In low range RFID systems, it can be stated that the main task is proximity identification, as low range RFID tags can be identified only within a few meters (less than 3 meters) form a reader (ISO 14443, 2008; ISO 15693, 2006). As mentioned before, RFID systems are vulnerable to relay attacks. In the relay attack, the attacker can remotely impersonate a tag and pretend that the tag is in the proximity of the reader. The attacker relays the communication between the reader and a remote (authentic) tag without altering the communication messages or attempting to penetrate the authentication mechanism.

The relay attack can be launched against any other wireless systems, but the damage is more severe in RFID systems. This is mainly due to the minimalist design of RFID tags that cannot support sophisticated authentication mechanisms. Let us imagine that in a vicinity payment system, users can make a quick payment simply by holding an RFID-enabled payment card close to a reader. The RFID cards are activated remotely. The payer's identification number is read from the RFID card he/she is holding. The payer is authenticated – often using the secret key stored in the RFID card – and billed if authentication is successful. Using the relay attack, an attacker could surreptitiously pass the bills to another bystander with any valid RFID card. The attacker succeeds by relaying the authentication messages between the reader and the victim's card. Successful cases of the relay attack have been reported in SpeedPass RFID cards (RFIDJournal (2005)), where a thief could get free gas by passing the charges to another SpeedPass card holder nearby. To increase the reading range in the relay attack, the attacker puts a proxy close the authentic tag, as shown in Figure 1. The proxy intercepts the signals from the tag and sends them over to a mole interfacing the authorized reader. The mole masquerades as the authentic tag's identity by reflecting its responses.

In the relay attack, the attacker is considered slower than the authentic tag in the authentication process, as it has to wait for the response from the tag and relay it to the reader. If the authentic tag is within reading range of the reader, it will oper-

Table 9. Comparing a PUF function with standard hash functions

Algorithm	MD4	MD5	SHA-256	AES	Yuksel (2004)	PUF
Number of gates	7,350	8,400	10,868	3,400	1,701	545

Figure 1. The relay attack: a proxy and mole are placed to increase the range of the attack

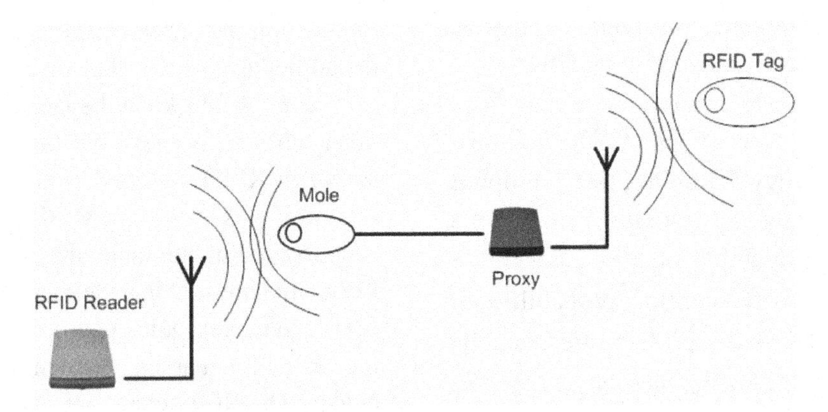

ate faster than the attacker and the anti-collision mechanism at the reader usually picks the first responder, i.e. the authentic tag. Therefore, the relay attack is effective when the authentic tag is physically far from the reader and its signal cannot reach the reader.

Physical distance has been used extensively in RFID systems as a countermeasure to thwart the relay attack (Kim & Avoine, 2009; Hancke & Kuhn, 2005; Munilla & Peinado, 2008; Singelée & Preneel, 2007). *Distance-bounding* refers to a suite of mechanisms that use the physical characteristics of the communication system to enforce an upper bound on the maximum distance an RFID tag can be placed from a reader.

Some physical characteristics of the communication channel have been used in distance-bounding methods, such as signal strength, angle of arrival and time of fly. The received signal strength is inversely proportional to the distance of a transmitter. Therefore, it is possible to estimate the distance of the transmitter on an RFID chip from a reader by measuring the power of the received signal (Hancke & Drimer, 2008). This approach, however, is not effective against a relay attack, as the attacker can easily increase the transmission power to fake an authentic transmitter in proximity. The angle of an arriving signal can also be used to determine the location of a transmitter (Hancke & Drimer, 2008). This approach fails against sophisticated attackers that can reflect signals from various directions. Time of fly measures the time elapsed for a message exchange from a reader to a tag and then back to the reader. By knowing the propagation speed in the communication channel, one could easily find the distance of the reader from the tag. This approach has received great attention in the security community (Hancke & Kuhn, 2005; Hancke, 2006; Munilla & Peinado, 2008; Munilla & Peinado, 2010; Kim & Avoine, 2009; Reid, Nieto, Tang, & Senadji, 2007), as it provides some level of protection against relay attacks.

A major contribution in distance-bounding protocols for RFID systems is by Hancke and Kuhn (2005). In their protocol, a random challenge is sent to the tag as a series of bits. The reader measures the time between sending each bit of the challenge and receiving the corresponding response. Measuring the time of fly between sent and received bits, the reader can estimate the distance to the transmitter in this protocol. The exchange of challenge and response bits is performed over a fast Ultra Wide Band (UWB) channel that provides a fine resolution for measuring the distance. Hancke and Kuhn's protocol is shown in Algorithm 6.

The RFID transponder starts the protocol by sending a random vector t of length n to the reader. The reader also generates two random vectors, d of length n and c of length l. The former d is transmitted to the transponder, but c is kept as the challenge. The reader and transponder both can calculate $f_k(t,d)$, where f is a known hash function and k is the shared secret key between the tag and the reader participating in the authentication protocol. The output of the hash function is *2l* bits long. The output split into two $l-$bit substrings m_1 and m_2,. The reader then sequentially transmits the bits c_i of the challenge c for all $i = 1,\cdots,l$. The transponder responds

Algorithm 6. Hancke-Kuhn's distance-bounding protocol

1. T : pick randomly t of length n
   $T : t \rightarrow R$
   R : pick randomly d of length n
   $R : d \rightarrow T$
5. T : compute $\{m_1, m_2\} = f_k(t,d)$
   R : pick randomly c of length l
   R : also find $\{m_1, m_2\} = f_k(t,d)$
   R : FOR $i = 1,\cdots,l$ DO
   $c_i \rightarrow T$
   T : set $r_i = m_{1,i}$ if $c_i = 0$,
   otherwise set $r_i = m_{2,i}$
10. R : check r_i END

to every challenge bit with $r_i = m_{1,i}$ if $c_i = 0$ and $r_i = m_{2,i}$ if $c_i = 1$, where $m_{1,i}$ and $m_{2,i}$ denote the $i-$th bit of m_1 and m_2, respectively. The reader measures the round-trip time between sending each c_i bit to the RFID tag and receiving the corresponding response bit r_i. By measuring the time of fly between sent and received bits, the reader can estimate the distance of the RFID transmitter from itself. The exchange of challenge c_i and response r_i bits is performed very quickly, as the response bits are already calculated by the RFID transponder. Discarding the time it takes to select the correct bit at the tag, the time of fly is inversely proportional to distance of the tag from the reader.

Time of fly has been used extensively in many other distance bounding protocols (Hancke & Kuhn, 2005; Hancke, 2006; Munilla & Peinado, 2008; Munilla & Peinado, 2010; Kim & Avoine, 2009; Reid, Nieto, Tang, & Senadji, 2007). The proposed distance-bounding scheme relies heavily on the precise timing of the bits exchanged between the RFID tag and reader. Delays and noise in returning the response bits will disturb the measured time and will therefore increase the false rejection rate. The performance of the distance-bounding protocols that are based on measuring the time of fly has yet to be tested in practice. A small tolerance of a few micro seconds in measuring the time of fly will result in a few hundred meters difference in permitted distance. Therefore, the distance bounding protocols are not strictly bordering the range of reading distance in relay attacks.

CONCLUSION

There are many unique technicalities and challenges in the design of a secure RFID authentication system. Many parameters, such as the operating frequency, size of memory and computing resources and operational power available at the RFID tag, directly affect the design and they have to be fully considered in the design process. Therefore, many different RFID authentication systems have been proposed in the literature. Authentication schemes vary from one another depending on the number of entities being authenticated, i.e. tag or reader only or both. The performance of each authentication scheme highly depends on the cryptographic functions being used in the scheme. Cryptographic functions often differ drastically from one another in the size of cryptographic keys, the method by which they share the keys, the complexity of encryption and decryption algorithms and the power consumed by the algorithms.

We have divided the existing authentication schemes for RFID systems into four main categories: The first category consists of the solutions that use standard, symmetric cryptographic systems, such as hash/MAC functions and symmetric encryption algorithms. Standard, symmetric cryptographic systems have a security record that can be trusted when applied to RFID systems. They are very efficient for hardware implementation and therefore desirable in small devices. However, the key management in symmetric systems prevents them from being used in large applications. The second category is based on asymmetric cryptographic solutions using standard algorithms that can handle the key management problem in large, distributed systems. However, most asymmetric algorithms based on standard cryptography require operations with large numbers and complex circuitries in implementation. In the specially designed category, new cryptographic solutions have been proposed to satisfy all the requirements of RFID systems. These solutions are designed to be efficient and have a small footprint in hardware, in order to be utilized in low-cost RFID tags. Finally, non-cryptographic schemes have also been designed for RFID systems to provide security in areas where cryptographic solutions have been unsuccessful.

RFID systems are heterogeneous systems that are comprised of various components, including tags, readers, and servers. Therefore, no single algorithm can provide a comprehensive defense mechanism against all existing attacks on all RFID systems. For a complete security solution, a mix of these solutions has to be used. For example, a specially designed code-based cryptosystem can be used inside RFID tags for the purpose of authentication. RFID readers might also be authenticated via a symmetric-key, MAC-based scheme and the cryptographic keys in the entire RFID system can be managed by an asymmetric system such as NTRU. To secure the authentication scheme against the relay attack, a distance-bounding mechanism should also be added to the security system.

RFID systems are used in numerous applications, and the level of security in the authentication process highly depends on the underlying application. The service of the RFID system varies greatly from one application to another. It can simply be matching the identity of the tag to a price in a retail store or making a payment via a credit card. To build a fully secure system, the entire RFID system must be protected at its various parts including the wireless link, tag, reader and server (computer support system). Due to the restrictions on the RFID tags' price and size, the main challenge is often to design light-weight schemes that meet the security requirements in low-cost RFID tags. In this chapter, we have given the specific requirements in RFID systems and the challenges that arise in designing a secure identification and authentication wireless system.

REFERENCES

Barasz, M., Boros, B., Ligeti, P., Loja, K., & Nagy, D. (2007*a*). *Breaking LMAP*. RFIDSec07.

Bárász, M., Boros, B., Ligeti, P., Lója, K., & Nagy, D. A. (2007*b*). *Passive attack against the M2AP mutual authentication protocol for RFID Tags*. In EURASIP International Workshop on RFID Technology.

Batina, L., Mentens, N., Sakiyama, K., Preneel, B., & Verbauwhede, I. (2006). Low-cost elliptic curve cryptography for wireless sensor networks. In *Workshop on Security and Privacy in Ad-Hoc and Sensor Networks: ESAS06, Lecture Notes in Computer Science, vol. 4357*, (pp. 6–17). Springer-Verlag.

Berger, T. P., Cayrel, P.-L., Gaborit, P., & Otmani, A. (2009). Lecture Notes in Computer Science: *Vol. 5580. Reducing key length of the McEliece cryptosystem. Advances in Cryptology (AFRICACRYPT09)* (pp. 77–97). Springer-Verlag.

Blake, I. F., Seroussi, G., & Smart, N. P. (Eds.). (2005). *Advances in elliptic curve cryptography*. Cambridge University Press. doi:10.1017/CBO9780511546570

Bolotnyy, L., & Robins, G. (2007). Physically unclonable function-based security and privacy in RFID systems. In *Fifth IEEE International Conference on Pervasive Computing and Communications* (PERCOMP07), (pp. 211–220). IEEE Computer Society.

Boneh, D., & Franklin, M. (2003). Identity-based encryption from the Weil pairing. *SIAM Journal on Computing*, *32*(3), 586–615. doi:10.1137/S0097539701398521

Castro, J. C. H., Estevez-Tapiador, J. M., Peris-Lopez, P., & Quisquater, J.-J. (2008), *Cryptanalysis of the SASI ultralightweight RFID authentication protocol with modular rotations*. In CoRR labs.

Chien, H.-Y. (2007). SASI: A new Ultralightweight RFID authentication protocol providing strong authentication and strong integrity. *IEEE Transactions on Dependable and Secure Computing*, *4*(4), 337–340. doi:10.1109/TDSC.2007.70226

Driessen, B., Poschmann, A., & Paar, C. (2008). Comparison of innovative signature algorithms for WSNs. In *The First ACM Conference on Wireless Network Security*, ACM, (pp. 30–35).

Federal Information Processing Standards Publication 197. (2001). *Advanced encryption standard (AES)*. Retrieved from http://www.itl.nist.gov/fipspubs/

Feldhofer, M., Dominikus, S., & Wolkerstorfer, J. (2004). Strong authentication for RFID systems using the AES algorithm. In *Workshop on Cryptographic Hardware and Embedded Systems (CHES04), Lecture Notes in Computer Science, vol. 3156*, (pp. 357–370). Springer-Verlag.

Feldhofer, M., & Rechberger, C. (2006). Lecture Notes in Computer Science: *Vol. 4277. A case against currently used hash functions in RFID protocols. OTM 2006 Workshops* (pp. 372–381). Springer-Verlag.

Feldhofer, M., & Wolkerstorfer, J. (2007). Strong crypto for RFID tags – A comparison of low-power hardware implementations. In *IEEE International Symposium on Circuits and Systems (ISCAS07)*, (pp. 1839–1842).

Feldhofer, M., Wolkerstorfer, J., & Rijmen, V. (2005). AES implementation on a grain of sand. In *IEE Proceedings of Information Security: IS05, Vol. 152*, (pp. 13–20).

Finkenzeller, K. (2003). *RFID handbook: Fundamentals and applications in contactless smart cards and identification*. Wiley.

Gilbert, H., Robshaw, M. J., & Sibert, H. (2005). An active attack against HB+: A provably secure lightweight authentication protocol. *IEE Electronics Letters*, *41*(21), 1169–1170. doi:10.1049/el:20052622

Gilbert, H., Robshaw, M. J., & Seurin, Y. (2008). HB#: Increasing the security and efficiency of HB+. In Advances in Cryptology (Eurocrypt08), Lecture Notes in Computer Science, vol. 4965, (pp. 361–378). Springer.

Girault, M., Poupard, G., & Stern, J. (2006). On the fly authentication and signature schemes based on groups of unknown order. *Journal of Cryptology*, *19*, 463–487. doi:10.1007/s00145-006-0224-0

Global, E. (2010). *Class-1 generation-2 UHF air interface protocol standard*. Retrieved from http://www.epcglobalinc.org/standards/.

Gura, N., Patel, A., Wander, A., Eberle, H., & Shantz, S. C. (2004). *Comparing elliptic curve cryptography and RSA on 8-bit CPUs*. In CHES04.

Hancke, G. P. (2006). Practical attacks on proximity identification systems. In *IEEE Symposium on Security and Privacy,* (pp. 328-333).

Hancke, G. P., & Drimer, S. (2008). Secure proximity identification for RFID. In *Security in RFID and sensor networks* (pp. 170–194). CRC Press.

Hancke, G. P., & Kuhn, M. G. (2005). An RFID distance bounding protocol. In *Security and Privacy for Emerging Areas in Communications Networks (SecureComm05)* (pp. 67–73). IEEE. doi:10.1109/SECURECOMM.2005.56

Hancke, G. P., & Kuhn, M. G. (2008). Attacks on time-of-flight distance bounding channels. In *First ACM Conference on Wireless Network Security: WiSec08*, ACM, (pp. 194–202).

Hankerson, D., Menezes, A. J., & Vanstone, S. (2004). *Guide to elliptic curve cryptography*. Springer-Verlag.

Hoffstein, J., Howgrave-Graham, N., Pipher, J., Silverman, J. H., & Whyte, W. (2003). NTRU sign: Digital signatures using the NTRU lattice. In *Topics in Cryptology – The Cryptographers Track at the RSA Conferenc* (*Vol. 2612*, pp. 122–140). Lecture Notes in Computer Science Springer-Verlag.

Hoffstein, J., Pipher, J., & Silverman, J. H. (1998). NTRU: A ring-based public key cryptosystem. In *Algorithmic Number Theory: ANTS III* (*Vol. 1423*, pp. 267–288). Lecture Notes in Computer Science Springer-Verlag. doi:10.1007/BFb0054868

Hoffstein, J., Pipher, J., & Silverman, J. H. (2001). NSS: An NTRU lattice-based signature scheme. In *Advances in Cryptology (EUROCRYPT01), Vol. 2045*, (pp. 211–228).

Hoffstein, J., & Silverman, J. H. & report 012, W. W. N. (2003). *Estimated breaking times for NTRU lattices*. Technical Report 12, NTRU Cryptosystems, Inc.

Hutter, M. (2009). *RFID authentication protocols based on elliptic curves: A top-down evaluation survey*. In *International Conference on Security and Cryptography: SECRYPT09*, (pp. 101–110). Retrieved from https://online.tu-graz.ac.at/tug_online/voe_main2.get%volltext?pDocumentNr=106022

ISO/IEC 15693. (2006). *Identification cards – Contactless integrated circuit cards – Vicinity cards.*

ISO/IEC 14443. (2008). *Identification cards – Contactless integrated circuit cards – Proximity cards*

ISO. IEC 9798-2. (1993). Information technology security techniques entity authentication mechanisms- Part 2: Entity authentication using symmetric techniques. ISO/IEC.

Juels, A., & Weis, S. A. (2005). Authenticating pervasive devices with human protocols. In *Advances in Cryptology (Crypto05)* (*Vol. 3126*, pp. 198–293). Lecture Notes in Computer Science Springer.

Kaps, J.-P. (2006). *Cryptography for ultra-low power devices*. PhD thesis, ECE Department of Worcester Polytehcnic Institute.

Keller, M., & Marnane, W. (2007). Low power elliptic curve cryptography. In N. Azémard & L. J. Svensson (Eds.), *Integrated Circuit and System Design, 17th International Workshop on Power and Timing Modeling, Optimization and Simulation (PATMOS 2007), Lecture Notes in Computer Science, vol. 4644*, (pp. 310–319). Gothenburg, Sweden: Springer-Verlag.

Kim, C. H., & Avoine, G. (2009). RFID distance bounding protocol with mixed challenges to prevent relay attacks. In *International Conference on Cryptology and Network Security: CANS09, Lecture Notes in Computer Science*, (pp. 119–133). Berlin, Germany: Springer-Verlag.

Kim, S., Kim, Y., & Park, S. (2007). RFID security protocol by lightweight ECC algorithm. In *Sixth International Conference on Advanced Language Processing and Web Information Technology* (ALPIT07), IEEE Computer Society, (pp. 323–328).

Lenstra, A. K., & Verheul, E. R. (2000). The XTR public key system. In *Advances in Cryptology (CRYPTO 2000), Lecture Notes in Computer Science, vol. 1880*, Springer-Verlag, (pp. 1–19).

Lenstra, A. K., & Verheul, E. R. (2001). Selecting cryptographic key sizes. *Journal of Cryptology: The Journal of the International Association for Cryptologic Research, 14*(4), 255–293.

Li, T., & Deng, R. (2007). Vulnerability analysis of EMAP - an efficient RFID mutual authentication protocol. In *International Conference on Availability, Reliability and Security*, (pp. 238–245).

Li, T., & Wang, G. (2007). Security analysis of two ultra-lightweight RFID authentication protocols. In *IFIP-SEC07*, (pp. 14–16).

Lim, D., Lee, J., Gassend, B., Suh, G., Dijk, M., & Devadas, S. (2005). Extracting secret keys from integrated circuits. *IEEE Transactions on Very Large Scale Integration (VLSI). Systems, 13*(10), 1200–1205.

Majzoobi, M., Koushanfar, F., & Potkonjak, M. (2009). Techniques for design and implementation of secure reconfigurable PUFs. *ACM Transactions on Reconfigurable Technology and Systems, 5*(2), 1–33. doi:10.1145/1502781.1502786

Menezes, A. J., van Oorschot, P. C., & Vanstone, S. A. (1996). *Handbook of applied cryptography*. CRC Press. doi:10.1201/9781439821916

Molnar, D., & Wagner, D. (2004). Privacy and security in library RFID: Issues, practices, and architectures. In *Conference on Computer and Communications Security* (CCS04), ACM, (pp. 210–219).

Munilla, J., & Peinado, A. (2008). Distance bounding protocols for RFID enhanced by using void-challenges and analysis in noisy channels. *Wireless Communication and Mobile Computing, 8*(9), 1227–1232. doi:10.1002/wcm.590

Munilla, J., & Peinado, A. (2010). Enhanced low-cost RFID protocol to detect relay attacks. In *Wireless Communications and Mobile Computing*, Vol. 10, (pp. 361–371).

Okamoto, T. (1992). Provably secure and practical identification schemes and corresponding signature schemes. In *Advances in Cryptology (CRYPTO92)* (*Vol. 740*, pp. 31–53). Lecture Notes in Computer Science Springer.

Peris-Lopez, P., Castro, J. C. H., Estévez-Tapiador, J. M., & Ribagorda, A. (2006a). EMAP: An efficient mutual authentication protocol for low-cost RFID tags. In *OTM Workshops On The Move to Meaningful Internet Systems* (*Vol. 4277*, pp. 352–361). Lecture Notes in Computer Science Springer-Verlag. doi:10.1007/11915034_59

Peris-Lopez, P., Castro, J. C. H., Estévez-Tapiador, J. M., & Ribagorda, A. (2006*b*). *LMAP: A real lightweight mutual authentication protocol for low-cost RFID tags*. In Workshop on RFID and Lightweight Cryptography.

Peris-Lopez, P., Castro, J. C. H., Estévez-Tapiador, J. M., & Ribagorda, A. (2009). Advances in ultra-lightweight cryptography for low-cost RFID tags: Gossamer protocol. In *International Workshop Information Security Applications: WISA08, Lecture Notes in Computer Science, vol. 5379*, (pp. 56–68). Springer.

Peris-Lopez, P., Hernandez-Castro, J. C., Estevez-Tapiador, J. M., & Ribagorda, A. (2006). M2AP: A minimalist mutual-authentication protocol for low-cost RFID tags. In *UIC06* (*Vol. 4159*, pp. 912–923). Lecture Notes in Computer Science Springer-Verlag. doi:10.1007/11833529_93

Pramstaller, N., Mangard, S., Dominikus, S., & Wolkerstorfer, J. (2005). Efficient AES implementations on ASICs and FPGAs. In *Advanced Encryption Standard (AES05)* (*Vol. 3373*, pp. 98–112). Lecture Notes in Computer Science Springer. doi:10.1007/11506447_9

Reid, J., Nieto, J. M. G., Tang, T., & Senadji, B. (2007). Detecting relay attacks with timing-based protocols. In *2nd ACM Symposium on Information, Computer and Communications Security: ASIACCS07*, (pp. 204 – 213).

Research, C. (2000). *Standards for efficient cryptography - SEC2: Recommended elliptic curve domain parameters*. Retrieved from http://www.secg.org/download/aid-386/sec2_final.pdf

RFIDJournal. (2005). *Attack on a cryptographic RFID device*. Retrieved from http://www.rfidjournal.com/article/view/1415/1/82

Schnorr, C. (1989). Efficient identification and signatures for smart cards. In *Advances in Cryptology (CRYPTO89)* (*Vol. 435*, pp. 239–252). Lecture Notes in Computer Science Springer.

Singelée, D., & Preneel, B. (2007). Distance bounding in noisy environments. In *European Workshop on Security and Privacy in Ad-Hoc and Sensor Networks: ESAS07, Lecture Notes in Computer Science, vol. 4572*, (pp. 101–115). Springer-Verlag.

Stinson, D. R. (2002). *Cryptography, theory and practice* (2nd ed., pp. 73–116). Chapman and Hall/CRC.

Yuksel, K. (2004). *Universal hashing for ultra-low-power cryptographic hardware applications*. Master's thesis, Worcester Polytechnic Institute, Worcester, Massachusetts, USA.

Section 4
Privacy

Chapter 8
Privacy Issues in RFID

Boyeon Song
Korea Institute of Science and Technology Information, Republic of Korea

ABSTRACT

The chapter first investigates privacy issues in RFID systems, namely information privacy threats and location privacy threats. RFID systems should be able to resist tag information leakage and tag tracking attacks. Next, the author presents a few formal models in which the notion of privacy in RFID systems is defined. To measure the privacy level of various RFID protocols, a formal privacy definition is needed. Formal models for RFID systems are continually being developed. Here, the chapter describes definitions of RFID systems, adversaries, experiments, and privacy in the most popular models so far: the Avoine model, the Juels-Weis model, and the Vaudenay model.

8.1 INTRODUCTION

Radio frequency identification (RFID) is a technology for automatic identification and data capture of objects such as products, animals or persons, using radio frequency. An RFID system typically consists of tags, readers and a back-end server. A tag is an electronic device that consists of an antenna for receiving and transmitting a radio-frequency signal, and an integrated circuit for modulating and demodulating the signal and storing and processing information (Juels, 2006; Song & Mitchell, 2009; 2011). When a reader emits a radio-frequency signal via its antenna, any tag within range of the signal responds with certain stored data, such as a tag identifier (Juels, 2006; Song & Mitchell, 2009; 2011). The reader then passes the received tag data to the back-end server for further processing, including tag identification and information retrieval (Song & Mitchell, 2009; 2011).

Key features of RFID tags include that it has a unique identifier and that it can be scanned efficiently by RFID readers out of the line of sight, without requiring physical contact (Juels, 2006; Song & Mitchell, 2009; 2011). Thus, this identification technology has been applied in a wide variety of fields including product management, animal supervision, transportation payments, library book administration, and entry access

DOI: 10.4018/978-1-4666-3685-9.ch008

control (Ayoade, 2007; Dimitriou, 2008; Garfinkel, Juels, & Pappu, 2005; Juels, 2006; Song & Mitchell, 2009; 2011).

However, the use of such tags gives rise to security and privacy concerns for the following reasons. RFID readers and RFID tags communicate via a wireless channel using radio-frequency (Song, 2008). Thus, interactions between readers and tags are susceptible to eavesdropping (Song, 2008). Also, each RFID tag has a unique value that is used for identification (Song, 2008). If a tag emits its fixed value to every reader that queries it, then the location of the tag can be tracked by an attacker, and thus the privacy of the tag holder could be invaded (Song, 2008). Moreover, an RFID tag is typically designed to be inexpensive for mass distribution. Such a low-cost tag has limited memory capacity and processing ability, and its memory is typically not tamper-resistant (Juels, 2006; Song, 2008).

In this chapter, we focus on privacy issues, including the possibilities of user information leakage and location tracking. If hidden malicious readers can scan a tag in reading range, they may be able to retrieve information stored in the tag and trace the tag's location, or the location of the tag's holder, without his or her knowledge, thereby leading to a violation of user privacy and tracking of individuals by the tags they carry (Dimitriou, 2008). Thus, for wide usage of RFID technology, the protection of privacy is a necessary requirement. That is, only authorised readers should be able to read the tag and get information related to the tag.

In the following sections, we first describe privacy concerns in RFID systems and then introduce a few formal definitions of RFID privacy. Next, we suggest future work and summarise this chapter.

8.2 RFID PRIVACY ISSUES

This section describes why user privacy is the main concern of users of RFID systems, what kind of privacy issues exist and what privacy requirements in RFID systems are.

Privacy itself is defined in dictionaries as "a state in which one is not observed or disturbed by others"(Oxford dictionary), "someone's right to keep their personal matters and relationships secret" (Cambridge dictionary), "the quality or state of being apart from company or observation: freedom from unauthorised intrusion" (Merriam-Webster dictionary).

It can be asked: How can the privacy of RFID tag users be threaten? Can users of tags be free from observation or disturbance by unauthorised entities?

For the answer, we will first look at why RFID systems are useful. RFID tags have the following key features (Juels, 2006):

- **Unique identification:** Each RFID tag has its unique identifier or serial number that enables it to be distinguished from many millions of identically manufactured objects. The unique identifers in RFID tags can act as pointers to database entries containing information for individual items.
- **Automation:** RFID tags are readable without line-of-sight contact and do not require careful physical positioning on a RFID reader. RFID readers can scan tags at rates of hundreds per second.

These properties of RFID tags allow accuracy and efficiency in managing products, animals or people, leading to speediness, convenience and productivity in many possible applications: for example, warehouse management and logistics, product tracking in a supply chain, transport payments, livestock identification, vehicle identification, library book administration, patient medical care, etc. (Ayoade, 2007; Dimitriou, 2008; Garfinkel, Juels, & Pappu, 2005; Juels, 2006).

Along with such benefits, however, the features of RFID tags can pose privacy threats. The communication between RFID tags and readers is wireless. Identifying of tags is automatic, without physical contact and out of sight. An RFID tag does not typically know whether authorised RFID readers read it, or unauthorised readers do. That is, tags in reading range of a reader will automatically respond to the reader's query (Juels, 2006), and the interactions will occur without alerting the tags' users. Thus, if an RFID tag replies with its unique identifier whenever queried, or a reader can distinguish a tag's replies from random numbers, the reader may be able to trace the tag (Juels, 2006). If an RFID tag replies to a query with its stored data, then an unauthorised reader may be able to do clandestine scanning and get information from the tag (or information about the item attached to the tag) (Juels, 2006). In such cases, users carrying RFID tags may broadcast the tags' identifiers to nearby readers, allowing tag location tracking and/or tag information leakage. The threats to privacy are worse when a tag identifier is linked to its user information.

Figure 1 presents a good scenario to show how personal privacy can be invaded by RFID tags. Suppose that tags are attached to individual items, the tags carry information about the items, and unauthorised readers can read the tags and get the information about the items. Then an adversary can silently scan what items a person carries and harvest the person's information: what types of credit cards she carries, what sort of clothes she wears, what she bought, where she is going, and further what her interests, tastes and preferences are. That is, a person carrying tagged items is subject to clandestine inventorying, if she does not care about such privacy invasion. Even if users care about privacy, it is difficult to remove RFID tags, because tags are tiny and may be hidden or embedded inside a product (Hunt, Puglia, & Puglia, 2007).

Hunt, Puglia, and Puglia (2007) point out the following reasons for RFID privacy problems:

- Hidden placement of tags in documents and products which cannot be detected by individuals,
- Unique identifiers for all objects worldwide which leads to linking of purchaser or owner at the point of sale or transfer,
- Massive data aggregation which could lead to linking with personal identifying data,
- Hidden readers which could read tags from a distance without the individual knowing,

Figure 1. Privacy invasion example

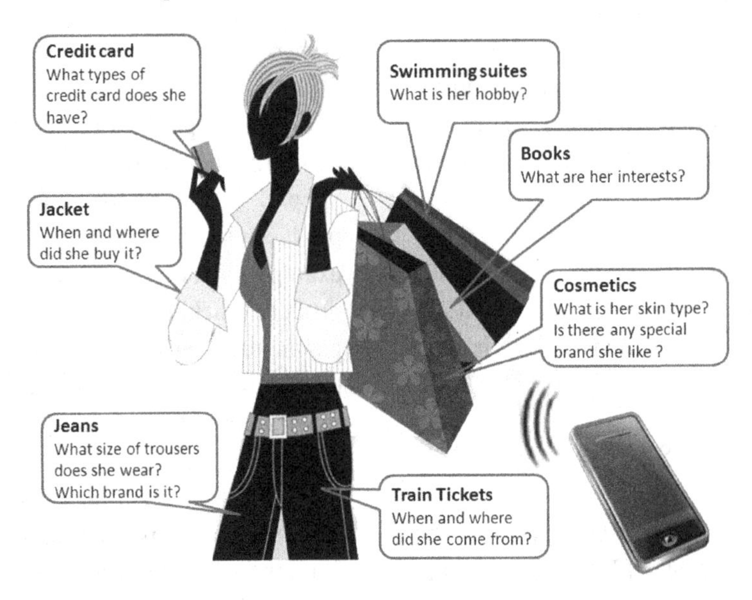

- Individual tracking and profiling which is possible as individuals could be mapped to their RFID tags without their knowledge or consent.

We define two major privacy issues as the following (Avoine, 2005; Juels, 2006; Ohkubo, Suzuki, & Kinoshita, 2003; Ohkubo, Suzuki, & Kinoshita, 2005; Weis, Sarma, Rivest, & Engels, 2003; Song & Mitchell, 2009; 2011):

- **Tag Information Leakage:** In a typical RFID system, when a reader queries a tag, the tag responds with its identifier. If unauthorised entities can also obtain a tag identifier, then they may be able to request and obtain the private information related to the tag held in the server database. For example, if the information associated with a tag attached to a passport, ID-card or medical record could be obtained by any reader, then the damage would be very serious.
- **Tag Tracking:** If the responses of a tag are linkable to each other or distinguishable from those of other tags, then the location of a tag could be tracked by multiple collaborating unauthorised entities. For example, if the response of a tag to a reader query is a static ID code, then the movements of the tag can be monitored, and the social interactions of an individual carrying a tag may be available to third parties without him or her knowing.

8.2.1 Privacy Requirements

RFID systems have the following privacy requirements in line with the two privacy threats described above.

- **Tag Information Privacy:** RFID systems should be able to resist tag information leakage.
- **Tag Location Privacy:** RFID systems should be able to resist tag tracking attacks.

To protect against privacy threats, RFID systems need to be controlled so that only authorised entities are able to access a tag and the information associated with it. A variety of technical solutions for privacy protection have been proposed, including killing, sleeping, blocking or hiding tags (Ayoade, 2007; Dimitriou, 2008; Garfinkel, Juels, & Pappu, 2005; Juels, 2006; Langheinrich, 2009; Ohkubo, Suzuki, & Kinoshita, 2005).

One approach is to use a tag authentication scheme with encryption techniques in which a tag is both identified and verified in a manner that does not reveal the tag identity to an eavesdropper (Song & Mitchell, 2011). A large number of tag authentication protocols of this type have been proposed. Also, if messages from tags are anonymous to each reader query, then the problem of tag tracking by unauthorised entities can be avoided. The concept of this anonymity includes *indistinguishability* (Ohkubo, Suzuki, & Kinoshita, 2003) in which a tag's outputs must be indistinguishable from truly random values; moreover, they should be unlinkable to an identifier of the tag (Ohkubo, Suzuki, & Kinoshita, 2003). To achieve such anonymity, many RFID protocols use pseudonyms based on random numbers; whenever a tag is queried, it responds with a different cryptographically derived pseudonym (see for example Dimitriou, 2008; Lim & Kwon, 2006; Ohkubo, Suzuki, & Kinoshita, 2005; Song & Mitchell, 2008; 2009; 2011; Weis, Sarma, Rivest, & Engels, 2003).

8.3 RFID PRIVACY MODELS

In this section we introduce formal definitions of privacy in RFID systems. A formal definition of privacy is necessary to measure the privacy level of various RFID protocols. To define privacy in RFID systems, the followings should be formalised: what privacy threats are possible, in other words, what potential adversaries exist, and what kind of RFID systems the adversaries are against. In general, a formal RFID privacy model defines RFID systems, capabilities of adversaries in a system, and the notion of privacy. The privacy level of RFID protocols is defined according to the abilities of the adversary.

In cryptography, a formal model usually takes the form of an *experiment*, a program that intermediates communications between a model adversary, characterised by a probabilistic algorithm (or Turing machine), and a model runtime environment containing system components (often called oracles) (Juels & Weis, 2007). The characterisation of the adversary is specified by the actions that she is allowed to perform (i.e. the oracles she can query), the goal of her attack (i.e. the game she plays) and the way in which she can interact with the system (i.e. the rules of the game) (Vaudenay, 2007).

Recently, some noteworthy formal RFID models have been designed. Here, we give a brief description of three main models among them.

8.3.1 The Avoine Model

Avoine (2005) proposed a formalism for traceability suited to RFID protocols, and defined the notions of existential and universal untraceability.

In the Avoine model, definitions of RFID systems and adversaries in a system are as follows:

- **RFID systems:** An RFID system typically consists of entities, i.e. a back-end server, readers and tags, and communication channels. In the model, a back-end server and readers are considered a single entity, named 'reader'. A tag $\mathcal{T}$ and a reader $\mathcal{R}$ participate in an RFID protocol $\mathcal{P}$. Each of them can run several instances of $\mathcal{P}$. Tag instances are denoted by $\pi_{\mathcal{T}}^{i}$ and reader instances by $\pi_{\mathcal{R}}^{j}$, where i and j are integers. The communication channels are divided into two types: the channel from a reader to a tag, called the forward channel, and the channel from a tag to a reader, called the backward channel. The former is generally longer than the latter, and thus the former is easier to eavesdrop on than the latter, because a reader emits more power than a tag.
- **Adversaries:** The means of the adversary are represented using oracles as follows:
  - **Query**$(\pi_{\mathcal{T}}^{i}, m_1, m_3)$: This query models $\mathcal{A}$ sending request m_1 to $\mathcal{T}$ through the forward channel and subsequently sending it message m_3 after having received its answer.
  - **Send**$(\pi_{\mathcal{R}}^{j}, m_2)$: This query models $\mathcal{A}$ sending message m_2 to $\mathcal{R}$ through the backward channel and receiving its answer.
  - **Execute**$(\pi_{\mathcal{T}}^{i}, \pi_{\mathcal{R}}^{j})$: This query models $\mathcal{A}$ executing an instance of $\mathcal{P}$ between $\mathcal{T}$ and $\mathcal{R}$, obtaining the messages exchanged on both the forward and the backward channels.
  - **Execute*** $(\pi_{\mathcal{T}}^{i}, \pi_{\mathcal{R}}^{j})$: This query models $\mathcal{A}$ executing an instance of $\mathcal{P}$ between $\mathcal{T}$ and $\mathcal{R}$, but obtaining the messages exchanged on the forward channel only.
  - **Reveal** $(\pi_{\mathcal{T}}^{i})$: This query models $\mathcal{A}$ obtaining the content of $\mathcal{T}$'s memory channel. This query can be used only once and Query, Send, Execute, and Execute* cannot be used after it.

It is said that *a protocol is resistant to attacks* $\mathcal{A}-\mathcal{O}$ or that *it is* $\mathcal{A}-\mathcal{O}$ if it is resistant to an attack $\mathcal{A}$ when the adversary has access to the oracles of $\mathcal{O} \subset \{Q, S, E, E^*, R\}$, where Q, S, E, E^* and R represent, respectively, the oracles Query, Send, Execute, Execute* and Reveal.

- **Privacy Experiment:** An interaction is a set of executions on the same tag at a time when the adversary is in a position to physically identify it. It is more precisely defined by $\Omega_I(\mathcal{T}) = \{\omega_i(\mathcal{T}) \mid i \in I\} \cup \{\text{Send}(\pi_*^i, *) \mid i \in J\}$ where $J \subset \mathbb{N}$, I is a sub-interval of $\mathbb{N}$, and $w_i(\mathcal{T})$ is the result of the application of an oracle Q, E, E^* or R on a tag $\mathcal{T}$: $w_i(\mathcal{T}) \in \{\text{Query}(\pi_\mathcal{T}^i, m_1, m_3)$, Execute $(\pi_\mathcal{T}^i, \pi_\mathcal{R}^j)$, Execute* $(\pi_\mathcal{T}^i, \pi_\mathcal{R}^j)$, Reveal $(\pi_\mathcal{T}^i)\}$.

Oracle allows us to simulate the set of oracles to which the adversary has access; *Oracle* calls the oracles of $\mathcal{O} \subset \{Q, S, E, E^*, R\}$ according to the model chosen, and sends back $\hat{\Omega}_I(\mathcal{T})$, being given $\mathcal{T}$ and I, where $\hat{\Omega}_I(\mathcal{T})$ is the interaction which maximise the adversary's advantage. The *Challenger* supplies two tags to the adversary, one of which is the target tag.

- **Definition of Privacy:** The notion of untraceability is defined as follows:
  - **Existential Untraceability:** (Parameters: $l_{ref}, l_{chal}, \mathcal{O}$)
    - $\mathcal{A}$ queries the *Challenger* thus receiving her target $\mathcal{T}$. $\mathcal{A}$ chooses I and calls *Oracle*$(T, I, \mathcal{O})$ where $|I| \leq l_{ref}$ and l_{ref} is a given parameter, then receives $\hat{\Omega}_I(\mathcal{T})$.
    - $\mathcal{A}$ queries the *Challenger* thus receiving her challenge $\mathcal{T}_1$ and $\mathcal{T}_2$. $\mathcal{A}$ chooses I_1 and I_2 such that $|I_1| \leq l_{chal}, |I_2| \leq l_{chal}$, and $(I_1 \cup I_2) \cap I = \varnothing$, where l_{ref} and l_{chal} are given lengths. $\mathcal{A}$ calls *Oracle*$(\mathcal{T}_1, I_1, \mathcal{O})$ and *Oracle*$(\mathcal{T}_2, I_2, \mathcal{O})$, then receives $\hat{\Omega}_I(\mathcal{T}_1)$ and $\hat{\Omega}_I(\mathcal{T}_2)$. $\mathcal{A}$ decides which of $\mathcal{T}_1$ or $\mathcal{T}_2$ is $\mathcal{T}$, then outputs her guess $\mathcal{T}'$.
  - **Universal Untraceability:** (Parameters: $l_{ref}, l_{chal}, \mathcal{O}$)
    - $\mathcal{A}$ queries the *Challenger* thus receiving her target $\mathcal{T}$. $\mathcal{A}$ chooses I and calls *Oracle*$(T, I, \mathcal{O})$ where $|I| \leq l_{ref}$, then receives $\hat{\Omega}_I(\mathcal{T})$.
    - $\mathcal{A}$ queries the *Challenger* thus receiving her challenge $\mathcal{T}_1, \mathcal{T}_2, I_1$ and I_2. $\mathcal{A}$ calls *Oracle*$(\mathcal{T}_1, I_1, \mathcal{O})$ and *Oracle*$(\mathcal{T}_2, I_2, \mathcal{O})$, then receives $\hat{\Omega}_I(\mathcal{T}_1)$ and $\hat{\Omega}_I(\mathcal{T}_2)$. $\mathcal{A}$ decides which of $\mathcal{T}_1$ or $\mathcal{T}_2$ is $\mathcal{T}$, then outputs her guess $\mathcal{T}'$.

The advantage of $\mathcal{A}$ for a given protocol $\mathcal{P}$ is defined by:

$$\text{Adv}_\mathcal{P}^{\text{UNT}}(\mathcal{A}) = [2\Pr(T' = T) - 1]$$

where the probability space is over all the random tags. If $\mathcal{A}$'s advantage is negligible with the parameters l_{ref}, l_{chal} and $\mathcal{O}$, $\mathcal{P}$ *is said to be UNT* $_{l_{ref}, l_{chal}}-\mathcal{O}$, usually simply denoted by UNT-$\mathcal{O}$.

8.4.2 The Juels-Weis Model

Juels & Weis (2007) proposed a simple formal definition of strong privacy, which is similar in flavour to the Avoine model described above. Avoine captures a range of adversarial abilities, while Juels and Weis characterise a very strong adversary with a relatively simple definition. Strong privacy is a strong notion, because the adversary is given a lot of power; she can corrupt any number of tags and read their contents, and can eavesdrop and schedule the tag/reader communication any way she wants (Damgård & Østergaard, 2006). In contrast to the Avoine model, the Juels-Weis model has no difference in the ability of an adversary to eavesdrop on the forward channel and the backward channel. They use the term *functionality* to describe tags and readers, instead of using *oracle*.

RFID systems: An RFID system comprises a single reader $\mathcal{R}$ and a set of n tags $\mathcal{T}_1, \cdots, \mathcal{T}_n$. Each party is a probabilistic interactive Turing machine with an independent source of randomness and unlimited internal storage. Tags and readers are modeled as ideal functionalities. Functionalities have a well-defined interface, may receive messages, and may respond with messages of their own. Tags and readers may maintain internal states, are randomised, and may adaptively change their output.

Tag Functionalities: Each tag functionality $\mathcal{T}_i$ stores an internal secret key and a session identifier sid. SETKEY messages make a tag assign a new key and disgorge its current key. The message (TAGINIT, sid) makes a tag set to a new value sid and delete information associated with an existing sid. In other words, a tag may be involved in only one protocol session at a time. When a tag receives a challenge c_j, it returns a response r_j, which may be null.

Reader Functionalities: The reader functionality initialises a new session upon receipt of a message of the form READERINIT. When receiving a READERINIT message, $\mathcal{R}$ generates a fresh session identifier sid and the first challenge of an interactive challenge-response protocol c_0. For each READERINIT received, the reader creates a new internal entry of the form $(sid, \text{`open'}, c_0)$. Any responses containing sid are appended to that entry, as well as subsequent challenges, or any other auxiliary data. This entry is marked as 'closed' and becomes read-only when the reader ultimately accepts or rejects a session.

Functionality Interaction: The sid values generated by passing READERINIT messages to $\mathcal{R}$ may be passed as a TAGINIT message to some $\mathcal{T}$. Once a tag is initialised with a sid, it can respond to challenge c_j with response r_j. This response can be a function of the key, sid, the $j-1$ previous challenge-response pairs, or locally generated randomness. The tag may internally log both the latest received c_j and the response r_j.

Similarly, reader functionality $\mathcal{R}$ responds to messages of the form (sid, r_j) by first searching for a $(sid, \text{`open'}, c_0, r_0, \cdots, c_j)$ session entry, if any exists. At this point, $\mathcal{R}$ performs a verification step by computing a function over its entire internal state, including all open and closed sessions and any internal key material. $\mathcal{R}$ can output an 'accept' or a 'reject' and mark the session entry as $(sid, \text{`open'}, c_0, r_0, \cdots, c_j, r_j)$. Alternatively, $\mathcal{R}$ can compute the next challenge c_{j+1} and append it to the corresponding session entry.

Adversaries: An adversary, $\mathcal{A}$ can issue its own SETKEY, TAGINIT, READERINIT, challenge, and response messages. The number of READERINIT messages sent by $\mathcal{A}$ is parameterised by r, the number of computational steps it performs by s, and the number of TAGINIT message sent by t. Any tag that receives a SETKEY message from the adversary is considered 'corrupted'.

Privacy Experiment: The idea is that an RFID protocol may be considered private for some parameter values if no adversary has a significant advantage in this experiment. The goal of the adversary in this experiment is to distinguish between two different tags within the limits of its computational power and functionality-call bounds. The privacy experiment for an RFID system is denoted by $\mathbf{Exp}^{priv}_{\mathcal{A},\mathcal{S}}[k, n, r, s, t]$. Here,

$\mathcal{S} = (GEN, \mathcal{R}, \{\mathcal{T}_i\})$, where $\{\mathcal{T}_i\}$ contains n tags, and k is a security parameter. Adversary $\mathcal{A}$ with parameters r, s, and t is denoted by $\mathcal{A}[r,s,t]$, where r, s and t are the respective parameters for reader initialisations, computation steps, and tag initialisations.

8.4.2.1 Experiment $\mathbf{Exp}_{\mathcal{A},\mathcal{S}}^{priv}[k,n,r,s,t]$:

- Setup
  - $\text{GEN}(1^k) \to (key_0, \cdots, key_n)$
- Initialise $\mathcal{R}$ with $(key_0, \cdots, key_n)$.
- Set each $\mathcal{T}_i$'s key to key_i with a SETKEY call.
  - **Phase 1 (Learning):** $\mathcal{A}$ may do the following in any interleaved order; i.e. calls of the same type need not be consecutive.
- Make READERINIT calls, without exceeding r overall calls.
- Make TAGINIT calls, without exceeding t overall calls.
- Make arbitrary SETKEY calls to any $(n-2)$ tags.
- Communicate and compute, without exceeding s overall steps.
  - Phase 2 (Challenge)
- $\mathcal{A}$ selects two tags, $\mathcal{T}_i$ and $\mathcal{T}_j$ to which it has not sent SETKEY messages.
- Let $\mathcal{T}_0^* = \mathcal{T}_i$ and $\mathcal{T}_1^* = \mathcal{T}_j$ and remove both of these from the current tag set.
- Let $b \xleftarrow{R} \{0,1\}$ and provide $\mathcal{A}$ access to $\mathcal{T}_b^*$.
- $\mathcal{A}$ may do the following in any interleaved order:
  - Make READERINIT calls, without exceeding r overall calls.
  - Make TAGINIT calls, without exceeding t overall calls.
  - Make arbitrary SETKEY calls to any tag in the current tag set except $\mathcal{T}_b^*$.
  - Communicate and compute, without exceeding s overall steps.
- $\mathcal{A}$ outputs a guess bit $b = b'$.
  - Exp succeeds if $b = b'$
  - Definition of Privacy

A protocol run within an RFID system $\mathcal{S} = (GEN, \mathcal{R}, \mathcal{T}_i\})$ is defined to be *private* if no adversary $\mathcal{A}[r,s,t]$ has a non-negligible advantage in successfully guessing b in the experiment above.

Definition (RFID (r,s,t) – Privacy): A protocol initiated by $\mathcal{R}$ in an RFID system $\mathcal{S} = (GEN, \mathcal{R}, \{\mathcal{T}_i, \cdots, \mathcal{T}_i\})$ with security parameter k is (r,s,t) – private if: $\forall \mathcal{A}[r,s,t] \Pr[\mathbf{Exp}_{\mathcal{A},\mathcal{S}}^{priv}[k,n,r,s,t]$ succeeds in guessing $b] \leq 1/2 + 1/poly(k)$

The variables r, s and t can be functions of system parameters like k, if desired. The notation $poly(k)$ is used to represent any polynomial function of k.

8.3.3 The Vaudenay Model

Vaudenay (2007) proposed a formal model for identification schemes, giving definitions of strong security and strong privacy. He defined the privacy property as the ability to resist adversaries who aim at identifying, tracing, or linking tags. The model focuses on passive tags which are exempt from side channel attacks except by full corruption.

A function in terms of a security parameter k is said to be *polynomial* if there exists a constant n such that it is $O(k^n)$. A function is said to be *negligible* if there exists a constant k such that it is $O(x^{-k})$.

RFID Systems: An RFID scheme is composed of:

- **SetupReader** (1^k) : A setup scheme which generates a private/public key pair, (K_S, K_P) for the reader dependent on a security parameter k. The key K_S is stored in the reader backend. The key K_P is publicly released.
- **SetupTag** $_{K_P}$ (ID): A polynomial-time algorithm which returns (K, S), the tag specific secret K and the initial state S of the tag. The pair (ID, K) is stored in the reader backend if the tag is legitimate.
- A polynomial-time interactive protocol between a reader and a tag in which the reader ends with a tape Output.

Adversaries: In the model, at every step, a tag can either be a free tag or a drawn tag. Drawn tags are the ones within visual contact of the adversary so that she can communicate while being able to link communications. Free tags are all the other tags.

An adversary is a polynomial-time algorithm which takes a public key K_P as input and runs by using the eight following oracles:

- **CREATETAG**^{b} (ID): This oracle creates a free tag, either legitimate $(b = 1)$ or not $(b = 0)$, with unique identifier ID. This oracle uses the SetupTag $_{K_P}$ algorithm to set up the tag and (for $b = 1$ only) to update the system database. By convention, b is implicitly 1 when omitted.
- **DRAWTAG:** This oracle moves from the set of free tags to the set of drawn tags, a set of tags chosen at random following the probability distribution distr(which is specified by a polynomially bounded sampling algorithm). The oracle returns a vector of fresh identifiers. Drawing tags already drawn or non-existent provoke the oracle to return in place of the respective virtual tag. Further, this oracle returns bits telling whether the drawn tags are legitimate or not. This oracle keeps a hidden table **T** such that $T(vtag)$ is the ID of $vtag$.
- **FREE** $(vtag)$: This oracle moves the virtual tag $vtag$ back to the set of free tags. This makes $vtag$ unreachable.
- **LAUNCH** $\rightarrow \pi$: This oracle makes the reader launch a new protocol instance of π.
- **SENDREADER** $(m, \pi) \rightarrow m'$ (resp. SENDTAG $(m, vtag) \rightarrow m'$) : This oracle sends a message m to a protocol instance of π for the reader (resp. to the virtual tag $vtag$) and receives the answer m'.
- **RESULT** $(\pi) \rightarrow x$: When π is complete, this oracle returns 1 if Output $\neq \perp$, and 0 otherwise.
- **CORRUPT** $(vtag) \rightarrow S$: This query returns the current state S of a tag. If $vtag$ is no longer used after this oracle call, we say that $vtag$ is destroyed.

Privacy Experiment: The adversary uses the oracles following the rules of the game and produces an output. Depending on the rules, the adversary wins or loses.

Polynomial-time adversaries are classified as follows, depending on corruption:

- **STRONG:** The class of adversaries who have access to the oracles above.
- **DESTRUCTIVE:** The class of adversaries who never use $vtag$ again after a CORRUPT $(vtag)$ query (i.e. who destroy it).
- **FORWARD:** The class of adversaries in which CORRUPT queries can only be followed by other CORRUPT queries.
- **WEAK:** The class of adversaries who do no CORRUPT queries.
- **NARROW:** The class of adversaries who do no RESULT queries. That is, adversaries who do not have access to side channel information are called narrow adversaries.

Definition of Privacy: Privacy is defined in terms of the ability to infer non-trivial ID relations from protocol messages. This generalises the notions of anonymity (where the ID of a tag cannot be inferred) and untraceability (where the equality of two tags cannot be inferred).

Definition (Privacy): Adversaries are considered who start with an attack phase allowing oracle queries, then pursue an analysis phase with no oracle queries. In between phases, the adversary receives the hidden table **T** of the DRAWTAG oracle, then outputs either true or false. The adversary wins if the output is true. It is said that the RFID scheme is $P-private$ if all such adversaries which belong to class P are *trivial* following the definition described below.

Definition (Blinder, trivial adversary): A blinder $\mathcal{B}$ for an adversary $\mathcal{A}$ is a polynomial-time algorithm which sees the same messages as $\mathcal{A}$ and simulates the LAUNCH, SENDREADER, SENDTAG, and RESULT oracles to $\mathcal{A}$. The blinder does not have access to the reader tapes so does not know the secret key nor the database. A blinded adversary $\mathcal{A}^{\mathcal{B}}$ is itself an adversary who does not use the LAUNCH, SENDREADER, SENDTAG, and RESULT oracles. An adversary $\mathcal{A}$ is *trivial* if there exists a $\mathcal{B}$ such that $|\Pr[\mathcal{A}\text{wins}] - \Pr[\mathcal{A}^{\mathcal{B}}\text{wins}]|$ is negligible.

Informally, an adversary is trivial if it makes no effective use of protocol messages. Namely, these messages can be simulated without significantly affecting the probability of success of the adversary. This privacy notion measures the privacy loss in the wireless link but not through tag corruption; it is assumed that corrupting a tag always compromises privacy.

In the Vaudenay model, the eight privacy classes are defined, classified by different oracles collections and different natures on accessing CORRUPT$(vtag)$ according to the abilities of the adversary (Ng, Susilo, Mu and Safavi-Naini (2008)).

- **Weak privacy:** Weak privacy is a basic privacy class where access to all the oracles is allowed except CORRUPT$(vtag)$.
- **Forward privacy:** Forward privacy is less restrictive than Weak privacy since access to CORRUPT$(vtag)$ is allowed under the condition that after it has been accessed the first time, no other types of oracles can be accessed except for more CORRUPT $(vtag)$ (which can be used on different handles).
- **Destructive privacy:** Restrictions are further relaxed. Whenever CORRUPT$(vtag)$ is accessed, the handle $vtag$ cannot be used again (i.e. virtually destroyed). However, there are no further restrictions on any other handle.
- **Strong privacy:** This is even less restrictive than Destructive privacy. The condition for accessing CORRUPT$(vtag)$ is removed. It is the strongest defined privacy class.

Each of these privacy classes also has their Narrow counterparts, namely Narrow-Strong, Narrow-Destructive privacy, Narrow-Forward, and Narrow-Weak. These classes share the same definitions as their counterparts except that there is no access to RESULT(π).

8.3.4 Other Models

Lim & Kwon (2006) improved the Avoine model as a generic security model that can cover the weakest to the strongest possible security in RFID protocols and is more general and flexible by incorporating the various possible restrictions existing in RFID systems. They focused on the strongest privacy notion of untraceability: forward and backward untraceability.

Damgård & Østergaard (2006) proposed a simple extension of the Juels-Weis model, adding to completeness and soundness requirements, i.e., a reader should accept only valid tags. They provided a class of protocols offering a new range of tradeoffs between security and efficiency.

Ng, Susilo, Mu, and Safavi-Naini (2008) reduced the eight RFID privacy classes defined in the Vaudenay Model into three classes, i.e. strong, forward, and weak, under appropriate assumptions. They showed that the strongest privacy level is achievable, while Vaudenay postulated that achieving strong privacy in RFID is impossible.

In independent work, Burmester, van Le, and de Medeiros (2006) also proposed a universally composable security model for RFID systems based on the universal composability framework proposed by Canetti (2000). Their model is under the assumption of the existence of pseudo-random function families, without employing random oracles.

8.4 FUTURE RESEARCH DIRECTIONS

Privacy in RFID applications is a significant concern, as we have discussed above. As methods for privacy protection, a number of RFID protocols have been suggested. However, many of them do not provide security proofs in a formal model. Formal models for RFID systems are getting developed. So far, all existing work have their own limitations. The classes of protocols that achieve privacy in one model do not fit in another model. In addition, some RFID protocols are provably secure in formal models but fail to resist realistic attacks (Burmester, van Le, & de Medeiros, 2006). Thus, the main influential formal models accepted for RFID protocols are still under construction. The proposed models do not capture the full spectrum of real-world needs (Juels & Weis, 2007). Juels and Weis (2007) emphasised that "a fertile and essential area of investigation is definitions and protocols for RFID privacy that are weaker, but more practical and useful." Devising appropriate formalism for use in analysing the privacy level of RFID protocols remains a challenging and potentially fruitful topic.

CONCLUSION

In this chapter, we have discussed privacy issues in RFID systems. We have first described why privacy matters in usage of RFID tags, and have presented two privacy requirements, namely tag information privacy and tag location privacy. Next, we have introduced formal definitions of RFID privacy, especially three main formal models: the Avoine model, the Juels-Weis model and the Vaudenay model. We have also suggested future work.

REFERENCES

Avoine, G. (2005). *Cryptography in radio frequency identification and fair exchange protocols*. PhD thesis, Ecole Polytechnique Federale de Lausanne, Lausanne, Switzerland.

Avoine, G. (2005). *Adversary model for radio frequency identification* (Tech. Rep. LASEC-REPORT-2005-001). Swiss Federal Institute of Technology, Lausanne, Switzerland.

Ayoade, J. (2007). Roadmap to solving security and privacy concerns in RFID systems. *Computer Law & Security Report*, *23*(6), 555–561. doi:10.1016/j.clsr.2007.09.005

Burmester, M., van Le, T., & de Medeiros, B. (2006). Provably secure ubiquitous systems: Universally composable RFID authentication protocols. *Conference on Security and Privacy for Emerging Areas in Communication Networks – SecureComm 2006* (pp. 1-10). Baltimore, MD: IEEE Computer Society.

(2010). *Cambridge Dictionaries*. Cambridge University Press.

Canetti, R. (2000). *Universally composable security: A new paradigm for cryptographic protocols (Report 2000/067). Cryptology ePrint Archive.* International Association for Cryptologic Research.

Damgård, I., & Østergaard, M. (2006). *RFID security: Tradeoffs between security and efficiency (Report 2006/234). Cryptology ePrint Archive.* International Association for Cryptologic Research.

Dimitriou, T. (2008). RFID security and privacy. In P. Kitsos & Y. Zhang (Eds.), *RFID security: Techniques, protocols and system-on-chip design* (pp. 57-79). Springer Science+Business Media LLC.

Garfinkel, S. L., Juels, A., & Pappu, R. (2005). RFID privacy: An overview of problems and proposed solutions. [IEEE Computer Society.]. *IEEE Security and Privacy*, *3*(3), 34–43. doi:10.1109/MSP.2005.78

Hunt, D., Puglia, A., & Puglia, M. (2007). *RFID-A guide to radio frequency identification*. John Wiley & Sons, Inc. doi:10.1002/0470112255

Juels, A. (2006). RFID security and privacy: A research survey. *IEEE Journal on Selected Areas in Communications*, *24*(2), 381–394. doi:10.1109/JSAC.2005.861395

Juels, A., & Weis, S. (2007). Defining strong privacy for RFID. *International Conference on Pervasive Computing and Communications – PerCom 2007* (pp. 342-347). IEEE Computer Society.

Langheinrich, M. (2009). A survey of RFID privacy approaches. *Personal and Ubiquitous Computing*, *13*(6), 413–421. doi:10.1007/s00779-008-0213-4

Lim, C., & Kwon, T. (2006). Strong and robust RFID authentication enabling perfect ownership transfer. In P. Ning, S. Qing, & N. Li (Eds.), *International Conference on Information and Communications Security – ICICS'06, Lecture Notes in Computer Science: Vol. 4307*, (pp. 1-20). Springer.

Ng, C., Susilo, W., Mu, Y., & Safavi-Naini, R. (2008). RFID privacy models revisited. *13th European Symposium on Research in Computer Security – ESORICS 2008, Lecture Notes in Computer Science, 5283/2008*, 251-266. Springer.

Ohkubo, M., Suzuki, K., & Kinoshita, S. (2003). *Cryptographic approach to "privacy-friendly" tags*. Paper presented at RFID Privacy Workshop, MIT, Massachusetts, USA.

Ohkubo, M., Suzuki, K., & Kinoshita, S. (2005). RFID privacy issues and technical challenges. *Communications of the ACM*, *48*(9), 66–71. doi:10.1145/1081992.1082022

Song, B. (2008). *RFID tag ownership transfer.* Workshop on RFID Security - RFIDSec'08, Budapest, Hungary.

Song, B., & Mitchell, C. J. (2008). RFID authentication protocol for low-cost tags. In V. D. Gligor, J. Hubaux, & R. Poovendran (Eds.), *ACM Conference on Wireless Network Security – WiSec '08*, (pp. 140-147). Alexandria, VA: ACM Press.

Song, B., & Mitchell, C. J. (2009). Scalable RFID pseudonym protocol. *The 3rd International Conference on Network & System Security - NSS 2009*, (pp. 216-224). Gold Coast, Australia: IEEE Computer Society.

Song, B., & Mitchell, C. J. (2011). Scalable RFID security protocols supporting tag ownership transfer. [Elsevier.]. *International Journal Computer Communications*, *34*(4), 556–566. doi:10.1016/j.comcom.2010.02.027

Vaudenay, S. (2007). On privacy models for RFID. In Kurosawa, K. (Ed.), *Advances in Cryptology – Asiacrypt 2007* (*Vol. 4833*, pp. 68–87). Lecture Notes in Computer Science Springer. doi:10.1007/978-3-540-76900-2_5

Weis, S., Sarma, S., Rivest, R., & Engels, D. (2003). Security and privacy aspects of low-cost radio frequency identification systems. In D. Hutter, G. Müller, W. Stephan, & M. Ullmann (Eds.), *International Conference on Security in Pervasive Computing – SPC 2003, Lecture Notes in Computer Science, Vol. 2802*, (pp. 201-212). Springer-Verlag.

KEY TERMS AND DEFINITIONS

Back-End Server: It has a database containing information associated with the RFID tags which it manages, and can retrieve the detailed tag information using the tag response as a key.

Formal RFID Privacy Model: A model in which RFID systems, capabilities of adversaries in a system, and the notions of privacy and security are defined.

Privacy: A state in which one is not observed by unauthorised entities, and keeps his or her personal matters secret.

Radio Frequency Identification (RFID): A technology for automatic identification and data capture of objects such as persons, products or animals, using radio frequency.

RFID Reader: A device that can recognise the presence of RFID tags and read the information supplied by them. It emits radio frequency signals via its antenna to interrogate an RFID tag.

RFID Tag: An electronic device that consists of an antenna and an integrated circuit.

Tag Information Privacy: A state in which a tag user is free from leakage of information related to the tag.

Tag Location Privacy: A state in which a tag user is free from being tracked.

Chapter 9
DoS Attacks on RFID Systems:
Privacy vs. Performance

Dang Nguyen Duc
Auto-ID Lab Korea, KAIST, Republic of Korea

Kwangjo Kim
Auto-ID Lab Korea, KAIST, Republic of Korea

ABSTRACT

In this chapter, the authors discuss the impact of providing tag privacy on the performance of an RFID system, in particular the complexity of identifying the tags being queried at the back-end server. A common technique to provide tag privacy is to use pseudonyms. That is, for each authentication session, a tag uses a temporary and random-looking identifier so that it is infeasible for attackers to relate two authentication sessions. A natural question which should arise here is how the server can identify a tag given that the tag's identity is changing all the time. This problem becomes even more serious when the shared secret key between a tag and the server is updated after every authentication session to provide forward privacy. In the first part of this chapter, the authors review different techniques to deal with this problem. They then point out that most of the existing techniques lead to vulnerability of the back-end server against Denial-of-Service (DoS) attacks. They illustrate some of these attacks by describing methods which attackers can use to abuse the server's computational resources in several popular RFID authentication protocols. Finally, the authors discuss some techniques to address the privacy vs. performance dilemma so that DoS attacks can be prevented while keeping tag identification efficient.

9.1 INTRODUCTION

RFID is an emerging technology which promises many powerful applications in supply chain management, smart home appliances and ubiquitous computing, *etc*. The key idea is to attach to each and every object a low-cost, wirelessly readable RFID tag (by the so-called RFID readers). Each RFID tag carries a unique string, serving as an object identifier so that this identifier can be used as a pointer to look up detailed information on the corresponding object (at some database servers or simply referred to as a back-end server where all the detailed information on tagged objects are

DOI: 10.4018/978-1-4666-3685-9.ch009

stored and managed). The identifier that a tag carries is usually called the *Electronic Product Code* (EPC). Unfortunately, this very core operation of an RFID tag also causes serious security concerns. These concerns are two-fold.

- **Counterfeiting product:** When RFID is widely deployed, we will come to depend on it to recognize surrounding objects, especially merchandized objects. However, the object identifier stored in an RFID tag can be read by any compatible RFID reader. In addition, wireless communication is inherently insecure against eavesdropping attacks which might help attackers capture the object identifiers without querying the tags. Malicious parties can collect legitimate tag identifiers and create tags that emit the same identifiers. We call these tags *cloned tags*. The cloned tags can be placed on counterfeiting products to avoid being detected.
- **Consumer privacy:** The uniqueness and availability of object identifiers raise another side effect for consumers, their privacy. With the vision of RFID tags being everywhere, it will become common for a person to carry with him/her several RFID-tagged objects. Therefore, a malicious hacker equipped with a compatible RFID reader, can identify objects carried by RFID holders as well as trace their movements. For the rest of this chapter, we will use consumer privacy and tag privacy as two equivalent terms.

To deal with counterfeiting products, one can implement an authentication protocol between the reader and the tag so that fake tags are detected and malicious readers cannot collect useful information from legitimate tags. This can be done easily, even on low-cost hardware like RFID tags. Many RFID authentication protocols (Ohkubo et al. (2004)) which employ just a cryptographic hash function or other lightweight cryptographic primitives have been proposed so far. However, dealing with privacy issues is much more difficult. This is not because privacy is a new problem in cryptography. Rather, it is due to the cost of providing privacy which in some cases could require the back-end server to go through the whole tag database in order to identify a single tag. A common technique for providing privacy is to use pseudonyms for RFID tags instead of their true identifiers. That is, for each authentication session, a tag uses a temporary, random-looking identifier (thus, it is called pseudonym) so that it is infeasible for attackers to relate two authentication sessions, even of the same tag. However, if a tag emits a different identifier whenever it is queried, the back-end server will have difficulty in identifying the tag. This problem becomes even more serious when forward privacy is required. Forward privacy is a security notion which specifies that the privacy of a tag is still partially preserved even if the secret key of the tag is exposed. By partially preserved, we mean that the querying sessions which happened before the secret key exposure remain anonymous (or sometimes referred to as unlinkable). In other words, the privacy of the tag is guaranteed up to the point of the secret key exposure. To provide forward privacy, a common practice is to refresh the secret key after every querying session. Unfortunately, this could add additional burdens on the back-end server when identifying and authenticating a tag because the server has to take care of situations where the secret keys shared between the server and tags were inconsistently refreshed (we refer to this problem as *de-synchronization of secret*).

In this chapter, we will demonstrate how malicious parties can exploit the privacy-performance dilemma in several privacy-preserving RFID authentication protocols to abuse the back-end server's computational resources. This means if an attacker can mount the attack on a large scale, it can cause denial-of-service attack on the back-end server. We then propose a few approaches to prevent this type of attack.

9.2 BACKGROUND

In this section, we briefly review several RFID authentication protocols with privacy-preserving property. Since protocols mentioned in this chapter use different approaches to achieve privacy-preserving property, we categorize them based on the complexity of looking up a tag in a tag database. The notations used to describe these protocols are depicted in Table 1.

It is important to mention some conventions which we will use to describe several different protocols in a consistent manner. For a tag being queried or an unknown tag, we do not use any subscripts in the notations related to the tag like its secret key and pseudonym. However, the same notations for tags in the database will have subscripts. By default, we assume that there are N tags in the back-end database and therefore the subscripts will range from 1 to N. We will also use some C-language conventions when describing how an entity processes information. For example, "return $\mathrm{ACCEPT}(i)$" means a tag is successfully authenticated and identified as the tag numbered i in the database while "return REJECT" implies a failed authentication.

Table 1. Notations

Notation	Description
S	Back-end server
D	Tag database
N	Number of tags in D
T	RFID tag
$e()$	A one-way trapdoor function
$f()$	A pseudo-random function
$h(), g()$	Two cryptographic hash functions
PRNG(.)	A pseudo-random number generator
||	Bit string concatenation

Note that, when describing each RFID authentication protocol in Figure 1, we omit the role of RFID readers. We assume that an RFID reader just acts as an intermediate party which relays messages between the back-end server and an RFID tag. This assumption does not affect the point we are going to make in this chapter. However, when the role of RFID readers is required, we will mention it specifically.

9.2.1 Exhaustive-Search Protocols

The most popular exhaustive-search protocol is the OSK protocol by Ohkubo et al. (2004). The protocol assumes that a tag can compute two cryptographic hash functions, $g()$ and $h()$. A tag is given an initial identifier s^1 which is also stored in the database at the back-end server. After each interrogation session, the tag identifier is updated in a chaining fashion, that is $s^{k+1} = h(s^k)$ after the $k-$th authentication session.

During the $k-$th authentication session, a tag computes its authentication token r^k as the hash of its current identifier, *i.e.,* $r^k = g(s^k)$. To verify a tag's authentication token, the server tries to match the tag being queried with all tags in the server's database. In particular, for an initial identifier of the i-th tag in the database, the server computes $g(s_i^1), g(s_i^2), \cdots$ for $i = 1, 2, \cdots, N$ until a match is found. We illustrate the OSK protocol in Figure 1 where M is the maximum number of times the back-end server will try to match a tag being queried with another tag in the database.

As we can see, the back-end server in the OSK protocol needs to perform $O(NM)$ hash operations to identify one tag in the database. This is the worst overhead for the sake of providing privacy among all RFID privacy-preserving authentication protocols and therefore we call the OSK protocol a brute-force protocol.

9.2.2 Tree-Based Protocols

Tree-based protocols are among the first attempts to reduce the cost of privacy suffered by the OSK protocol. We review here a protocol by Molnar et al. (2005) which achieves $O(\log_2(N))$ instead of $O(NM)$ for the cost of identifying a tag. In order to achieve such a reduced computational complexity, a tree-based protocol employs a binary tree data structure to store the secret keys that the server shared with tags in the tag database. We call this protocol the Molnar-Soppera-Wagner protocol.

In the Molnar-Soppera-Wagner protocol, the tree of secrets is prepared as follows: First, a complete binary tree of depth d_1, in which each node is labeled with an independently and randomly chosen secret key, is generated. In this tree, each leaf node corresponds to a tag in the tag database; each tag is uniquely identified by a list of d_1 secret keys collected from the nodes along the path from the root to the leaf node representing the tag. A tag is initialized with these d_1 secret keys when it is admitted to the system. The tree is then expanded with $N = 2^{d_1}$ complete binary subtrees, each is of depth d_2 and rooted at a node corresponding to one tag. Each node in these subtrees is also assigned a secret key which is not chosen at random but computed from a deterministic pseudo-random number generator seeded with the secret key of the parent node. For example, let K be the secret key associated with the parent node and PRNG(.) be the pseudo-random number generator, then the secret keys associated with the left and the right child nodes are computed as the first and the second halves of the output of PRNG(K), respectively. It is easy to see that knowing the secret key associated with a node at depth d_1 is sufficient to compute all secret keys associated with this node's descendants. Each leaf node in a subtree is also assigned with a d_2 – bit number as there are 2^{d_2} such leaf nodes per subtree; For our convenience, we assign numbers to leaf nodes in increasing order from left to right, *e.g.*, the left-most leaf

Figure 1. The k – th authentication session of a tag in OSK protocol

S(D)		**T**(s^k)
	$\xrightarrow{Request}$	
		$r^k \leftarrow g(s^k)$ $s^{k+1} \leftarrow h(s^k)$
	$\xleftarrow{r^k}$	
for $i = 1$ to N for $j = 1$ to M $r_i^j \leftarrow g(s_i^j)$ if $r^k = r_i^j$ return ACCEPT(i) $s_i^{j+1} = h(s_i^j)$ return REJECT		

node's number is 0 and the right-most leaf-node's number is $2^{d_2} - 1$, *etc.* It is also convenient to view the number assigned to a leaf node as a one-to-one correspondence to the path from the root of the subtree to the leaf node. Indeed, a tag pseudonym is computed from d_2 secret keys associated with nodes along such a path and d_1 secrets associated with nodes along the path from the root of the expanded tree to the root of the subtree. It is easy to see that each tag has at most 2^{d_2} pseudonyms.

Without loss of generality, assume that a tag is initialized with d_1 secret keys $(K_1, K_2, \cdots, K_{d_1})$ and an appropriate counter c. The tag database D contains the whole tree of secrets, whose depth is $d_1 + d_2$. A detailed description of the Molnar-Soppera-Wagner protocol is given in Figure 2.

Note that the Molnar-Soppera-Wagner tree-based protocol does not aim to provide authentication. Instead, its goal is to provide privacy and identification of tags.

9.2.3 Constant-Cost Key-Lookup Protocols

In this section, we will review several privacy-preserving RFID authentication protocols which claim to achieve constant cost of key lookup. These schemes represent the most advanced techniques in achieving privacy without degrading performance.

O-FRAP Protocol

O-FRAP, which stands for *Optimistic Forward secure RFID Authentication Protocol*, is one of the first protocols to claim constant-cost key-lookup proposed by Le et al. (2007). O-FRAP achieves privacy by giving each tag a random pseudonym (denoted by r_{tag}) and updating this pseudonym after every authentication session. The pseudonym is maintained at both the tag and the back-end server which also uses the pseudonym to index the tag database. Let call $D.query()$. the procedure to look up a tag in the tag database D given the tag pseudonym r_{tag}. Each tag and the back-end server also share a secret key (denoted by k_{tag}). This secret key is also updated after every authentication to provide forward privacy. To deal with the de-synchronization of the shared secret key, the server keeps two versions of the secret key for each tag. One version is the secret key used in the last *normal* authentication session (denoted by k_{tag}^{prev}) and another version is the current secret key (denoted by k_{tag}^{cur}). The server updates this pair of secret keys by executing the $D.update()$ procedure as follows:

- If the tag is authenticated with k_{tag}^{cur}, *i.e.,* a normal authentication session, the server does: $k_{tag}^{prev} = k_{tag}^{cur}$ and $k_{tag}^{cur} = k_{tag}^{new}$ where k_{tag}^{new} is a newly generated key.
- If the tag is authenticated with k_{tag}^{prev}, *i.e.,* an abnormal authentication session, the server preserves k_{tag}^{prev} and lets $k_{tag}^{cur} = k_{tag}^{new}$. The reason that the server does not update k_{tag}^{prev} is to prevent further de-synchronization-of-secret attacks on this particular tag from de-legitimating the tag because this tag may no longer has the secret key that is actually kept in the back-end database.

In addition to two versions of the secret key for each tag, the server also keeps two corresponding versions of the tag pseudonym. Let $Prev_j$ and Cur_j denote two pairs $\langle$ pseudonym, secret key $\rangle$ kept in the back-end database for one of previous session and the current session, respectively. A detailed description of O-FRAP is given in Figure 3.

The O-FRAP protocol requires additional costs to update the database index as the tag pseudonym is updated after every authentication. For this

Figure 2. Molnar-Soppera-Wagner tree-based protocol

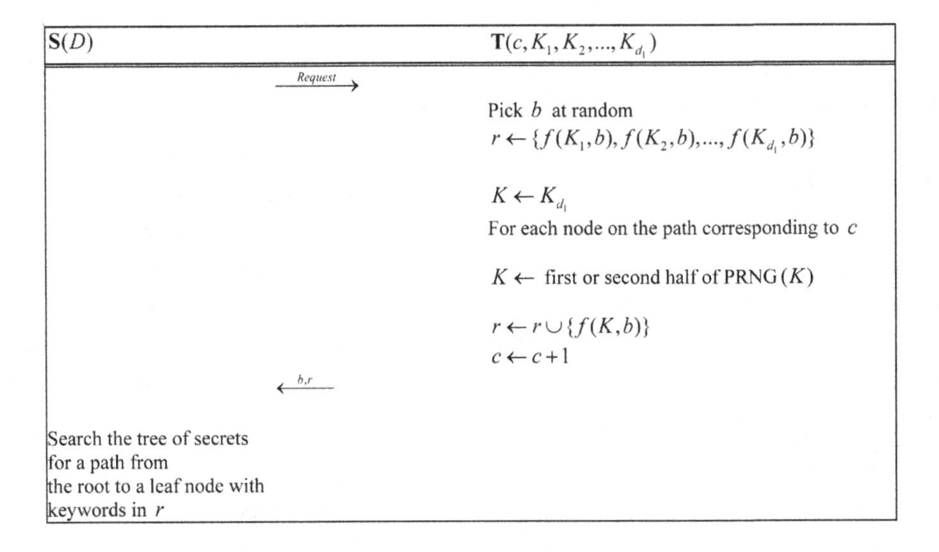

Figure 3. The O-FRAP protocol

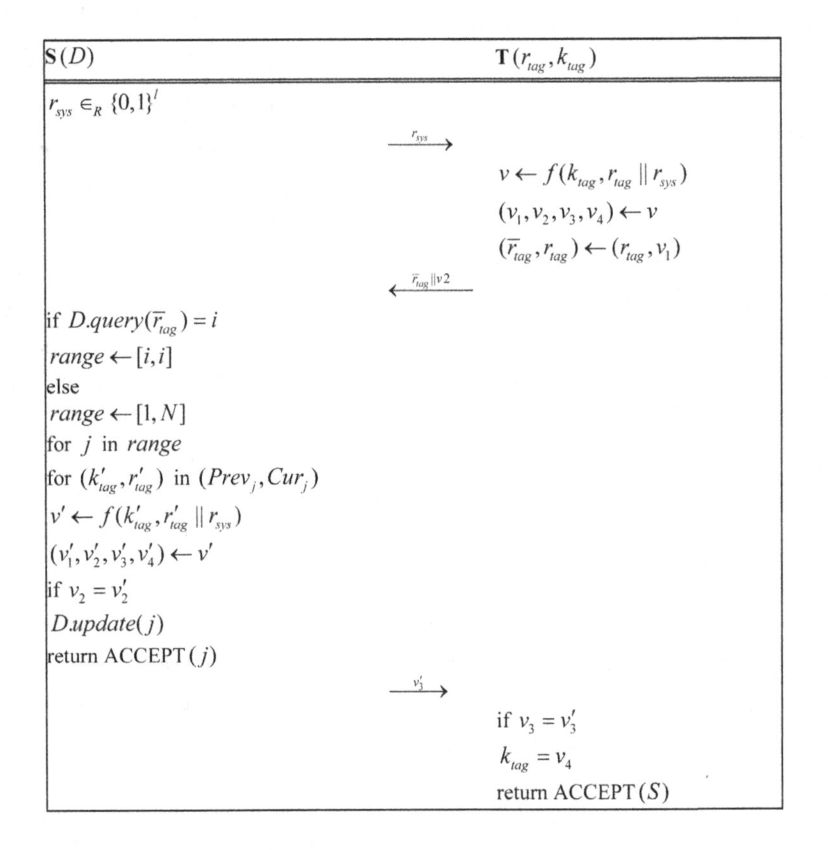

reason, we place O-FRAP below two other protocols in the category of constant-cost key-lookup schemes.

Ryu-Takagi Protocol

Ryu & Takagi (2009) proposed a unique approach to addressing the cost-of-privacy problem. The pseudonyms for a tag are created by probabilistically encrypting the tag identifier so that when presented with a pseudonym, the back-end server can recover the unique and fixed tag identifier and quickly look-up the tag in the database. We call this protocol the Ryu-Takagi protocol. In the Ryu-Takagi protocol, a trusted party prepares tags and populates the tag database as follows: Each tag is assigned a unique identifier ID and a secret key K; The tag identifier is then encrypted m times to produce a set of encrypted tag pseudonyms $\Delta = \{r_1, r_2, \cdots, r_m\}$ where r_i is a ciphertext of ID; Lastly, the pair $\langle \Delta, K \rangle$ is stored in the tag's memory and the tag database D is populated with N pairs $\langle ID, K \rangle$ for N tags. A detailed description of Ryu-Takagi protocol is illustrated in Figure 4.

Note that, according to Duc et al. (2010), the Ryu-Takagi protocol does not satisfy the privacy-preserving property as claimed by its authors. The reason is that malicious parties can collect the list of m pseudonyms of a tag by repeatedly querying the tag. Duc et al. (2010) also suggested an improved protocol which uses the same idea that we will present later in this chapter.

Burmester-Medeiros-Motta Protocol

Burmester-Medeiros-Motta protocol proposed by Burmester et al. (2008) is not a new protocol per se. It is a generic method to convert any secure challenge-response protocol to a secure privacy-preserving RFID authentication protocol with constant-cost key lookup. We describe Burmester-Medeiros-Motta construction by using a secure challenge-response authentication protocol depicted in Figure 5.

The key idea of Burmester et al. (2008) is somewhat similar to that of Ryu-Takagi protocol. More specifically, the tag identifier is encrypted to create tag pseudonyms. However, the different part here is that the tag is the one to encrypt its identifier, not the back-end server. As a result, a tag does not need to store a list of its pseudonyms because it can generate these pseudonyms on-the-fly. The encryption algorithm should work like a public-key cryptosystem so that the tag can encrypt its identifier and only the server can decrypt the tag pseudonym to get back the tag identifier. A symmetric cipher would be better suited for low-cost tags but there is a risk of leaking the secret key used to encrypt the tag identifier once a tag is corrupted. The question remained is how to design an efficient encryption scheme (or more technically, a one-way trapdoor function where the trapdoor can be used to invert/decrypt the function) that fits on the limited hardware of low-cost RFID tags. Fortunately, there is one candidate for such a scheme called SQUASH suggested by Shamir (2008). SQUASH is designed as a hash function and simply a modular squaring operation with modulus n as the product of two large prime numbers. The one-way property of SQUASH is equivalent to the security of Rabin's encryption scheme and the trapdoor information is the two prime numbers. As estimated in Burmester et al. (2008), when adapting SQUASH to some RFID authentication protocols, the cost of hardware should be less than 3,000 Gate-Equivalents for computation and storage.

Let $e()$ be a one-way trapdoor function and t be the trapdoor information, we describe the privacy-preserving RFID authentication protocol with constant-cost key-lookup converted from a basic challenge-response authentication protocol in Figure 6.

Figure 4. Ryu-Takagi protocol

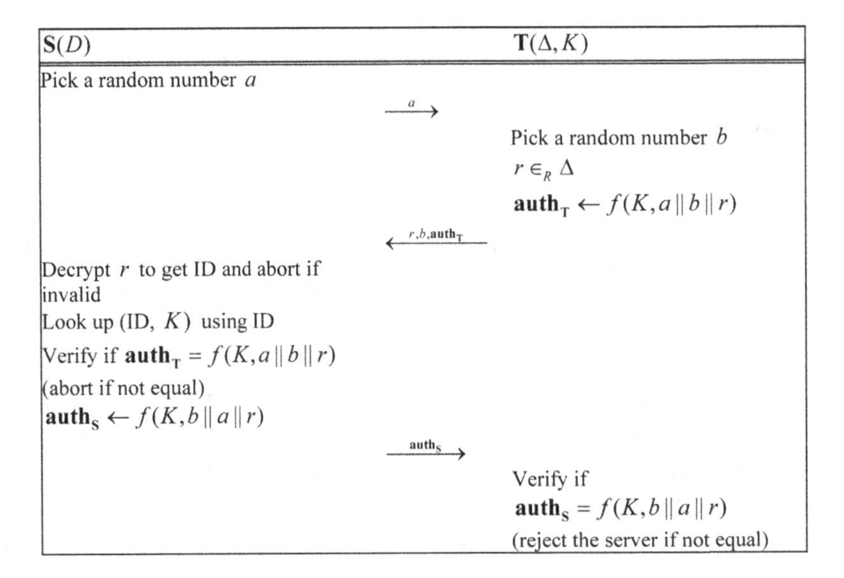

Figure 5. A basic challenge-response authentication protocol

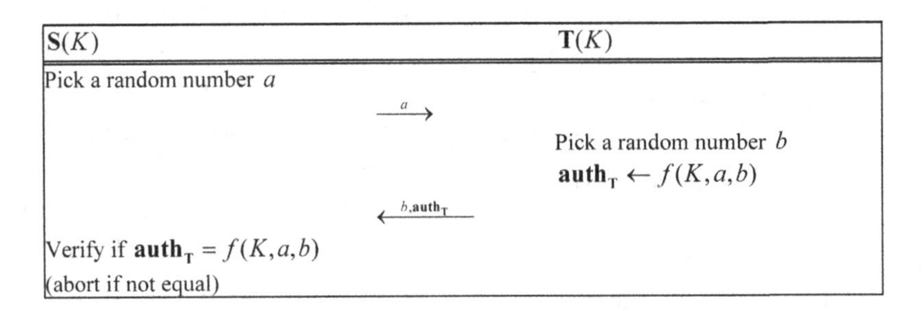

9.3 DENIAL-OF-SERVICE ATTACKS ON PRIVACY-PRESERVING RFID AUTHENTICATION PROTOCOLS AND COUNTERMEASURES

9.3.1 Denial-of-Service Attacks on Privacy-Preserving RFID Authentication Protocols

We now describe how malicious parties can abuse the back-end server's computational resources by exploiting the server's behavior in handling the identification of tags being queried.

First of all, in the OSK protocol, malicious parties can abuse the back-end server's computational resources as follows:

- Malicious parties send query requests to a large number of tags in order to make these tags update their identifiers. When these tags are actually scanned by the server, the server will have to compute a longer hash chain to find a match. Note that, if a tag is queried by malicious parties more than M times, this tag will no longer be identified by the server because the server only com-

Figure 6. A privacy-preserving RFID authentication protocol with constant-cost key-lookup

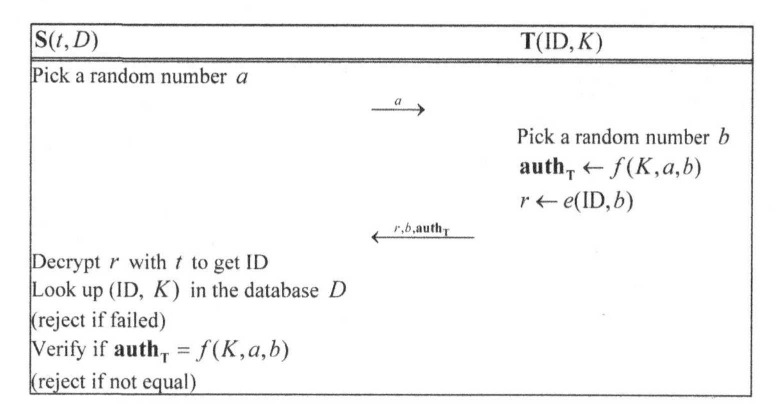

pute at most M values per hash chain. This constitutes a valid denial-of-service attack on the RFID system.

- Malicious parties themselves deploy a large number of fake tags. These tags simply backscatter random numbers when interrogated by the server. Since it is unlikely that the random numbers would match with some hash values, the server will have to go through all N hash chains for the N tags, each with M hash values, in the database before rejecting these tags. In other words, malicious parties can make the server run into the worst-case scenario and therefore use the most of its computational resources.

In the case of the Molnar-Soppera-Wagner tree-based protocol, malicious parties can also mount similar attacks to abuse the server's computational resources. The difference here is that the attacker needs to eavesdrop a valid pseudonym and replay it to the server so that the server will waste most of its computational resources. Simply sending random pseudonyms is not effective because the server would stop the search early when there is invalid information on the pseudonym.

In Duc & Kim (2011), the authors presented two methods that malicious parties can use to abuse the back-end server's computational resources in the O-FRAP protocol. We summarize their observations here. In the O-FRAP protocol, the server tries match a tag being queried with every tag in the tag database $(range \leftarrow [1, N])$ when it fails to find a tag given a tag pseudonym $(\overline{r}_{tag})$. The reason for this behavior is that the tag pseudonym is updated by a tag whenever it receives a query request (r_{sys}). As a result, the tag pseudonym is an easy target for a de-synchronization attack which leads to failure in looking up a tag by its pseudonym. To identify tags that have been subject to de-synchronization, the server needs to go through the whole tag database. Similar to DoS attacks on the OSK protocol, malicious parties can mount the attack either by querying a large number of tags or deploying fake tags. We illustrate the two attack scenarios in Figure 7.

It is much harder to mount DoS attacks on Ryu-Takagi and Burmester-Medeiros-Motta protocols because the server in these two protocols never has to go through the whole tag database. Nevertheless, the server still has to decrypt the tag pseudonym each time it interrogates a tag. Note that the server does not verify the integrity of tag pseudonyms backscattered by tags at all. This poses a potential weakness for malicious parties to exploit. For example, malicious parties can deploy a large number of fake tags at sensitive locations where the responsiveness of scanning activities are critical so that the server has a significant load to handle including decryption of tag pseudonyms in a short period of time. In such a scenario, it is really beneficial if fake tags are detected and filtered out before arriving at the back-end server. An issue with the Burmester-Medeiros-Motta protocol is the cost of implementing SQUASH on RFID tags, especially passive tags.

9.3.2 Countermeasures

We observe that a common problem leading to abuse of the server's computational resources is the lack of integrity checking on tag pseudonyms. That is, the server has no idea whether a pseudonym backscattered by a tag is valid or not. Therefore, we come to the conclusion that the server should be able to verify the validity of a tag pseudonym before actually using it for searching the tag in the database. More importantly, if this task can be done before the tag pseudonym reaches the server (*e.g.*, by the RFID readers who actually do the interrogation of tags), then the malicious party's attempt to push an illegitimate load on the back-end server will be stopped before the attacking traffic arrives at the server. To be able to verify a tag pseudonym even before identifying the tag, we propose that the server shares a common secret key K_S with all the tags in the database. The reasons are two-fold.

- Before a tag is identified, the server does not know the secret shared with the tag and therefore cannot use this secret key to verify the tag pseudonym.
- The key, K_S can also be used by the tag to verify the server before sending its pseudonym. This can prevent malicious parties from harvesting tag pseudonyms. This is how Duc et al. (2010) addressed the privacy issue of the Ryu-Takagi protocol.

We call our approach *two-phase authentication* because a tag is authenticated and identified in two phases:

- In the first phase, the tag pseudonym is authenticated so that the server is guaranteed that this particular tag is actually in the server's tag database. This phase can be done by the RFID readers provided that the server gives the RFID readers the secret key shared with all tags.
- In the second phase, the secret key shared between the server and only this tag is lo-

Figure 7. DoS attack on O-FRAP

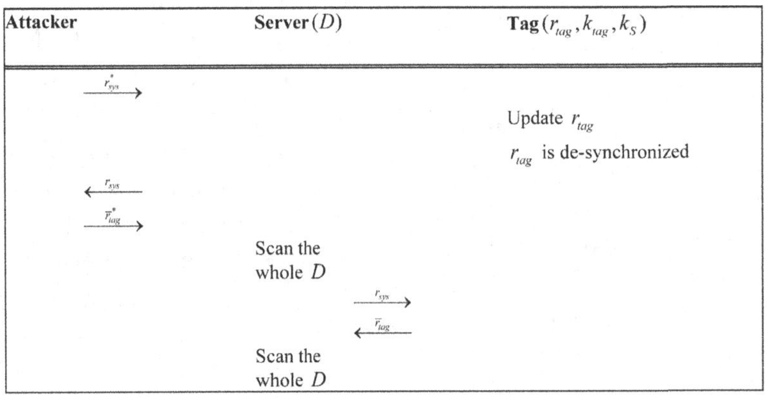

Attacker	**Server** (D)		**Tag** (r_{tag}, k_{tag}, k_S)
$\xrightarrow{r^*_{sys}}$			
			Update r_{tag}
			r_{tag} is de-synchronized
$\xleftarrow{r_{sys}}$			
$\xrightarrow{\overline{r}^*_{tag}}$			
	Scan the whole D		
		$\xrightarrow{r_{sys}}$	
		$\xleftarrow{\overline{r}_{tag}}$	
	Scan the whole D		

cated in the tag database. At this point, the tag is identified and verified by this private secret key.

We apply this approach to a privacy-preserving RFID authentication protocol with constant-cost key-lookup in Figure 6. We make the following modification: the server now shares with all tags a common secret key K_S; Before sending its response, a tag computes an authentication token for its pseudonym r as $\mathbf{auth}_r = f(K_S, a, r)$; Upon receiving the tag's response, the server (or the RFID readers in practice) first verifies the validity of the tag pseudonym r using $\mathbf{auth}_r$ and proceeds only if $\mathbf{auth}_r$ is correctly verified. The resulting protocol is illustrated in Figure 8.

This same idea can be applied to the O-FRAP and Ryu-Takagi protocols as we can see in Duc & Kim (2011) and Duc et al. (2010), respectively. Indeed, the two-phase authentication approach can be used by most RFID authentication protocols which require detection of illegitimate tags at the reader level.

FUTURE RESEARCH DIRECTIONS

Our proposed two-phase authentication approach is not without drawbacks. In particular, the same secret key has to be stored on every tag in the tag population. As a result, if one tag is corrupted and the secret key is leaked, it will affect all other tags in the system. Technically speaking, our two-phase authentication is not secure against key exposure since there is no way the key K_S can be updated and synchronized in all tags and the server. While security against the risk of key exposure is an overkilled feature in many applications, this is still a valid concern and needs to be addressed. We propose the following directions for further investigation:

- One may use some trusted computing primitives to protect the secret keys against leakage.
- The tag database can be partitioned into different parts and all tags in each part will share a different secret key. Then, when a tag is interrogated, the server and/or

Figure 8. Our proposed privacy-preserving RFID authentication protocol with constant-cost key-lookup

$\mathbf{S}(t, K_S, D)$		$\mathbf{T}(\mathrm{ID}, K_S, K)$
Pick a random number a		
	$\xrightarrow{a}$	
		Pick a random number b $\mathbf{auth}_T \leftarrow f(K, a, b)$ $r \leftarrow e(\mathrm{ID}, b)$ $\mathbf{auth}_r \leftarrow f(K_S, a, r)$
	$\xleftarrow{r, b, \mathbf{auth}_r, \mathbf{auth}_T}$	
Verify if $\mathbf{auth}_r = f(K_S, a, r)$ (reject if not equal) Decrypt r with t to get ID Look up (ID, K) in the database D (reject if failed) Verify if $\mathbf{auth}_T = f(K, a, b)$ (reject if not equal)		

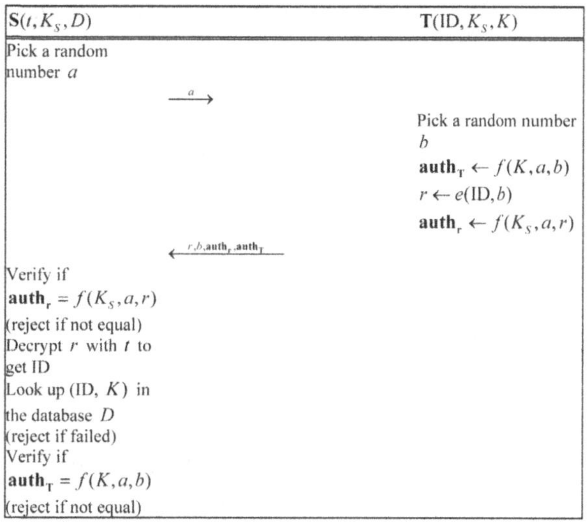

the RFID readers are required to identify which partition of the tag database the tag belongs to.

Another issue with the current situation in RFID security is the lack of research on cryptographic primitives that are suitable for low-cost RFID tags. For example, SQUASH is the only efficient one-way trapdoor function designed specifically for RFID tags. We need even more efficient schemes for really low-cost tags like passive tags.

CONCLUSION

In this chapter, our goal is to show that privacy and performance are often two conflicting goals in RFID system. We reviewed different privacy-preserving RFID authentication protocols and pointed out that they suffer from different levels of vulnerability against DoS attacks. We then presented a generic method to address DoS attacks on privacy-preserving protocols, which we called two-phase authentication. The goal of two-phase authentication is to detect and filter out illegitimate tags at the reader level so that the server does not waste computational resources in processing these tags. We believe that privacy-vs-performance is a central issue in RFID security and much more research effort is needed in order to find an optimal solution.

REFERENCES

Burmester, M., de Medeiros, B., & Motta, R. (2008). Anonymous RFID authentication supporting constant-cost key-lookup against active adversaries. *International Journal of Applied Cryptography*, *1*(2), 79–90. doi:10.1504/IJACT.2008.021082

Duc, D. N., & Kim, K. (2011). Defending RFID authentication protocols against DoS attacks. *Elsevier's Journal of Computer Communications*, *34*(3), 384–390. doi:10.1016/j.comcom.2010.06.014

Duc, D. N., Yeun, C. Y., & Kim, K. (2010). Reconsidering Ryu-Takagi RFID authentication protocol. In *the 2nd International Workshop on RFID/USN Security and Cryptography (RISC)*, (pp. 1–6).

Le, T. V., Burnmester, M., & de Medeiros, B. (2007). Universally composable and forward secure RFID authentication and authenticated key exchange. In *the Proceedings of the 2nd ACM Symposium on Information, Computer and Communications Security*, (pp. 242-252).

Molnar, D., Soppera, A., & Wagner, D. (2005). A scalable, delegatable pseudonym protocol enabling ownership transfer of RFID tags. In *the Proceedings of Selected Areas in Cryptography'05, LNCS 3897,* (pp. 276-290). Springer-Verlag.

Ohkubo, M., Suzuki, K., & Kinoshita, S. (2004). Efficient hash-chain based RFID privacy protection scheme. In *the Proceedings of International Conference on Ubiquitous Computing, Workshop Privacy*.

Ryu, E.-K., & Takagi, T. (2009). A hybrid approach for privacy-preserving RFID tags. *Journal of Computer Standards and Interfaces*, *31*, 812–815. doi:10.1016/j.csi.2008.09.001

Shamir, A. (2008). SQUASH - A new MAC with provable security properties for highly constrained devices such as RFID tags. *In the Proceedings of Fast Software Encryption 2008* [Springer-Verlag.]. *LNCS*, *5086*, 144–157.

ADDITIONAL READING

Avoine, G. (2010). *RFID Security & Privacy Lounge*. Retrieved from http://www.avoine.net/rfid/

Avoine, G., Dysli, E., & Oechslin, P. (2005). Reducing time complexity in RFID system. In *the Proceedings of Selected Areas in Cryptography'05*, LNCS 3897, (pp. 291-306). Springer-Verlag.

Avoine, G., & Oechslin, P. (2005). A scalable and provably secure hash-based RFID protocol. In *the Proceedings of Workshop on Pervasive Computing and Communications Security - PerSec'05*, (pp. 110-114).

Burnmester, M., Le, T. V., De Medeiros, B., & Tsudik, G. (2009). Universally composable RFID identification and authentication protocols. *ACM Transactions on Information and System Security, 12*(4), 1–33. doi:10.1145/1513601.1513603

Duc, D. N. (2010). *A study on cryptographic protocols for RFID tags*. Doctoral dissertation, KAIST, Republic of Korea. Retrieved from http://caislab.kaist.ac.kr/publication/thesis_files/2010/Duc_Thesis.pdf

Juels, A. (2006). RFID security and privacy: A research survey. *Journal of Selected Areas in Communication, 24*(2), 381–395. doi:10.1109/JSAC.2005.861395

Juels, A. (2007). The vision of secure RFID. *Proceedings of the IEEE, 95*(8), 1507–1508. doi:10.1109/JPROC.2007.900324

Juels, A., Rivest, R., & Szydlo, M. (2003). The blocker tag: Selective blocking of RFID tags for consumer privacy. In *the Proceedings of the 10th ACM conference on Computer and communications security*, (pp. 103-111). ACM Press.

Kohno, T. (2008). An interview with RFID security expert Ari Juels. *IEEE Pervasive Computing / IEEE Computer Society [and] IEEE Communications Society, 7*(1), 10–11. doi:10.1109/MPRV.2008.4

Vaudenay, S. (2007). On privacy models for RFID. In *the Proceedings of ASIACRYPT'2007*, LNCS 4833, (pp. 68-87). Springer-Verlag.

Weis, S. (2003). *Security and privacy in radio-frequency identification devices*. Master thesis, MIT, USA. Retrieved from http://saweis.net/pdfs/weis-masters.pdf

Weis, S., Sarma, S., Rivest, R., & Engels, D. (2003). Security and privacy aspects of low-cost radio frequency identification systems. In *the Proceedings of Security in Pervasive Computing, LNCS 2802*, (pp. 201-212). Springer-Verlag.

KEY TERMS AND DEFINITIONS

Authentication: Authentication is a process of proving one's identity (the prover) to another party (the verifier) over an insecure communication channel. A typical approach in designing an authentication protocol is to use a challenge-response protocol. That is, the verifier sends a random challenge and the prover replies with a response which is computed as a function of the verifier's challenge and secret information, *e.g.*, a secret key shared in advance between the prover and the verifier. The response can then be checked by the verifier in such a way that if the response is correctly verified, the verifier is confident that the prover indeed knows the secret information used to compute the response.

Denial-of-Service (DoS) Attack: DoS attack refers to a type of attack in which the victim's computational resources are abused to the point where no legitimate activities can be served. In RFID, the DoS attack is usually targeted at the back-end server where all the detailed information about tagged objects are stored and queries are processed. The unavailability of the back-end server would mean a total breakdown of an RFID system.

De-Synchronization Attack: De-synchronization attack refers to the acts by malicious parties which cause inconsistency in updating shared secret keys and sometimes pseudonyms between RFID tags and the back-end server. For example, an attacker may interfere in an authentication session to cause the server to update the shared secret key but the RFID tag. As a result, the tag might be no longer accepted by the server since the twos possess two different secret keys.

Electronic Product Code (EPC): A unique number stored in an RFID tag's memory. The EPC serves as the identifier for a product on which the tag is attached. An EPC can be retrieved via radio communication by RFID readers.

Forward Privacy: Forward privacy is a security notion which addresses the key exposure problem. As an RFID tag is generally a low-cost hardware, it is an easy target to be captured and dissected to reveal the secret key stored in the tag's memory. Forward privacy refers to the ability to preserve the privacy of a tag up to the point of the secret key exposure. A common method to achieve forward privacy is to update the secret key and/ or the tag pseudonym after every authentication session.

Privacy: Privacy has a broad meaning in cryptography. In case of RFID, privacy refers to the privacy of RFID tags which in turn means the privacy of people carrying tagged items. Intuitively speaking, the privacy of an RFID is guaranteed if its identity, movements and sometimes even its existence are hidden from unauthorized parties.

Pseudonym: Pseudonym is a temporary name used by an entity instead of its real name. The goal of using pseudonym is to preserve the anonymity of the entity when communicating over an insecure channel or to possibly other un-trusted entities.

Unlinkability: Unlinkability is a frequent interpretation of privacy in RFID. Unlinkability usually means that it is infeasible for unauthorized parties to decide whether two arbitrary authentication sessions involve a same tag. In other words, to the eyes of an outsider, an authentication session could be of any tag and therefore the outsider cannot learn any useful information and violate the privacy of tags.

Untraceability: Untraceability is an alternative interpretation of tag privacy. In simply means the impossibility to trace a tag by unauthorized parties.

Chapter 10
Malware Protection on RFID-Enabled Supply Chain Management Systems in the EPCglobal Network

Qiang Yan
Singapore Management University, Singapore

Yingjiu Li
Singapore Management University, Singapore

Robert H. Deng
Singapore Management University, Singapore

ABSTRACT

As RFID-enabled technology is becoming pervasive in enterprise systems and human life, it triggers significant concerns over the malware that can infect, damage, and even destroy RFID-enabled network systems. RFID malware can spread malicious codes or data quickly to a large number of RFID systems via RFID read and write, which are pervasive operations on RFID tags that are transported from one RFID system to another. To address this concern, this chapter uses RFID-enabled supply chain management systems in the EPCglobal network as a case study to demonstrate the important issues in RFID malware protection. This case study shows that although there are fundamental difficulties in preventing RFID malware from entering the systems, the behaviors of RFID malware resemble traditional malware after it enters the systems. Based on this characteristic, the security threats of RFID malware can be effectively controlled.

DOI: 10.4018/978-1-4666-3685-9.ch010

10.1 INTRODUCTION

RFID is an automated data collection technology that uses radio frequency waves to transfer data between a reader and an RFID tag to identify, track or locate the physical item to which the tag is attached. Since RFID technology improves the automation processes significantly, it works its way into many enterprise systems like supply chain management systems. The market value of RFID technology keeps growing even under the current global financial crisis. According to a market report from IDTechEx (Das & Harrop (2009)), the value of the entire RFID market was $5.56 billion in 2009, up from $5.25 billion in 2008; by 2019, the market value of RFID technology is predicted to grow over five times to exceed $27 billion, and the number of RFID tags sold yearly will increase over ten times to exceed 100 billion. If we further consider the trends of integrating RFID systems over the Internet ("The Internet of Things"), the future large-scale deployments of RFID-enabled network systems are very likely to become attractive targets for malware developers.

As an emerging threat to high value RFID applications, RFID malware got the attention of the industry and the public after the first proof-of-concept RFID malware was reported in 2006 (Rieback, Crispo & Tanenbaum (2006)). It was classified as a long-term threat in the latest IT security threat report from Gartner (Pescatore, Young, Allan, Girard, Feiman & MacDonald (2008)). Compared to other long-term threats, the RFID attack was placed at the earliest position, which means that the threats of RFID attacks were the most imminent. Gartner further identified the biggest risk associated with RFID applications as:

RFID systems, especially readers, were developed without security in mind(Pescatore, Young, Allan, Girard, Feiman & MacDonald (2008), p. 20)

Hence, reading RFID data from ubiquitous acquisition points without proper security protection will incur high risks and pose serious threats to enterprise IT security. In addition, as RFID readers are integrated into personal devices such as PDAs or mobile phones, and more RFID applications are developed on these devices, RFID malware will eventually penetrate into our daily lives.

These potential threats have motivated us to provide a comprehensive investigation of RFID malware protection. We use RFID-enabled supply chain management systems in the EPCglobal network as a case study. This kind of systems is one of the most important enterprise applications of RFID technology. RFID technology has been widely envisioned to have significant impact on modern supply chain management as an inevitable replacement of barcodes in the near future. EPCglobal network[1], the de-facto industry standard for RFID-based trading systems, further integrates RFID technology with Internet and networking technology, to enable contactless information collection, integration, sharing and querying in real time over the Internet.

To demonstrate the important issues on RFID malware protection for RFID-enabled supply chain management systems, we describe and analyze a demo security system, RFscreen. It is designed to detect and filter out generic RFID malware by protecting critical points on each layer of an RFID-enabled supply chain management system. Our analysis on RFscreen shows that there are fundamental difficulties in preventing RFID malware from entering the systems. But after it enters the systems, it behaves similarly to traditional malware whose threats can be effectively detected and prevented.

This chapter aims to provide the necessary background and a comprehensive analysis of RFID malware for RFID-enabled supply chain management systems. The background section describes the EPCglobal network architecture, the capabilities of current RFID infrastructure, and the basic characteristics of RFID malware. The remainder of this chapter is devoted to the analysis of the potential security threats of RFID malware and the design of corresponding countermeasures.

10.2 BACKGROUND

This section first introduces the background knowledge of RFID-enabled supply chain management systems specified by the EPCglobal network. After that, we describe the basic characteristics of RFID malware.

10.2.1 EPCglobal Network Architecture

A typical RFID-enabled supply chain management system in the EPCglobal network is a layered system which consists of the following components:

- **RFID tag:** Is a microchip combined with an antenna. It stores the information about the physical item to which it is attached. An RFID tag can be passive or active. A passive RFID tag cannot emit an RF (*radio frequency*) signal until it gains enough energy from the external RF signals emitted by a RFID reader. An active tag is able to emit RF signals without relying on the energy gained from the external RF signals, as it is equipped with a battery. A variance of active RFID tags is the BAP (*battery-assisted passive*) RFID tag. A BAP RFID tag is equipped with a battery, but behaves like a passive RFID tag, as it emits a RF signal only after receiving a valid external RF signal. The battery of a BAP RFID tag is mainly designed to enable advanced cryptographic functions that usually require much more energy than that harvested from the external RF signals. Among these three types of RFID tags, passive RFID tags are most widely used, as they are the cheapest.
- **RFID reader:** Is an interrogator for RFID tags. It emits an RF signal by its antenna, and RFID tags respond by sending back their data after receiving valid requests. RFID readers usually are not simply peripherals, but are also integrated with embedded operating systems. So most RFID readers are able to preprocess the collected data before sending them to back-end enterprise systems. This capability is also widely used in RFID protocol design (Juels (2006), Ma, Li, Deng & Li (2009)).
- **RFID middleware:** Is the middle layer that links RFID readers and back-end enterprise systems by routing the RFID data among them. Besides managing and transferring RFID data, business logic and user interfaces are also usually implemented by RFID middleware. RFID middleware can be integrated with individual RFID readers, but more often, it operates as a separate component, which is essential to enable central management to aggregate RFID data from multiple RFID readers.
- **EPC information service (EPCIS):** Is the local back-end database repository (EPCglobal (2007)). It is designed to store the detailed supply chain information of the physical items labeled by scanned RFID tags. EPC information service receives RFID data processed by RFID middleware, and also provides query interfaces for RFID middleware to implement required business logic. To enable global data sharing, EPC information service uploads the index information to the EPCglobal network core services to make information visible to other supply chain partners.

- **EPCglobal network core services:** Are the global data repositories that enable real-time sharing and querying of supply chain information on a global scale. The major core services include 1) Object Naming Service (ONS) (EPCglobal (2008*a*)) that provides the location of the information service of the manufacturer of the item; 2) EPC Discovery Service (EPC DS)[2] that lists the locations of the information services of supply chain partners who once took over the item. From these core services, any user is able to query the real time tracking information of a specific item, and retrieve detailed supply chain information from the information services whose locations are contained in the response from core services.

The interactions among these components follow the protocol stack specified by the EPCglobal network architecture, which are shown in Figure 1.

In this protocol stack, RFID data are aggregated from lower layers to upper layers. Upper layer data may also flow back to lower layers, and eventually update the data on a large number of RFID tags. This figure shows the information flow in the EPCglobal network, and also indicates the potential propagation paths of RFID malware. If no proper protection mechanisms are deployed to sanitize these propagation paths, RFID malware will be able to cause serious damage and infect a large number of victim systems when it compromises a component in a higher layer.

Figure 1. Protocol stack of EPCglobal network

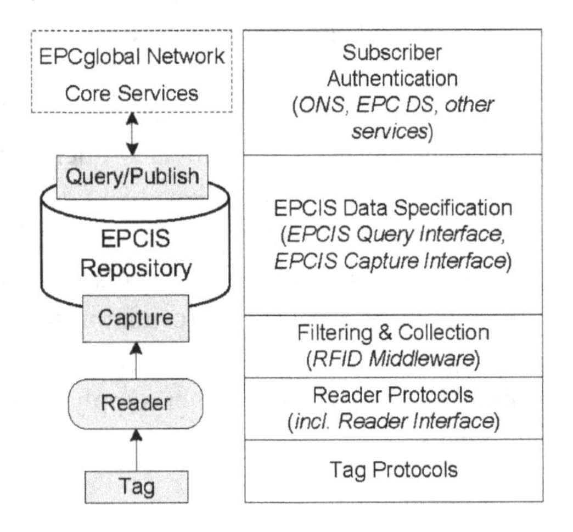

10.2.2 Capabilities of Current RFID Infrastructure

The low capabilities of RFID tag was one of the major reasons that many people believe that RFID malware is a myth. The EPC Class 1 Gen 2 tag (EPCglobal (2008*b*)) is the cheapest and most common type of RFID tag in RFID-enabled supply chain systems. It is a passive device with only 96 bits of writable memory. When an EPC Class 1 Gen 2 tag receives a request from an RFID reader, it responds with the data stored in its memory, its EPC code (*Electronic Product Code*). The data in its memory can be updated and it can be disabled by accepting a correct pin number. There are no more operations that can be performed by this tag. However, even for a RFID tag with such low capabilities, it is still possible to implant RFID malware into it (Rieback, Crispo & Tanenbaum (2006)). Even more serious threats come from the trend that more and more RFID tags with higher capabilities are being deployed as they are becoming cheaper. According to IDTechEx (Das & Harrop (2009)), active RFID tags accounted for about 10% of all RFID expenditure in 2008. This ratio would have been even higher if it were not depressed by the huge number of passive RFID tags used in Chinese national ID card scheme. Compared to passive RFID tags, these active tags have much more memory and some of them are even equipped with powerful embedded processors to support advanced cryptographic functions or other complex computational tasks. These extra capabilities are often required by military and other organizations that enforced high security requirements. Table 1 lists the capacities of representative

RFID tags for passive and battery-assisted passive RFID tags. Fully active RFID tags are not given in this table as they are currently not common in supply chain systems due to their high prices.

This table shows that the current newly developed RFID tags already have enough memory to store malware (4KB is enough to hold common PC malware). Certain active RFID tags like LIME tag are even able to execute malware on the tag. So the capabilities of future RFID tags are unlikely to be a major obstacle for the development of RFID malware. The capability enhancement of RFID tags seems inevitable as we have to use these enhanced capabilities to finish complex tasks required by high security requirements.

Another type of RFID infrastructure is the RFID reader. Table 2 lists the capabilities of representative RFID readers.

This table shows that current RFID readers work on mainstream platforms or as a peripheral of a mainstream platform. The integration of RFID readers with mainstream platforms means RFID systems will not only enjoy the benefits but also suffer from the vulnerabilities of these platforms. Since these platforms are the major targets of malware developers, it makes RFID systems vulnerable by reusing attack techniques in RFID malware design. The wireless interfaces equipped by certain models of RFID readers further create more attack vectors for RFID malware.

Therefore, the capabilities of RFID infrastructure are already powerful enough to sustain RFID malware.

Table 1. Capabilities of representative RFID tags

Type	Model	Memory			Other Capacities
		Tag ID/bits	EPC ID/bits	User Memory/bytes	
Passive	RI-UHF-00001-01[3]	32	96	N/A	N/A
Passive	Higgs-3[4]	64	96(up to 496)	64	N/A
Battery-assisted Passive	ALB-2484[2]	Unknown	96	4K	N/A
Battery-assisted Passive	LIME Tag[5]	128	96	512K	8051 processor with 2KB system RAM and 32KB System Flash, and reprogrammable 1M gate FPGA

Table 2. Capabilities of representative RFID readers

Model	Platform Support	Interface	Storage
ALR-9900 reader[6]	Linux, Java & .Net	LAN TCP/IP, RS232	64M RAM, 64M Flash
MC9090-G handheld reader[7]	Windows Mobile 5.0	802.11a/b/g, Bluetooth v1.2	128M RAM, 64M Flash
UHF Interrogator CF Card[8]	Windows and Linux	CF Type 1	N/A

10.2.3 Basic Characteristics of RFID Malware

RFID malware can be generally defined as malicious software infiltrating RFID systems. In the narrow sense, it is specific to malicious software carried by RFID tags. It is a new member of the malware family. Until now, no RFID malware has been detected in the wild, and only a few proof-of-concept examples of RFID malware proposed by researchers (Rieback, Crispo & Tanenbaum (2006), Rieback, Simpson, Crispo, & Tanenbaum (2006), Sulaiman, Mukkamala & Sung (2008), Shankarapani, Sulaiman & Mukkamala (2009)) are available.

From the existing proof-of-concept RFID malware, similar characteristics among RFID malware and generic PC malware can be observed. The first proof-of-concept RFID malware proposed by (Rieback, Crispo & Tanenbaum (2006)) exploited traditional attack vectors including SQL-injection, service side includes, and buffer overflow. They presented the malware examples, from the simple denial-of-service attacks like shutting down the back-end database services, to complex attacks like opening a network port as a backdoor. These examples show the feasibility of RFID malware, but certain complex attacks like self-replication still required a single RFID tag with a large memory to hold several hundred bytes of malicious data. Later research (Shankarapani, Sulaiman & Mukkamala (2009)) proposed an improved malware design to solve this memory space problem. They partitioned a malware into multiple fragments of smaller size and stored each fragment in a RFID tag with large enough memory to hold it. These fragments were reconstructed later to form a complete RFID malware, and then the malware was activated. All these proof-of-concept examples of RFID malware were proposed to directly attack the back-end enterprise systems, especially database systems. Thus, after they entered the system, they behaved similarly to existing PC malware, targeting the same back-end enterprise systems.

In essence, RFID malware shares the same basic characteristics as the malware targeting other systems, which is to infiltrate the target system by exploiting its vulnerabilities. But RFID malware also has its unique characteristics, which makes it different from other malware. These characteristics include its specific propagation behaviors, specific strategies of attacking special devices like RFID readers, and specific survival skills under the restrictions of RFID systems. These unique characteristics were not fully explored by prior research (Rieback, Crispo & Tanenbaum (2006), Rieback, Simpson, Crispo. & Tanenbaum (2006), Sulaiman, Mukkamala & Sung (2008), Shankarapani, Sulaiman & Mukkamala (2009)), as their focus was to show the feasibility of RFID malware. We investigate them in the context of RFID-enabled supply chain management systems in the next section. Based on this investigation, we develop countermeasures to build our demo security system, RFscreen.

10.3 RFSCREEN

This section proposes a demo security system, RFscreen. This demo security system is used to demonstrate the important issues of RFID malware protection. We first describe its threat model, and then present its architecture, and finally give the detailed design and analysis for malware protection on each layer of RFID-enabled supply chain management systems in the EPCglobal network.

10.3.1 Threat Model

To provide a broad view of RFID malware, RFscreen is designed to defend against generic RFID malware for RFID-enabled supply chain management systems. Since RFID malware is a new family of malware that does not appear in the wild, we have to use artificial RFID malware as the adversary in our threat model, as researchers have done in prior work (Rieback, Crispo & Tanenbaum (2006), Rieback, Simpson, Crispo. & Tanenbaum (2006), Sulaiman, Mukkamala & Sung (2008), Shankarapani, Sulaiman & Mukkamala (2009)). We design them by carefully analyzing the critical intrusion points of each layer and the information flow among related layers in target systems. The details of these artificial attacks of RFID malware will be discussed along with their countermeasures later.

10.3.2 Architecture

A well-known fact against malware protection is that it is only a matter of time to develop new malware to breach any existing malware protection system (NSA (n.d.)). It is also true for our demo security system, especially under attack by a new rising malware, RFID malware. Fortunately, we can borrow the existing "best practices" strategies to minimize the impact of such potential attacks. Therefore, we designed the architecture of RFscreen by following the defense-in-depth strategy (Cohen (1992), NSA (n.d.)). This strategy can help RFscreen to detect and sanitize RFID malware before it gets too deep into the system. The main guidelines of the defense-in-depth strategy are listed as follows:

- Protect all critical points of the systems to defend against intrusion from both outsiders and insiders.
- Use layered protection mechanisms to resist intrusion and to buy precious response time when one protection layer is breached.
- Collect and analyze information to detect intrusion and correlate the results and countermeasures.

Figure 2 shows the architecture of RFscreen, which is organized as a layered structure as suggested by these guidelines. This layered structure can be naturally incorporated into any RFID-enabled supply chain management system in the EPCglobal network. RFscreen is divided into three abstract layers (*tag layer, reader layer, and back-end system layer*), according to their capability and position in the RFID-enabled supply chain management systems. Detailed analysis and description of each layer will be given in the following subsections. For each layer, we analyze the potential security threats of RFID malware during two basic phases:

- **Attack Phase:** The attacker intends to attack the current layer.
- **Propagation Phase:** The attacker has already compromised the current layer.

For each phase, we discuss the specific attack strategies, survival skills, and propagation behaviors of RFID malware, respectively. And then we propose the potential countermeasures to these malicious behaviors. Toward the end, we provide a global security threat map to summarize these potential security threats and their countermeasures.

10.3.3 Malware Protection on Tag Layer

The tag layer is the lowest layer in a RFID-enabled supply chain system. It is also the most vulnerable layer due to the lack of capability to support strong self-protection mechanisms and other operational risks. One of the most serious operational risks comes from the trait of supply chain systems that a large number of RFID tags are transported between multiple supply chain partners. Due to the ownership transfer behavior of supply chain systems and the contactless feature of RFID tech-

Figure 2. Architecture of RFscreen

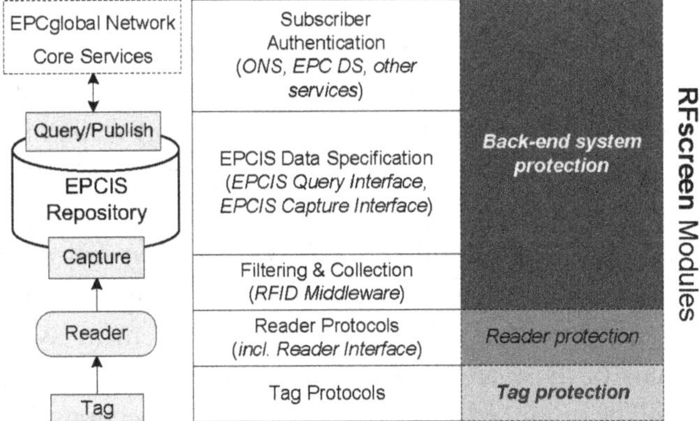

nology, it is very difficult to ensure that none of the RFID tags are infected by RFID malware (see Figure 3). These are the *fundamental difficulties* of preventing RFID malware from entering the supply chain systems. To address these difficulties, the detailed analysis and design of malware protection on the tag layer are presented below.

Attack Phase

Security threats:

- Confidential data stored in RFID tags are disclosed. These disclosed data could be
  - Enterprise secrets, or
  - Hints for subsequent attacks including secrets stored in RFID tags.
- Data stored in RFID tags are rewritten to store malicious data. The malicious data could be a complete malware, or a fragment of malware (that can be reconstructed to form a complete malware later), or an input of malware (that can trigger a certain activity of malware).

Attack strategies:

- Use a malicious RFID reader to read or write RFID tags if strong on-tag access control mechanisms are not deployed.

Figure 3. Attack phase of RFID malware on the tag layer

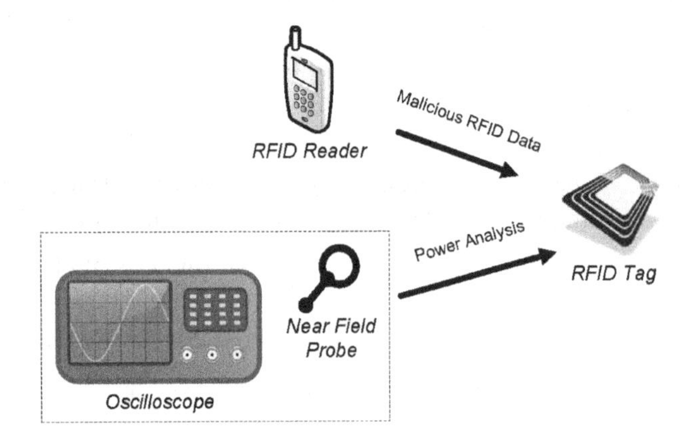

The most common type of RFID tag, EPC Class 1 Gen 2, only supports a 32-bit access password to guard the data stored in it (EPCglobal (2008*b*)).

- Use advanced hardware techniques to recover the secrets stored in the tag. Correlation power analysis is one such technique, which is widely addressed in RFID research areas (Brier, Clavier & Olivier (2004), Kasper, Oswald & Paar (2009)). It can even recover the secrets in high-cost RFID smartcards by analyzing the power leakage of physical implementation of a cryptographic algorithm with an oscilloscope and a near-field probe.

Survival skills:

- No special survival skills are required for RFID malware on the current layer. The environment can be considered under full control of the attacker as he has to be in the proximity of the target RFID tags to send RF signals to perform attacks. But it indeed requires certain skills for the attacker to hide himself without being detected.

RFscreen Design on Tag Layer

Available information for defense:

- Secrets stored in a RFID tag

Available defense capabilities:

- Check an access password (applicable for almost all RFID tags used in supply chain systems, including EPC Class 1 Gen 2 tags (EPCglobal (2008*b*))).
- Perform a cryptographic protocol (applicable for only certain high-cost active RFID tags).

Countermeasures:

- Encrypt the data stored in RFID tags to protect the confidentiality, which avoids valuable information being disclosed in plaintext.
- Enforce a secure cryptographic protocol to protect the data if applicable; otherwise set an access password. This represents best efforts to prevent unauthorized access of RFID data.
- Sign the data with a digital signature to protect the integrity of the original data. This increases the difficulty to an outside attacker, as it is difficult to forge a correct signature for malicious data without knowing the secret keys for signature generation. But digital signature cannot stop an inside attacker who intends to steal information or destroy the enterprise systems of its competitor.
- Design a safe data format (see Figure 4 for an example) for data stored in RFID tags if the extra space costs of the format specification do not affect the normal operation of the supply chain systems. The safe data format is able to prevent the common script injection attacks by prohibiting common dangerous keywords. For example, 'shutdown' should not appear in a valid data format.

Propagation Phase

- **Local-area propagation:** Upload malicious data from multiple RFID tags to a near RFID reader. Legitimate RFID readers may be infected after reading malicious data stored in RFID tags. Countermeasures to this threat will be discussed in the next subsection.

Figure 4. An example of a safe format for RFID data

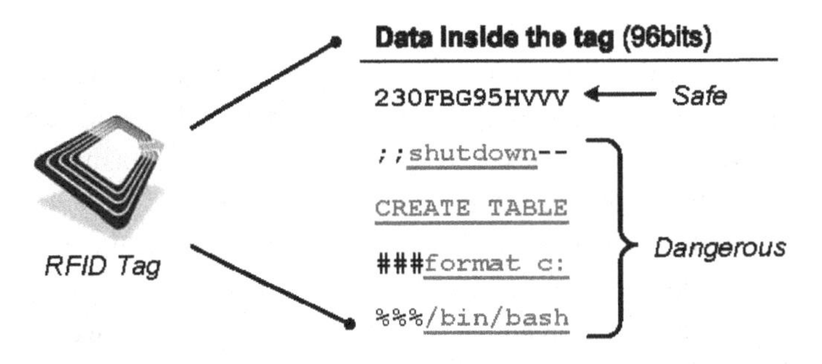

- **Local/Wide-area propagation:** Spread malicious data from one tag to nearby RFID tags. Uninfected RFID tags are re-written when receiving signals from an infected RFID tag. This tag-to-tag propagation is possible when the infected RFID tag is an active RFID tag and it is able to work in RFID reader mode. Luckily, the product with such capabilities seems not to be available on the current market. This could be the reason why this threat is not identified in prior literature. But there are no fundamental technical or economic difficulties in developing such a device, which can be implemented as a variant of RFID Guardian (Rieback, Gaydadjiev, Crispo, Hofman & Tanenbaum (2006)). The major threat of this propagation behavior is due to the malware only needing to compromise a RFID tag to spread itself, which is much easier than compromising a RFID reader. Moreover, these newly infected RFID tags may be transported to other supply chain partners causing wide-area propagation. The countermeasures for this propagation behavior are the same as those for the attack phase on the current layer. See Figure 5 for a visual representation of the propagation phase.

10.3.4 Malware Protection on Reader Layer

The reader layer is a boundary layer in a RFID-enabled supply chain system. Each RFID reader

Figure 5. Propagation phase of RFID malware on the tag layer

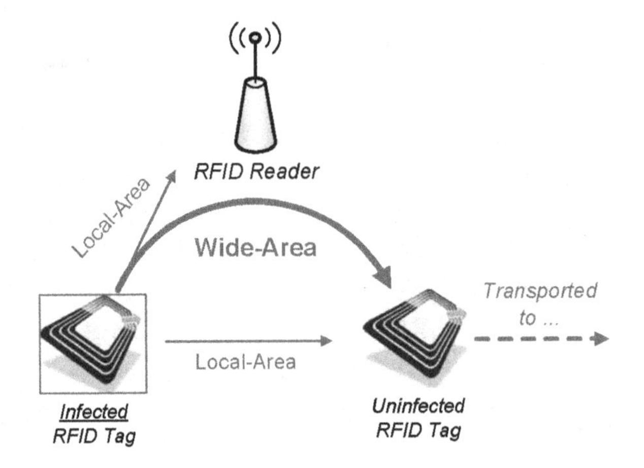

is an entry point of the enterprise systems of a supply chain partner. Stronger malware protection mechanisms should be deployed in this layer for the following reasons: 1) this is the lowest layer under full control of a single supply chain partner in the sense that the RFID infrastructure, the RFID reader, is supposed to be securely configured; 2) a RFID reader is more capable than a RFID tag, which is usually equipped with an embedded processor (ARM or XScale)[9] that is able to perform complex computational tasks which are unaffordable to a RFID tag; 3) More information is available to a RFID reader, as it is the first aggregation point for RFID data. Correspondingly, the threats from a compromised RFID reader will be more serious than an infected RFID tag, as its stronger capability is also available to RFID malware (see Figure 6). To mitigate these threats, we present our detailed analysis and design of malware protections on the reader layer as follows.

Attack Phase

Security threats:

- Confidential data stored in a RFID reader are disclosed. These disclosed data could be 1) enterprise secrets or 2) hints for subsequent attacks including plaintext RFID data and secret keys used in processing RFID data.
- RFID readers are compromised to perform malicious actions. The malicious actions could be uploading malicious data to upper layers, or writing malicious data to uninfected RFID tags, or accessing corporate networks for other malicious purposes.
- RFID readers remain intact, but malicious RFID data pass through them and reach upper layers.
- RFID readers remain intact, but they are disrupted by jamming signals such that they fail to read certain RFID tags.

Attack strategies:

- Use an infected RFID tag to feed malicious data to RFID readers. A RFID reader could be compromised by these data that exploit its vulnerabilities. Digital signatures and cryptographic protocols alone will not stop the intrusion when the vulnerabilities are exploited before finishing the signature checking and protocol execution. The malicious data may come not only from a normal RFID tag but also from a more power-

Figure 6. Attack phase of RFID malware on the reader layer

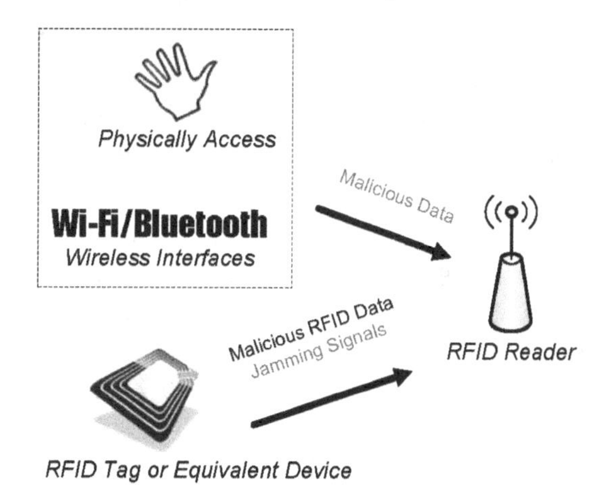

ful device like a RFID smart card that is able to adaptively feed malicious data.

- Physically access a RFID reader or access it from wireless remote configuration interfaces if these interfaces are available. This attack requires the attacker to be in the proximity of the target RFID reader.
- Send a jamming signal to (selectively) create collisions with the RF signal from a target RFID tag. The jamming signal will disrupt the collision resolution protocols (Rieback, Gaydadjiev, Crispo, Hofman & Tanenbaum (2006)) so that RFID data in the target RFID tag cannot be read by a RFID reader. Although the implementation of this denial-of-service attack has been reported (Rieback (2006)) and even patented[10], it is still not well recognized as a significant threat in the literature of security research on RFID-enabled supply chain systems.

Survival skills:

- Partition a complete malware into multiple fragments (Shankarapani, Sulaiman & Mukkamala (2009)). Each fragment will be stored in a separate RFID tag so as to evade detection based on the signature of a complete malware.
- Use polymorphic techniques (Spinellis (2003)) to evade detection based on exact pattern matching.
- Use packing techniques (Brosch & Morgenstern (2006)) to evade detection without corresponding unpacking functions.

RFscreen Design on Reader Layer

Available information for defense:

- Secrets stored in a RFID reader
- RFID data from RFID tags
- Event logs of reading RFID data from RFID tags
- Information provided by upper layers

Available defense capabilities:

- Affordable computational tasks by an embedded device (applicable for almost all standalone RFID readers)

Countermeasures:

- Avoid storing plaintext secrets in the storage of a RFID reader if possible. This reduces the loss when a RFID reader is compromised.
- Review the implementation of RFID readers to avoid simplistic assumptions. One common problematic assumption is that the RFID reader is supposed to read certain types of RFID tags, and the preset buffer size for RFID data is large enough for these types, without demanding further boundary checking. This assumption opens the door to RFID malware carried by a special tag with larger memory capacity than the preset buffer size.
- Lightweight malware protection mechanisms should be deployed on RFID readers, which are listed as follows.
    - **Data integrity protection technology:** Digital signature and safe data format mechanism introduced in the previous subsection provides basic protection against unauthorized modification of RFID data and limits the content of malicious data that can be placed in RFID tags.
    - **Execution integrity protection technology:** Current trusted comput-

ing technology (TCG (n.d.)) is able to efficiently ensure the execution integrity when a target system has simple behaviors and few states, which are applicable to RFID readers.
    - **Coarse-grained signature/anomaly-based malware detection technology (Bose, Hu, Shin & Park (2008), Liu, Yan, Zhang & Chen (2009)):** Fine-grained malware detection schemes usually generate many events that consume considerable resources to analyze them, which may not be affordable to RFID readers.
    - **Coarse-grained access control technology (Zhang, Aciicmez & Seifert (2009), Muthukumaran, Sawani, Schiffman, Jung & Jaeger (2008)):** Data read from RFID tags are labeled as tainted to track the data flow of RFID data. This prevents malware intrusion from traditional attack vectors like buffer overflow. All four types of technologies constitute a sandbox environment for applications run on RFID readers, which provide best-effort protection to avoid RFID readers being compromised and filter out malicious data in the current layer. A RFID reader may request the upper layers to analyze suspicious data, when it cannot decide on its own. The extra knowledge from upper layers further improves the effectiveness of the protection.
- Minimize the network access privilege of RFID readers to mitigate the threat of unauthorized access of corporate networks from a compromised RFID reader. The threat of accessing corporate networks could be more serious than any other threat as the attacker may now be able to access more valuable information besides RFID data.
- Analyze the RFID event logs to detect denial-of-service attacks by emitting jamming signals. Abnormal event patterns such as more than usual collision events are a sign of such attacks.

Propagation Phase

- **Local-area propagation:** Upload malicious data from a compromised RFID reader or multiple RFID tags to the back-end system layer. Back-end systems may be infected after reading these data. Countermeasures to this attack will be discussed in the next subsection.
- **Local/Wide-area propagation:** Use a compromised RFID reader to write malicious data to nearby RFID tags. There are usually no effective countermeasures to this propagation behavior, as the attacker is able to use all the secrets used in RFID communication protocols after he has fully compromised the RFID reader. These newly infected RFID tags may be transported to other supply chain partners to further cause wide-area propagation. See Figure 7 for a visual representation of the propagation phase.

10.3.5 Malware Protection on Back-End System Layer

The back-end system layer is the top layer including both local and global back-end systems. The local back-end systems of a supply chain partner consists of RFID middleware and EPC information services. RFID middleware provides the interfaces that link the front-end RFID readers and back-end data repository managed by EPC information

Figure 7. Propagation phase of RFID malware on the reader layer

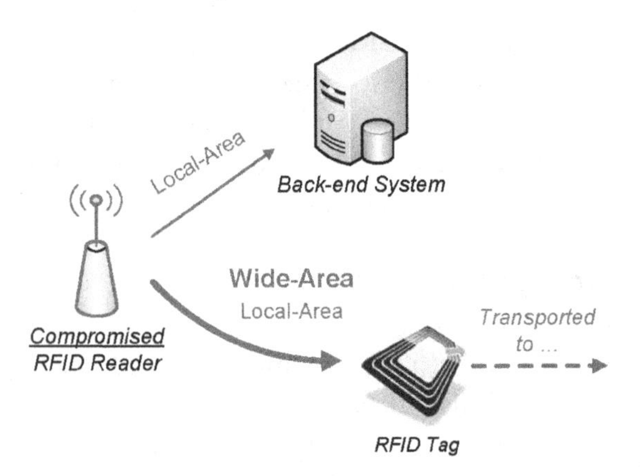

services. The global back-end systems are global services of the EPCglobal network, which enable real-time information sharing between multiple supply chain partners over the Internet. Unlike the previous layers, the capability of these back-end systems is *essentially equivalent* to that of general purpose computing systems, which is usually considered sufficient to support even heavyweight protection mechanisms. So malware protection on this layer can directly reuse the existing techniques designed for general purpose computing systems, which have been extensively investigated in prior literature (Newsome & Song (2005), Wang, Li, Xu, Reiter, Kil & Choi (2006), Martignoni, Christodorescu & Jha (2007), Lu, Park, Hu, Ma, Jiang, Li, Popa & Zhou (2007), Tan, Zhang, Ma, Xiong & Zhou (2008), Padioleau, Tan & Zhou (2009)). We incorporate these techniques into our design of RFscreen and highlight the major customizations for RFID-enabled supply chain management systems in the following analysis and discussion.

Attack Phase

Security threats:

- Confidential data stored in back-end systems are disclosed. These disclosed data could be 1) enterprise secrets or 2) hints for subsequent attacks including the information that is associated with RFID data but not stored in RFID tags and all the information that could be disclosed in the previous layers (plaintext RFID data and secret keys used in processing RFID data).
- Back-end systems are compromised to perform malicious actions (see Figure 8). The malicious actions could be sending malicious data from an infected back-end system to an uninfected back-end system, or dispatching malicious data to RFID readers that may write them into uninfected RFID tags later, or accessing corporate networks for other malicious purposes.
- Back-end systems remain intact, but they are disrupted by denial-of-service attacks and fail to function normally.

Attack strategies:

- Compromise RFID tags or RFID readers in the previous layers to feed malicious data to back-end systems. A back-end system could be compromised by these data, as it is very difficult to completely eliminate vulnerabilities in enterprise back-end systems due to their large code size and rapid development process under economic and time pressures. Even if a back-end system is able to resist the intrusion attempts after receiving these malicious data, it may still suffer denial-of-service attacks. The denial of service could be caused by receiving a large number of fake data or only a few elaborately crafted data to impose intensive computational or IO tasks to a back-end system.
- Exploit the vulnerabilities of the enterprise firewall to directly access a back-end system. This attack is usually much more difficult to perform than the previous one. To penetrate a securely configured enterprise firewall requires considerable effort and knowledge. Especially for an organization that maintains the global services of

Figure 8. Attack phase of RFID malware on the back-end system layer

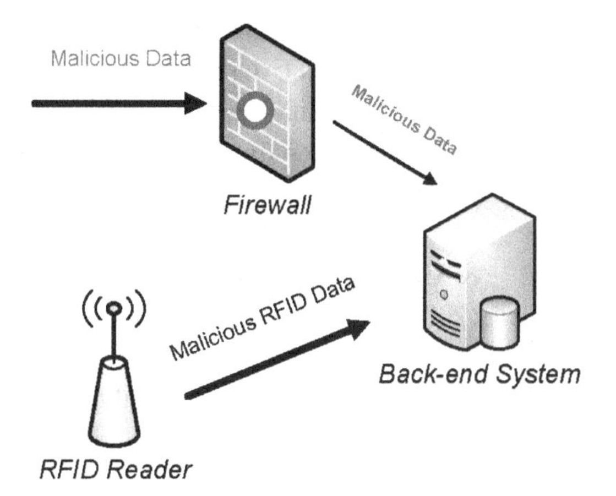

the EPCglobal network, the setup of a secure firewall is one of the basic capabilities that are required to gain the trust of supply chain partners; otherwise no supply chain partners will use these insecure services to share their confidential data.

Survival skills:

- There are no extra special survival skills besides those given for RFID malware on the reader layer. The effectiveness of these skills will be degraded on the current layer. This is because heavyweight malware detection mechanisms can now be deployed, which are not affordable on the previous layers. These mechanisms are better able to tear off the mask of RFID malware (Newsome & Song (2005), Wang, Li, Xu, Reiter, Kil & Choi (2006), Martignoni, Christodorescu & Jha (2007),Hu, Chiueh & Shin (2009)).

RFscreen Design on Back-End System Layer

Available information for defense:

- Secrets stored in a back-end system
- Aggregated RFID data and event logs from RFID readers
- Event logs of operations done in back-end systems
- Information provided by other back-end systems

Available defense capabilities:

- Affordable computational tasks by an enterprise server (usually considered sufficient to support any practical defense technologies)

Countermeasures:

- Avoid storing plaintext secrets in the storage of a back-end system if possible. This is not as easy as for RFID readers because certain secrets have to be cached in a back-end system to enable automated data processing. For example, RFID middleware must hold the authentication token (usually a pair of plaintext username and password) of a database server to access and upload RFID data. Trusted computing technology may mitigate the threat by storing secrets in tamper-resistant hardware. This tamper-resistant hardware ensures that the plaintext of the stored secrets is never revealed to the outside software system, so that the attacker is not able to know the secrets even if he can control the entire software system.
- Review the design and implementation of back-end systems before they are deployed. Common pitfalls, such as assigning the unnecessary privilege of database access to a back-end system, should be avoided. All the existing proof-of-concept RFID malware proposed in prior literature (Rieback, Crispo & Tanenbaum (2006), Rieback, Simpson, Crispo. & Tanenbaum (2006), Sulaiman, Mukkamala & Sung (2008), Shankarapani, Sulaiman & Mukkamala

(2009)) rely on these pitfalls to show their effectiveness. Fortunately, most of these pitfalls can be efficiently detected by manual or automated code review and bug analysis (Lu, Park, Hu, Ma, Jiang, Li, Popa & Zhou (2007), Tan, Zhang, Ma, Xiong & Zhou (2008), Padioleau, Tan & Zhou (2009)).

- Any practical malware protection mechanism can be deployed during the operation of back-end systems. Considering the endless potential of cloud computing technology[11], the capability of back-end systems will be technically sufficient to support even heavyweight mechanisms if they are worth the investment. The common types of protection mechanisms are given below, which are *essentially the same* as those given for the reader layer but not restricted to lightweight mechanisms.
    - **Data integrity protection technology:** Each data package should be associated with a digital signature if applicable. This increases the difficulty of attacks, as an attacker is first forced to compromise a device in order to generate correct signatures.
    - **Execution integrity protection technology:** Current trusted computing technology (TCG (n.d.)) may not be able to efficiently handle the entire enterprise back-end system due to the complex behaviors and large number of states. But it can still be used to protect the critical components of back-end systems, which are usually small enough.
    - **Practical signature/anomaly-based malware detection technology:** Malware detection techniques are now affordable to collect and analyze all available information. The information includes both fine-grained event information like system call sequences, and coarse-grained event information like the statistics on aggregated RFID data. This information can be analyzed by many well-developed detection algorithms used in prior research (Newsome & Song (2005), Wang, Li, Xu, Reiter, Kil & Choi (2006), Hu, Chiueh & Shin (2009)) or existing commercial products. For the same algorithm with a well-built signature/anomaly database, the effectiveness of detecting RFID malware will be similar to that of detecting other malware.
    - **Practical access control technology:** Any existing access control mechanisms can be easily adapted to monitor RFID data, by classifying RFID data into the lowest level of credibility and privilege when they are received from RFID readers. These four types of technologies together provide best-effort protection against RFID malware on the current layer. The overall protection effectiveness on a back-end system can be improved by cooperating with other back-end systems.
- Deploy intrusion resistant and backtracking mechanisms. These mechanisms minimize the loss of malware intrusion and the cost of identifying the source of the intrusion. One of the major intrusion resistant mechanisms is to create redundancy. The damaged data can be recovered from the last data backup. Besides recovering data, it is now also possible to rollback a damaged system to the last check point powered by virtualization technology (Borders, Weele, Lau & Prakash (2009)). These intrusion resistant mechanisms not only reduce the recovery cost, but also buy precious re-

sponse time for the enterprise. During this time, backtracking mechanisms will help to identify the source of RFID malware, which may eventually stop the intrusion if the source is identified in time. The most common backtracking mechanisms are to keep a log for sensitive system events like database accesses and to track the data flow within the systems. Digital signature technology is a practical anti-counterfeiting technique to label data source and route. Audit records consisting of the signatures of the devices that generate or process data will provide evidence to identify and isolate the source of malicious data.

Propagation Phase

Global propagation: Use a compromised back-end system to send malicious data to any other back-end systems or RFID readers of any supply chain partners. The consequences of this propagation behavior could be very serious as malicious data spread over the Internet may cause the global propagation of RFID malware. Certain statistical or behavioral patterns may expose such propagation, which can be used to design effective countermeasures after they are observed for specific RFID malware. See Figure 9 for a visual representation of the propagation phase.

10.3.6 Summary

This subsection summarizes the security threats of RFID malware and their countermeasures used in RFscreen design. Figure 10 is a conceptual map that captures the major security threats of RFID malware targeting RFID-enabled supply chain management systems in the EPCglobal network.

This map shows that the most significant threat of RFID malware comes from a new attack vector, malicious RFID data. Unlike traditional attack vectors, there are fundamental difficulties in eliminating malicious data from RFID tags: 1) *RFID is a wireless communication technology and a large number of RFID tags are transported between multiple supply chain partners*, which gives plenty of chances for an attacker to access some of them so as to implant RFID malware. 2) *The limited capability of RFID tags restricts the strength of self-protection mechanisms.* These characteristics create an easier path for an attacker to feed malicious data to a target system when compared to breaking through a heavily armed firewall. This is how RFID malware is different from other malware.

Figure 9. Propagation phase of RFID malware on the back-end system layer

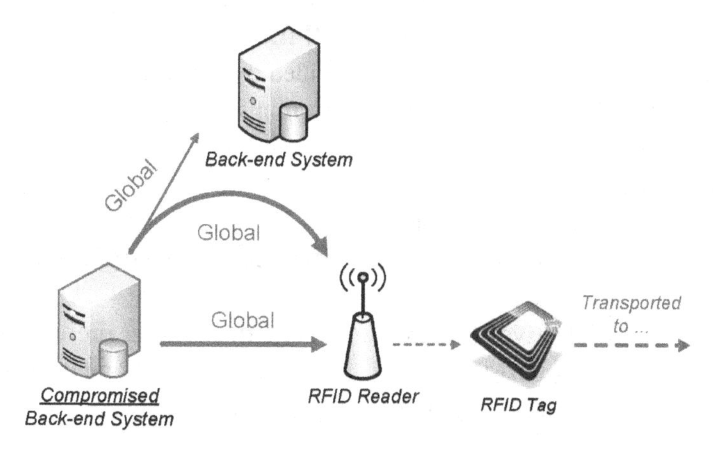

Figure 10. A global security threat map for RFID malware

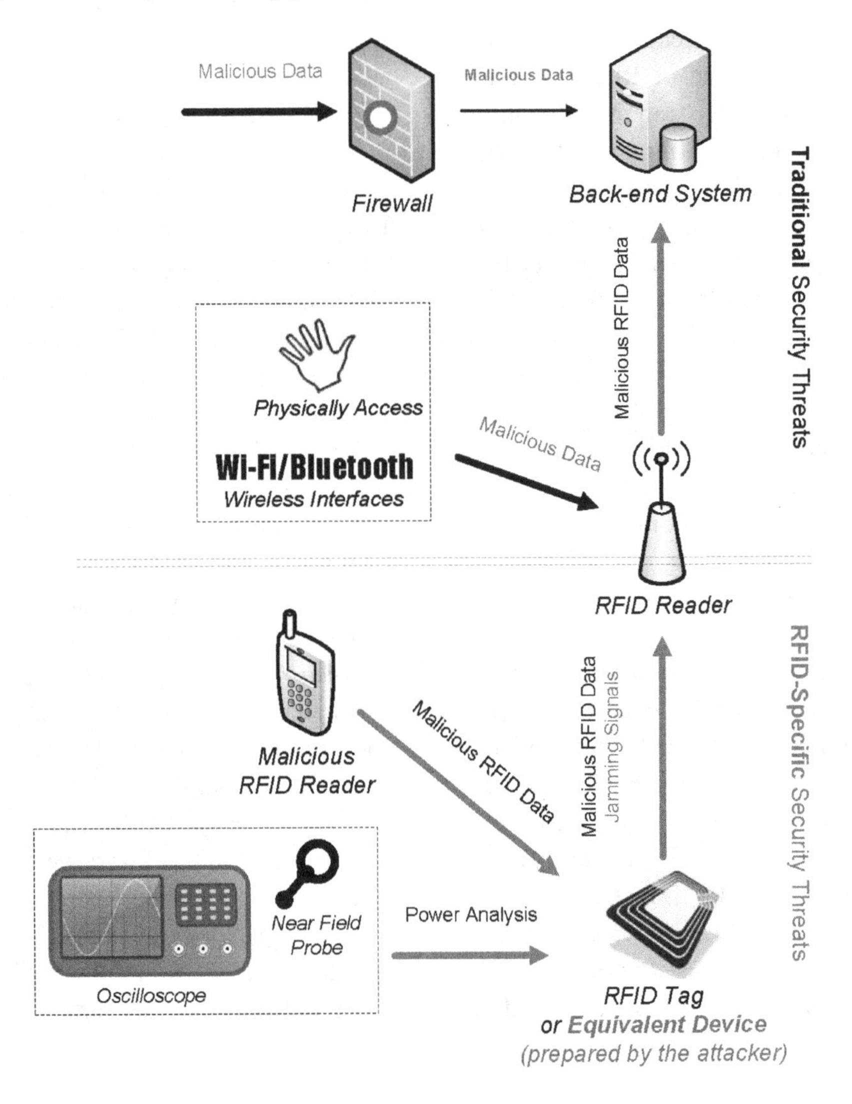

But after malicious data enter a back-end system, the characteristics of RFID malware constructed from these data become essentially the same as other malware, as we have analyzed in previous subsections. Correspondingly, the effectiveness of protection mechanisms against RFID malware will be on the same level as that against other malware, as long as we keep in mind that data read from pervasive RFID tags could be malicious. Therefore, it is an effective strategy for RFscreen to follow the existing best-practice architecture design and to reuse well-developed malware protection mechanisms. The intrusion of RFID malware can be effectively controlled, provided that the entire system is carefully designed to minimize valuable information disclosure in front-end systems.

FUTURE RESEARCH DIRECTIONS

While this chapter presents a high level analysis of RFID malware and its countermeasures in the context of RFID-enabled supply chain management systems, it is still too early to rigorously evaluate the effectiveness of any specific countermeasure with only the few artificial samples of RFID malware that are currently available. A future direction to address this problem is to carefully design more representative samples and to collect more wild samples. Another potential research direction is to investigate the hardware features of new RFID products including both RFID readers and tags, and to develop new attack or defense mechanisms that use these features under the capability limitations of the devices. Since RFID malware is more hardware-specific than traditional malware, this research direction is particularly meaningful in practice. Finally, a comprehensive analysis and prediction of future outbreak events is obviously important.

CONCLUSION

As a new rising threat, RFID malware is attracting more and more attention as RFID-enabled technology becomes pervasive in enterprise systems and human life. It is imperative to develop a comprehensive understanding of RFID malware. Regarding this threat, this chapter provided a high level analysis using RFID-enabled supply chain management systems as a case study. Our analysis shows that the most significant threat of RFID malware comes from malicious RFID data, which are difficult to detect and prevent by front-end systems. Once malicious RFID data enter a back-end system, RFID malware constructed from these data behaves similarly to traditional malware. Because of this feature, the security threats of RFID malware can be effectively controlled. Instructive guidelines were provided to develop specific countermeasures for each major security threat identified in this chapter. These guidelines will be valuable for building malware-resistant RFID systems in the future.

ACKNOWLEDGMENT

This work is partly supported by the Office of Research at Singapore Management University under Grant No. 11-C220-SMU-001.

REFERENCES

Abadi, M., Budiu, M., Erlingsson, U., & Ligatti, J. (2005). Control-flow integrity. In *Proceedings of the 12th ACM Conference on Computer and Communications Security*, (pp. 340–353) New York, NY: ACM.

Borders, K., Weele, E. V., Lau, B., & Prakash, A. (2009). Protecting confidential data on personal computers with storage capsules. In *Proceedings of the 18th USENIX Security Symposium, USENIX Association*, Berkeley, CA, USA, (pp. 367–382).

Bose, A., Hu, X., Shin, K. G., & Park, T. (2008). Behavioral detection of malware on mobile handsets. In *Proceeding of the 6th International Conference on Mobile Systems, Applications, and Services*, (pp. 225–238). New York, NY: ACM.

Brier, E., Clavier, C., & Olivier, F. (2004). Correlation power analysis with a leakage model. In *Proceedings of 6th International Workshop on Cryptographic Hardware and Embedded Systems*, (pp. 16–29). Berlin, Germany: Springer-Verlag.

Brosch, T., & Morgenstern, M. (2006). *Runtime packers: The hidden problem?* Presented at Black Hat USA, 2006.

Cohen, F. B. (1992). Defense-in-depth against computer viruses. *Computers & Security*, *11*(6), 563–579. doi:10.1016/0167-4048(92)90192-T

Cox, B., Evans, D., Filipi, A., Rowanhill, J., Hu, W., & Davidson, J. … Hiser, J. (2006). N-variant systems: A secretless framework for security through diversity. In *Proceedings of the 15th conference on USENIX Security Symposium*, (pp. 105–120). Berkeley, CA: USENIX Association.

Das, R., & Harrop, D. P. (2009). *RFID forecasts, players and opportunities 2009-2019.* Market report, IDTechEx. Retrieved from http://www.idtechex.com/research

EPCglobal. (2007). *EPC information services* (EPCIS) v1.0.1. Technical report, EPCglobal. Retrieved from http://www.epcglobalinc.org/standards

EPCglobal. (2008a). *EPCglobal object naming service (ONS) v1.0.1.* Technical report, EPCglobal. Retrieved from http://www.epcglobalinc.org/standards

EPCglobal. (2008b). *UHF class 1 gen 2 standard v1.2.0.* Technical report, EPCglobal. Retrieved from http://www.epcglobalinc.org/standards

Gao, D., Reiter, M. K., & Song, D. (2005). Behavioral distance for intrusion detection. In *Proceedings of the 8th International Symposium on Recent Advances in Intrusion Detection,* (pp. 63–81). Berlin, Germany: Springer-Verlag.

Hu, X., Chiueh, T.-C., & Shin, K. G. (2009), Large-scale malware indexing using function-call graphs. In *Proceedings of the 16th ACM Conference on Computer and Communications Security,* (pp. 611–620). New York, NY: ACM.

Juels, A. (2006). RFID security and privacy: A research survey. *IEEE Journal on Selected Areas in Communications, 24*(2), 381–395. doi:10.1109/JSAC.2005.861395

Kasper, T., Oswald, D., & Paar, C. (2009). *New methods for cost-effective side-channel attacks on cryptographic RFIDs.* In Workshop on RFID Security.

Kim, H., Smith, J., & Shin, K. G. (2008). Detecting energy-greedy anomalies and mobile malware variants. In *Proceeding of the 6th International Conference on Mobile Systems, Applications, and Services,* (pp. 239–252). New York, NY: ACM.

Liu, L., Yan, G., Zhang, X., & Chen, S. (2009). VirusMeter: Preventing your cellphone from spies. In *Proceedings of the 12th International Symposium on Recent Advances in Intrusion Detection,* (pp. 244–264). Berlin, Germany: Springer-Verlag.

Lu, S., Park, S., Hu, C., Ma, X., Jiang, W., Li, Z., et al. (2007). MUVI: Automatically inferring multi-variable access correlations and detecting related semantic and concurrency bugs. In *Proceedings of Twenty-First ACM SIGOPS Symposium on Operating Systems Principles,* (pp. 103–116). New York, NY: ACM.

Ma, C., Li, Y., Deng, R. H., & Li, T. (2009). RFID privacy: Relation between two notions, minimal condition, and efficient construction. In *Proceedings of the 16th ACM Conference on Computer and Communications Security,* (pp. 54–65). New York, NY: ACM.

Martignoni, L., Christodorescu, M., & Jha, S. (2007). OmniUnpack: Fast, generic, and safe unpacking of malware. In *Proceedings of the 23rd Annual Computer Security Applications Conference,* (pp. 431–441). Los Alamitos, CA: IEEE Computer Society.

Muthukumaran, D., Sawani, A., Schiffman, J., Jung, B. M., & Jaeger, T. (2008). Measuring integrity on mobile phone systems. In *Proceedings of the 13th ACM Symposium on Access Control Models and Technologies,* (pp. 155–164). New York, NY: ACM.

Newsome, J., & Song, D. (2005). Dynamic taint analysis for automatic detection, analysis, and signature generation of exploits on commodity software. In *Proceedings of the Network and Distributed System Security Symposium,* Internet Society, Reston, Virginia, USA.

NSA. (n.d.). *Defense in depth: A practical strategy for achieving information assurance in today's highly networked environments.* Technical report, National Security Agency. Retrieved from http://www.nsa.gov/ia/_files/support/defenseindepth.pdf

ODonnell. A. J., & Sethu, H. (2004). On achieving software diversity for improved network security using distributed coloring algorithms. In *Proceedings of the 11th ACM Conference on Computer and Communications Security,* (pp. 121–131). New York, NY: ACM.

Padioleau, Y., Tan, L., & Zhou, Y. (2009). Listening to programmers - Taxonomies and characteristics of comments in operating system code. In *Proceedings of the 31st International Conference on Software Engineering,* (pp. 331–341). Washington, DC: IEEE Computer Society.

Pescatore, J., Young, G., Allan, A., Girard, J., Feiman, J., & MacDonald, N. (2008). *Gartner 2008 IT security threat projection timeline.* Technical report, Gartner. Retrieved from http://www.gartner.com

Rieback, M. R. (2006). *A hackers toolkit for RFID emulation and jamming.* Presented at 23rd Chaos Communication Congress.

Rieback, M. R., Crispo, B., & Tanenbaum, A. S. (2006). Is your cat infected with a computer virus? In *Proceedings of the Fourth Annual IEEE International Conference on Pervasive Computing and Communications,* (pp. 169–179). Washington, DC: IEEE Computer Society.

Rieback, M. R., Gaydadjiev, G. N., Crispo, B., Hofman, R. F. H., & Tanenbaum, A. S. (2006). A platform for RFID security and privacy administration. In *Proceedings of the 20th conference on Large Installation System Administration,* (pp. 89–102). Berkeley, CA: USENIX Association.

Rieback, M. R., Simpson, P. N., Crispo, B., & Tanenbaum, A. S. (2006). RFID malware: Design principles and examples. *Pervasive and Mobile Computing, 2*(4), 405–426. doi:10.1016/j.pmcj.2006.07.008

Shankarapani, M. K., Sulaiman, A., & Mukkamala, S. (2009). Fragmented malware through RFID and its defenses. *Journal in Computer Virology, 5*(3), 187–198. doi:10.1007/s11416-008-0106-0

Spinellis, D. (2003). Reliable identification of bounded-length viruses is NP complete. *IEEE Transactions on Information Theory, 49*(1), 280–284. doi:10.1109/TIT.2002.806137

Sulaiman, A., Mukkamala, S., & Sung, A. (2008). SQL infections through RFID. *Journal in Computer Virology, 4*(4), 347–356. doi:10.1007/s11416-007-0075-8

Tan, L., Zhang, X., Ma, X., Xiong, W., & Zhou, Y. (2008). AutoISES: Automatically inferring security specifications and detecting violations. In *Proceedings of the 17th USENIX Security Symposium,* (pp. 379–394). Berkeley, CA: USENIX Association.

TCG. (n.d.). *Trusted computing group.* Retrieved from http://www.trustedcomputinggroup.org

Wang, X., Li, Z., Xu, J., Reiter, M. K., Kil, C., & Choi, J. Y. (2006), Packet vaccine: Black-box exploit detection and signature generation. In *Proceedings of the 13th ACM Conference on Computer and Communications Security,* (pp. 37–46). New York, NY: ACM.

Xie, Y., & Kim, H.-A. OHallaron, D., Reiter, M. K., & Zhang, H. (2004). Seurat: A pointillist approach to anomaly detection. In *Proceedings of the 7th International Symposium on Recent Advances in Intrusion Detection,* (pp. 238–257). Berlin, Germany: Springer-Verlag.

Yee, B., Sehr, D., Dardyk, G., Chen, J. B., Muth, R., & Ormandy, T. … Fullagar, N. (2009). Native client: A sandbox for portable, untrusted x86 native code. In *Proceedings of the 2009 30th IEEE Symposium on Security and Privacy,* (pp. 79–93). Washington, DC: IEEE Computer Society.

Zhang, X., Aciicmez, O., & Seifert, J.-P. (2009). Building efficient integrity measurement and attestation for mobile phone platforms. In *Proceedings of the First International Conference on Security and Privacy in Mobile Information and Communication Systems*, (pp. 71–82). Berlin, Germany: Springer-Verlag.

KEY TERMS AND DEFINITIONS

Attack Phase: A phase when RFID malware intends to attack.

Defense-in-Depth: A best-practices strategy for a security system design.

EPCglobal Network: The de-facto industry standard for RFID-based trading systems.

Propagation Phase: A phase when RFID malware has already compromised a device.

RFID: An automated data collection technology that uses radio frequency waves.

RFID Malware: Malicious software infiltrating RFID systems.

RFscreen: A demo security system against RFID malware.

Supply Chain System: A large-scale transportation system for asset supply.

ENDNOTES

1. EPCglobal Network, http://www.epcglobalinc.org
2. EPCglobal Discovery Services Standard (in development), http://www.epcglobalinc.org/standards/discovery
3. Texas Instruments, http://www.ti.com
4. Alien Technique, http://www.alientechnology.com
5. SecureRF, http://www.securerf.com
6. Alien Technique, http://www.alientechnology.com
7. Motorola, http://www.motorola.com
8. GAO RFID Inc, http://www.gaorfid.com
9. ALR-9900 Enterprise RFID Reader Family, http://www.alientechnology.com/docs/products/DS_ALR_9900+.pdf
10. Jamming device against RFID smart tag systems, United States Patent 7221900, http://www.freepatentsonline.com/7221900.html
11. Cloud computing, http://en.wikipedia.org/wiki/Cloud_computing

APPENDIX: ADDITIONAL RESOURCES

The following articles and websites give more details about RFID malware and RFID-enabled supply chain systems as discussed in this chapter.

Articles

- Rieback, Crispo & Tanenbaum (2006), Rieback, Simpson, Crispo. & Tanenbaum (2006), Sulaiman, Mukkamala & Sung (2008), and Shankarapani, Sulaiman & Mukkamala (2009) discuss RFID malware design and implementation in detail.
- Bose, Hu, Shin & Park (2008), Liu, Yan, Zhang & Chen (2009), Zhang, Aciicmez & Seifert (2009), Muthukumaran, Sawani, Schiffman, Jung & Jaeger (2008), Kim, Smith & Shin (2008), and Yee, Sehr, Dardyk, Chen, Muth, Ormandy, Okasaka, Narula & Fullagar (2009) show the recent progress on lightweight malware protection mechanisms designed for mobile devices and web browsers, which may provide insight into solving the capability limitation problem of RFID readers, especially for mobile RFID readers.
- (Abadi, Budiu, Erlingsson & Ligatti 2005), Newsome & Song (2005), Wang, Li, Xu, Reiter, Kil & Choi (2006), Martignoni, Christodorescu & Jha (2007), and Hu, Chiueh & Shin (2009) show the recent progress on traditional malware protection mechanisms.
- O'Donnell & Sethu (2004), Xie, Kim, O'Hallaron, Reiter & Zhang (2004), Gao, Reiter & Song (2005), and Cox, Evans, Filipi, Rowanhill, Hu, Davidson, Knight, Nguyen-Tuong & Hiser (2006) show the recent progress on system diversity, which could be useful to resist the intrusion of RFID malware for large-scale systems like RFID-enabled supply chain systems.

Websites

The following websites give more information on the experience of RFID malware:

- **RFID Viruses and Worms:** http://www.rfidvirus.org/
- **Hackers Clone E-Passports:** http://www.wired.com/science/discoveries/news/2006/08/71521
- **Kevin Warwick, a human infected with a computer virus:** http://www.kevinwarwick.com/ICyborg.htm and http://blog.jgc.org/2010/05/inside-rfid-virus.html

The following websites give more information on the standards and industrial communities related to RFID-enabled supply chain systems.

- **EPCglobal Network Standards:** http://www.epcglobalinc.org
- **RFID Journal:** http://www.rfidjournal.com
- **RFID CUSP:** http://www.rfid-cusp.org
- **RFID Security Alliance:** http://www.rfidsa.com

Chapter 11
Addressing Covert Channel Attacks in RFID-Enabled Supply Chains

Kirti Chawla
University of Virginia, USA

Gabriel Robins
University of Virginia, USA

ABSTRACT

RFID technology can help competitive organizations optimize their supply chains. However, it may also enable adversaries to exploit covert channels to surreptitiously spy on their competitors. We explain how tracking tags and compromising readers can create covert channels in supply chains and cause detrimental economic effects. To mitigate such attacks, the authors propose a framework that enables an organization to monitor its supply chain. The supply chain is modeled as a network flow graph, where tag flow is verified at selected key nodes, and covert channels are actively sought. While optimal taint checkpoint node selection is algorithmically intractable, the authors propose node selection and flow verification heuristics with various tradeoffs. The chapter discusses economically viable countermeasures against supply chain-based covert channels, and suggests future research directions.

11.1 INTRODUCTION

Radio Frequency Identification (RFID) technology enables the tracking of objects via attached tags that respond to radio signals from readers (Finkenzeller (2003); Sweeney (2003)). Organizations can use RFID technology to streamline their internal processes and can optimize various phases of the production cycle, including asset management, inventory control, production tracking, shipping, recalls, and warranties (Angeles (2005); Min and Zhou (2002)).

However, the use of RFID technology can also leak sensitive information to adversaries about the internal processes of a target organization (Mitrokotsa, Reiback and Tanenbaum (2008)). For example, an adversary can track and/or modify tags, insert duplicate tags into supply chains, and

DOI: 10.4018/978-1-4666-3685-9.ch011

even compromise the RFID readers in the supply chain of a target organization. Such attacks "taint" the information flow, resulting in covert channels in the supply chain of a target organization (Moskowitz and Kang (1994)).

These covert channels can surreptitiously reveal product flow patterns, site-specific inventories, delivery schedules, and other strategic information. An adversary can use this illicitly obtained sensitive information to gain an unfair (and not necessarily even illegal) strategic or economic advantage with respect to a target organization. Given such possible threats, it is important for a target organization to control and verify the information flow in order to detect the presence of covert channels and mitigate their effect (Chawla, Robins and Weimer (2010)).

In this chapter, we analyze threat sources in RFID-enabled supply chains and focus on four representative attacks which adversaries can use to track the supply chains of target organizations. We consider the ability of such attacks to affect market change by modeling supply chains as network flow graphs, where nodes represent sites and edges model product flow. We designate key "taint checkpoint" nodes in the supply chain flow graph to verify the product information flow, and note that selecting such "taint checkpoints" optimally is NP-Complete.

We develop taint check cover heuristics based on tradeoffs, including the desired coverage (i.e. the number and locations of the desired taint checkpoints). We describe algorithms that verify the flow of information in the supply chain, both locally and globally. These algorithms offer user-controlled tradeoffs between the strength of the verification results versus the time required to compute them. This enables post-detection actions to be taken by the target organization, either at a local site or along selected global paths. Finally, we evaluate these approaches using a supply chain simulator, and provide remedies that target organizations can utilize to mitigate the impact of covert channels.

11.2 BACKGROUND

Supply chains are a collection of organizational processes spanning multiple geographic sites for the purpose of transforming raw materials into finished products, and delivering them from producers to consumers. Due to the large size and complexity of a typical supply chain, it is difficult to track and maintain cross-site inventories. Furthermore, mishaps such as loss or theft of products can cause serious financial losses to the organizations operating these supply chains. Products affixed with RFID tags can be tracked and queried universally at any place and time. Thus, with the advent of RFID technology, supply chains are becoming more efficient in managing inventories and preventing theft (Niederman, Mathieu, Morley and Kwon (2005); Wilding and Delgado (2004)).

However, such flexibility also gives rise to novel spatial-temporal inferences through embedded covert channels that could reveal product flow patterns within supply chains. While this technology is critical to the smooth functioning of a target organization, it may also leak sensitive information to competitors who may unscrupulously leverage it to affect market changes. To remain economically viable, target organizations must actively mitigate the adversarial impact of covert channels on its profitability and economic competitiveness.

11.3 THREATS IN RFID-ENABLED SUPPLY CHAINS

To analyze potential threats to RFID-enabled supply chains, we present a threat model and focus on four possible supply chain attacks.

11.3.1 Threat Model

The proposed threat model uses a motivating example to highlight the underlying key assumptions. Consider two competing businesses that develop ubiquitous and interchangeable products (e.g. cellular phones). Such businesses use competitive pricing and features as key differentiating factors, tailoring their products according to brand loyalties and user preferences. Assume that both businesses are competing for the same consumer base in the same markets, and are striving to make their supply chains more efficient by optimizing their internal processes. Often the target business invests in new technology, such as RFID, after performing appropriate cost-benefit analyses. While the benefits may be immediately evident (e.g., efficient inventory control, real-time production tracking, speedy warranty authorizations, etc.,) the cost of utilizing such a technology may involve more than just the direct cost of installing RFID equipment and processes.

In order to remain competitive, the adversary business may also adopt RFID technology. Moreover, the adversary can also exploit this technology to fraudulently learn patterns of product flow in the supply chain of the target business. Such patterns can be used in time-sensitive ways to provide lower consumer prices, or flood the adversary's products into selected regions or stores while the products of the target business become scarcer.

Such practices can be construed as industrial or economic espionage and can significantly reduce the profitability of the target business. The resulting profit drop can be viewed as a hidden cost that would be difficult to anticipate or even identify by the target business. Moreover, recent advances in RFID technology and its wide-spread use can provide adversaries with selective "insider access" of target businesses' supply chains, without requiring direct access to their physical premises.

11.3.2 Possible Attacks

Supply chains of target organizations can unintentionally reveal product flow patterns to adversaries in a number of ways. We enumerate four representative attacks (see Figure 1), some of which have already received attention from the security research community (Avoine, Lauradoux and Martin (2009); Juels, A., Rivest, R.L., and Szydlo, M. (2003); Koscher, Juels, Brajkovic and Tadayoshi (2009); Mandel, Roach and Winstein (2004); Mitrokotsa, Reiback and Tanenbaum (2008); Weis, S. A. et al. (2004)). We explain the significance of these attacks when applied to a supply chain scenario, discuss the potential implications, and present possible ways to mitigate such attacks. While the proposed attacks can be executed over any given RFID standard, this chapter assumes the *EPC Gen2* RFID standard (EPCGlobal (2008); EPCGlobal (2011)). The following types of attacks create covert channels that are said to "taint" the supply chain of a target organization.

Attack 1: Tag Tracking

In this attack, an adversary tracks existing tags across the supply chain of a target business. Such tags can be applied at the product-level or case-level. We assume that a target business assembles the finished product at its factory, attaches the tags at the case-level, and then ships cases to geographically-separate warehouses. These cases are eventually organized into different batches and delivered to various retail stores. An adversary can learn the product flow patterns by copying the information stored in some case-level tags, and then querying them at different places in the supply chain (e.g. by deploying long-range readers). Such copied case-level tags constitute a covert channel, as they leak product flow information to the adversary, while traveling unobtrusively through the supply chain of the target business.

Attack 2: Tag Duplication

In this attack, an adversary copies the information stored in an existing tag and constructs a duplicate tag. Consequently, the adversary can attach this duplicate tag to a different case, enabling it to become part of the supply chain of the target

Figure 1. (a) Inserting duplicate tags, (b) modifying existing tags, and (c) compromising existing readers

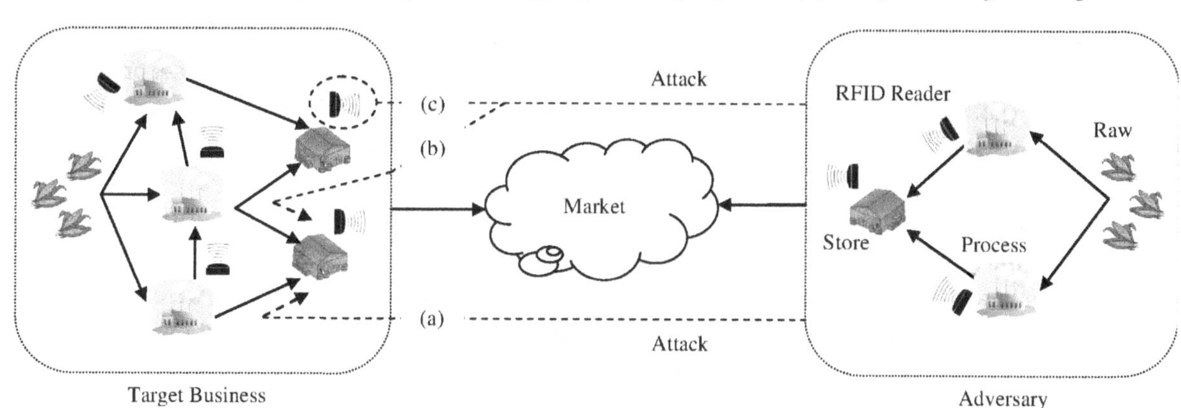

business and thus, a covert channel. The adversary then queries the cases at different points of the supply chain to determine the product flow (i.e. if the adversary sees two duplicate case-level tags at a warehouse then he can infer that they aggregated at that warehouse after originating from different locations). This attack scenario is more powerful than the previous tracking-only attack, since here the adversary can still track legitimate supply chain tags as well as its own surreptitiously inserted duplicated tags. An adversary can mount such an attack with modest effort, since tag duplication hardware is relatively inexpensive and easily available (Mandel, Roach and Winstein (2004)).

Attack 3: Tag Modification

In an *EPC Gen2*-compliant tag, there are four memory banks: *EPC*, *TID*, *User*, and *Reserved*. While the inventory process of a target business supply chain primarily uses the EPC portion of a tag's memory, the contents of the other memory banks are ignored. Therefore, an adversary can modify the information in the writable portions of other memory banks, which can then serve as a covert channel. Independently, it has been suggested that the unused portion of memory of a tag can be utilized to conceal information (Karygiannis, Phillips, and Tsibertzopoulos (2006); Mitrokotsa, Reiback and Tanenbaum (2008)). Such a vulnerability can be an attractive target to an adversary, due to its potentially large payoff versus the relatively low effort required to exploit it.

Attack 4: Reader Compromise

With rapid advances in RFID technology, RFID readers are available in a variety of form-factors, hardware/software combinations, and use-case scenarios (e.g., rack-mountable, battery-powered, etc.). Many of these readers are deployed in supply chains in a manner that enables an adversary to compromise them (e.g., snooping on a wireless reader transmission, compromising the on-board software or hardware of a mobile reader, etc.).

In Figure 2 (a), a compromised reader copies information stored in several case-level tags, and provides it to an adversary. In Figure 2 (b), an adversarial compromised reader selectively repudiates (i.e. intentionally fails to report) the presence of any duplicate or modified tags. Such compromised views from the readers enable covert channels to exist unobtrusively in the supply chains of target organizations.

We note that such a reader-compromise attack subsumes the tag duplication and modification attacks in terms of potential risks to the target organization. From the adversary's perspective, in order to ensure a successful attack, while at least one compromised tag at the case-level is necessary, it may not be sufficient, since that tag may fail or become undetectable. Thus, several compromised (i.e. duplicated and/or modified) tags should be used at the case-level (say, at least three tags per 100 cases). On the other hand, if too many compromised tags (i.e. half of the total) are deployed, the adversary's exposure risk also increases dramatically. Moreover, an adversary may not need to track product flow information at the product-level, since case-level tracking is sufficient for that purpose.

11.3.3 Market Change Scenarios

We examine two possible market scenarios to illustrate the potential harmful impact of the RFID-based attacks described above. A typical supply chain involves business-related variables such as inventory levels at factories and retailers, delivery schedules from raw material sites to warehouses, shipping capabilities, backlog of orders, etc. Leaks of such strategic business information can occur via attacks on supply chains, which can enable an adversary to engage in unfair competitive practices. Furthermore, an adversary can affect negative market changes by knowing the business practices of its competitors. We used the *Anylogic* supply chain model simulator (Anylogic (2009)) to generate economic impact projections and qualitatively outline possible outcomes of such attacks.

Case 1: Brand Loyalty Change

In the first scenario (Figure 3), we consider two businesses serving a population of 10,000 consumers with brands A and B, respectively. We assume that the two brands are interchangeable, and have the same retail price. The business with brand B is the target business, while the business with brand A is the adversary. Consumers must purchase either brand A or brand B per time unit (i.e. the product is a staple product).

In Figure 3 (a), consumers are projected to prefer brand B to brand A by 55% to 45% (i.e. whenever a consumer arrives at a store, he chooses a product at random from the set of available equivalent products, preferring brand B slightly over brand A). However, by carefully timing its production so that more brand A products are available at a time when fewer brand B products are available, an adversary can induce consumers to switch brands. In Figure 3 (b), the

Figure 2. Reader compromise attack: (a) copy case-level tags, and (b) repudiate the presence of a covert channel

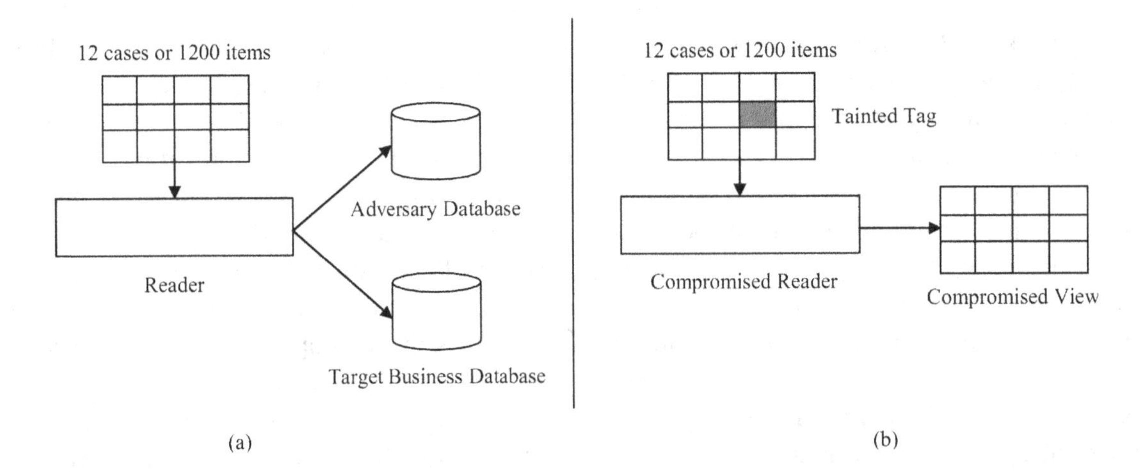

Figure 3. Market change projections using the anylogic supply chain simulator (Anylogic 2009): (a) consumers prefer brand B over brand A, (b) consumers switch from brand B to brand A, (c) brand B enjoys more demand than brand A, and (d) brand A demand increases, while brand B demand decreases

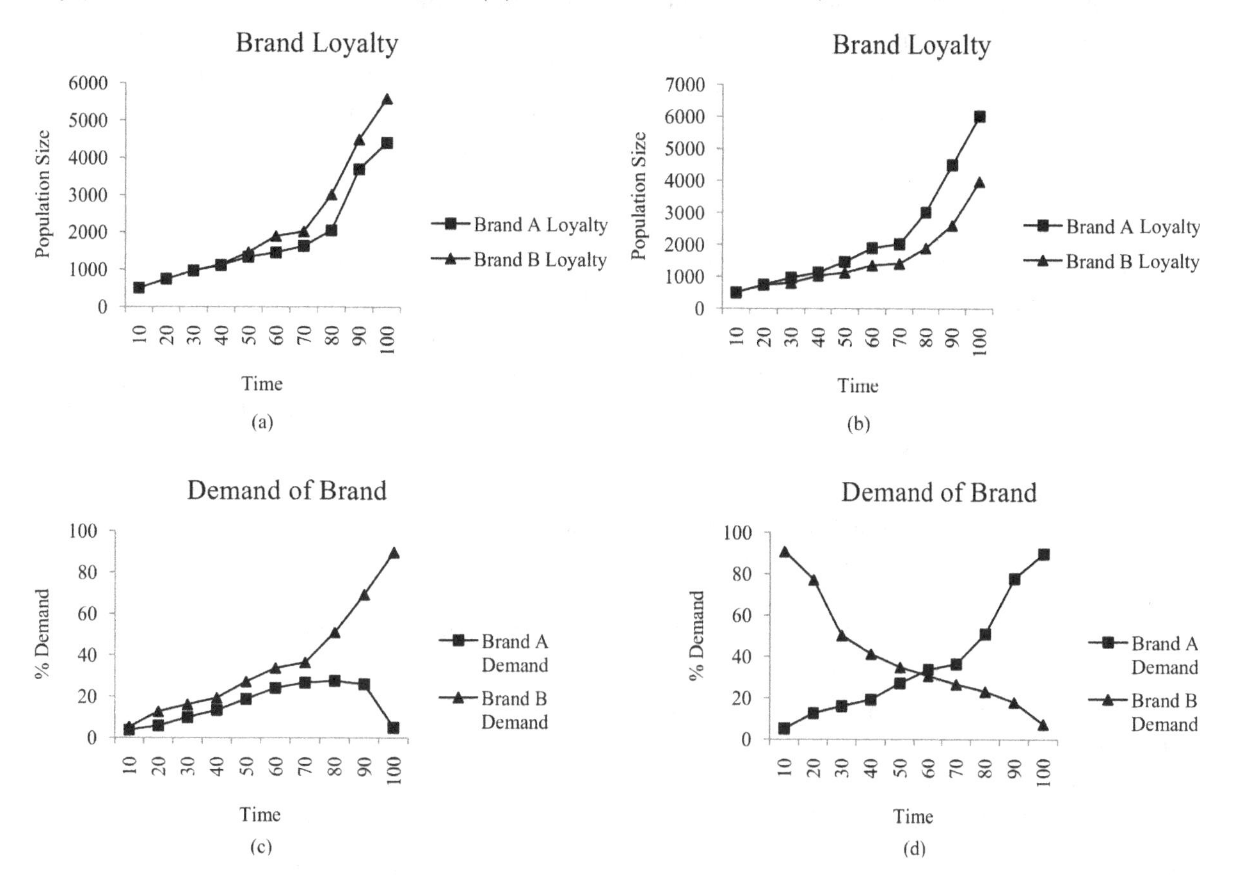

adversary has succeeded in inducing the consumers to switch brands, now favoring A over B by 57% to 43%.

Case 2: Brand Aversion

In the second scenario (Figure 3), we consider a neighborhood store served by two businesses A and B, as described above. Stores often stock products that enjoy consistent demand, in order to maintain profitability. Initially, the store stocks both products in equal amounts. However, at a later time, as shown in Figure 3 (c), brand B (i.e. product of the target business) is projected to have a higher demand than brand A by 89% to 5%. There is typically a demand threshold below which it will become non-profitable to stock a brand (i.e. leading to "brand aversion"). An adversary aiming to bolster its own shelf presence may resort to illegitimately acquiring sensitive supply chain information of the target business. Figure 3 (d) depicts such a scenario, where the adversary engages in supply chain attacks to obtain time-sensitive information about a target business, and uses that information to manipulate the market.

Observations for Market Change Scenarios

Augmenting a supply chain with RFID technology entails attaching RFID tags at the product-level or case-level, and then tracking them throughout the supply chain using RFID readers. The target business keeps track of products starting from the purchase phase (i.e. as raw materials) through the distribution phase (i.e. as finished products stored at warehouses or retail outlets). An adversary can use the possible attacks described above in order to learn vital strategic information, resulting in the projected market change scenarios, to the detriment of the target business.

If the potential benefits outweigh the incurred costs, adversaries have strong motivation (i.e. economic incentive) for perpetrating such attacks. Thus, such attacks are viable in an RFID-enabled supply chain given the potentially high payoff to an adversary, although specific occurrences of such attacks seem to have not yet been publicly reported. While we have argued that the exposure of only a few business variables to an adversary can result in an unfair (and not necessarily even illegal) marketplace economic advantage, it would be interesting to study more elaborate and detailed marketplace scenarios and projections. Such possible scenarios can stimulate further discussions regarding the associated risks as well as the effectiveness of possible solutions in RFID-enabled supply chains.

11.4 SUPPLY CHAIN MODEL

We now focus on the problem of modeling a supply chain, towards the goals of preventing covert channel attacks and mitigating their effects. A supply chain typically spans multiple geographically separate sites and involves numerous phases that include the sourcing of raw materials, processing and storing end-products, and delivering these products to markets and consumers. Supply chain models can be categorized as deterministic models, stochastic models, hybrid models, economic models, and IT driven models (Angeles (2005); Min and Zhou (2002); Swaminathan, Smith and Sadeh (1998)). While these models aim to capture many aspects of a supply chain in great detail, our goal is to construct a simpler model that will enable us to focus on the fundamental issues related to potential covert channel attacks.

In a supply chain, there are product flows between sites (i.e. raw materials moving among various locations). In an RFID-enabled supply chain, however, product flow between sites is analogous to "tag-flow", since RFID tags are attached to each product. The supply chain consists of multiple phases, wherein each phase is a collection of sites. Furthermore, to detect the presence of duplicate tags, modified tags, or compromised readers, we need mechanisms to track product flow between supply chain phases. With these three key observations in mind, we developed a model based on network flow graphs (Cormen, Leiserson, Rivest and Stein (2009)), called "supply chain flow graphs."

11.4.1 Logistical Phases

A supply chain typically spans three phases: the purchase phase, the production phase, and the distribution phase (e.g. sites associated with the production phase are involved primarily in manufacturing a product). Each phase of the supply chain is a collection of interconnected sites having product flows among them. We define the supply chain flow graph $G = (V, E)$ as a directed, connected graph, where a node p corresponds to a site, and an edge (p, q) models a product flow between the two sites. Each edge $(p, q) \in E$ has a positive product flow capacity $C(p, q) > 0$, while "non-edges" have no capacity (i.e. $\forall\ (p, q) \notin E$, $C(p, q) = 0$). There are two special nodes called the "source node" (S) and the "sink node" (T). We partition the supply chain flow graph into three sub-graphs, corresponding to the purchase phase, production phase, and distribution phase, respectively (other more specialized supply chains may contain additional phases).

Network flows are subject to the usual constraints on edge capacity and flow conservation at nodes (Cormen, Leiserson, Rivest and Stein

(2009)). We propose an additional property, namely the node maximal outgoing flow, which will enable us to address issues related to attacks. There are typically multiple paths for product flow in a supply chain. A "critical node" or "critical edge" may experience more product flow than those along other paths. We model such supply chain characteristics by keeping track of each node's maximum outgoing flow. If two nodes have the same maximal outgoing flow, we resolve the tie by giving precedence to the node having a predecessor with a higher flow value. Supply chain flow graphs having such critical paths facilitate reasoning about issues related to possible attack locations and product flow inspection sites.

11.4.2 Taint Checkpoints

A direct approach for detecting covert channel attacks may require looking for tainted RFID tags at every node of the supply chain. However, this would be prohibitively expensive and time consuming. Instead, we propose to select a subset of nodes, called "taint checkpoints", verify the product flow at these selected locations, and report the presence of any discovered covert channels in the supply chain flow graph. When RFID tags are attached to products by the target business in the early phases of the supply chain, the information onboard the tags is recorded in order to track the inventory.

In subsequent phases of the supply chain, this information is available to taint checkpoints for the purpose of inspection and verification. This verification process involves comparing the information present onboard a currently viewable RFID tag with the trusted and previously stored information. Any mismatch may indicate the presence of covert channels or other tampering. Figure 4 illustrates a supply chain flow graph, including several taint checkpoints where product flow is inspected.

11.4.3 Taint Checking and Verification Algorithms

Based on the above observations, we formulate the problem of optimally selecting taint checkpoints in the supply chain flow graph. We show that this problem is NP-Complete (i.e. computationally intractable), and suggest heuristics to generate good approximate solutions.

Problem Statement: Taint Check Cover

To ensure the absence of covert channels in the supply chain, the taint checkpoints should provide broad coverage for the entire graph. The related optimization problem is to select as few taint checkpoints as possible, while providing broader coverage for the entire supply chain flow graph. Thus, we seek a "taint check cover" V' of the

Figure 4. A supply chain flow graph with three taint checkpoints

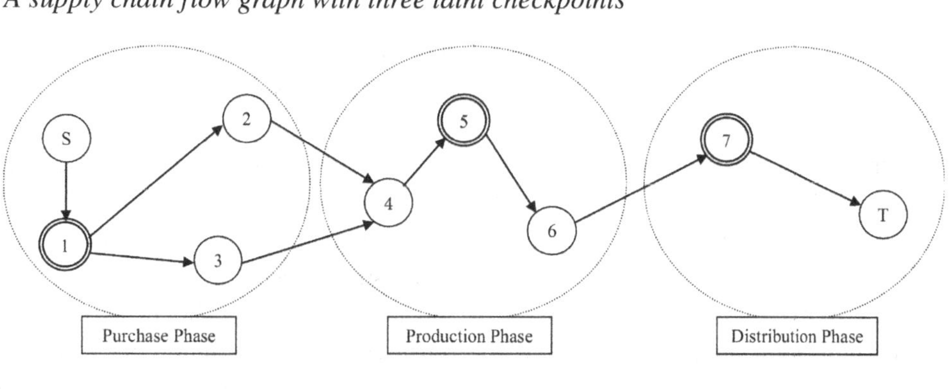

supply chain flow graph $G = (V, E)$, where $V' \subseteq V$, such that every edge of E has at least one of its end points (i.e. nodes) in V'. Note that we may choose to only cover some critical node subset of the supply chain flow graph rather than the entire graph. Either way, this objective corresponds to the classical graph vertex cover problem, which is known to be NP-complete (Cormen, Leiserson, Rivest and Stein (2009)).

Heuristic Taint Check Cover Generation

There is a known efficient heuristic for the vertex cover problem that produces solutions of size no worse than twice the optimal (Cormen, Leiserson, Rivest and Stein (2009)). This heuristic selects an arbitrary graph edge, adds its two end points to the growing vertex cover solution, eliminates this edge and its end points from the graph, and iterates until the entire graph is exhausted. To see that this scheme produces a worst-case twice-optimal solution, we observe that at least one of the two nodes of each removed edge must be present in any optimal solution.

Given the high degree of freedom in how edges (and thus nodes) are selected in constructing such a heuristic taint check cover solution, a target business may introduce different selection criteria, based on practical, economic, or strategic considerations. A target business may wish to limit the number of taint checkpoints, or balance the cost of its supply chain versus the coverage provided by the taint checkpoints. On the other hand, we may also consider taint checkpoint selection criteria based on the specific structure of the supply chain, or some other combination based on these considerations. To address these factors, we give several possible formulations of the supply chain flow graph coverage problem as follows:

- **Maximum Edge Cover:** Given a supply chain flow graph G and an integer K, find a taint check cover (node set) of size K having a maximum total number of adjacent edges.
- **Minimum Taint Checkpoint Cover:** Given a supply chain flow graph G and an integer J, find a minimum taint check cover (node set) with at least J total adjacent edges.

These formulations naturally generalize to scenarios where, as part of the input, we also designate a given subset of the nodes or edges (or both nodes and edges) that must be included in the taint check cover. When determining a taint check cover, the target business may choose from a continuous tradeoff between infrastructural costs and overall taint check coverage. This can be accomplished using greedy approaches such as the $2 \cdot \mathrm{OPT}$ node cover heuristic (Cormen, Leiserson, Rivest and Stein (2009)) discussed above, the techniques described in (Chen, Kanj and Jia (2001); Niedermeir and Rossmanith (1999)), or any other taint check cover heuristic.

Alternatively, nodes can be selected by decreasing order of maximal outgoing flow values, in order to give higher priority to high-flow nodes. Furthermore, we can utilize the classical min-cut-max-flow theorem (Cormen, Leiserson, Rivest and Stein (2009)) and preferentially select high-flow nodes along some minimum cut. This will tend to maximize the overall probability of detecting covert channels (which must cross any graph cut). Moreover, the nodes may be permuted in some other manner (e.g., by aggregate product value, time-criticality, geographical distribution, or even randomly), in order to capture topological or economic considerations. In summary, our algorithmic template is quite general and can utilize many alternative possible criteria to construct taint check covers.

Note that not every node and edge in the flow graph must necessarily be covered, since tainted tags that are missed at some points along the graph will likely be discovered at subsequent downstream locations. On the other hand, including any flow graph "cut" in the taint check cover ensures that every tainted tag will *eventually* be discovered in at least one location. Toward this goal, cuts containing smaller number of nodes are more efficient than larger node cuts (i.e. they will require fewer readers at taint checkpoints to achieve the same guaranteed taint checking coverage). Choosing small (or even optimal) graph cuts can be accomplished using known min-cut algorithms (Stoer and Wagner (1997)). Figure 4 illustrates a cut-based cover utilizing phase criteria for a sample supply chain flow graph.

Verification Algorithm

Each node (i.e. a taint checkpoint) in the taint check cover is responsible for inspecting and verifying the product flow passing through it. Each product in this flow has a unique RFID tag ID. If a taint checkpoint reads multiple counts of the same tag ID, or the system detects the same tag ID at two different places simultaneously, then a duplicate tag has been detected. By comparing the information present onboard each viewable tag with data stored a priori in a trusted backend database, modifications to tags can be detected at taint checkpoints. We note that such product flow verifications can be performed "locally" at a given taint checkpoint or "globally" across a given path or cut, as taint checkpoints accumulate, exchange, and compare tag information.

11.5 EVALUATION AND ANALYSIS

We used simulations to evaluate our proposed approaches. We randomly constructed a base supply chain flow graph configuration of 2000 nodes and selected among 10, 100, and 1000 nodes to be taint checkpoints. Each checkpoint is assumed to be able to verify 1000 cases of 100 units of product at each time interval. We assume each taint checkpoint has direct access to the trusted database implementing a tag lookup service. In our first simulation, we measured the relationship between the number of taint checkpoints and the cumulative time required to perform local verification (i.e. the maximum reading rate of the RFID readers).

Figure 5 (a) shows that as the number of taint checkpoints increases, there is a corresponding increase in the time to locally verify the product flow. Our second simulation evaluates the global verification algorithm, which collects local verification results from taint checkpoints. The cost of the collection process depends on the underlying bandwidth of the network links. Figure 5 (b) shows the simulated verification cost when the link cost is either constant (500 ms), or a variable (ranging from 2 to 1000 ms) time window proportional to the geographical distance of the taint checkpoints from the central database server.

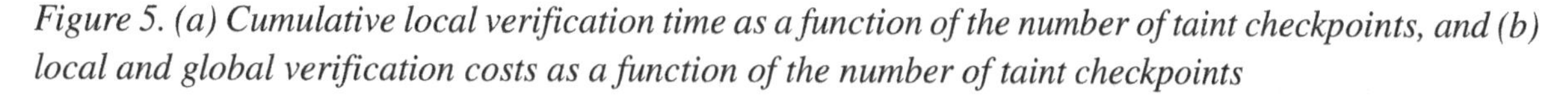
Figure 5. (a) Cumulative local verification time as a function of the number of taint checkpoints, and (b) local and global verification costs as a function of the number of taint checkpoints

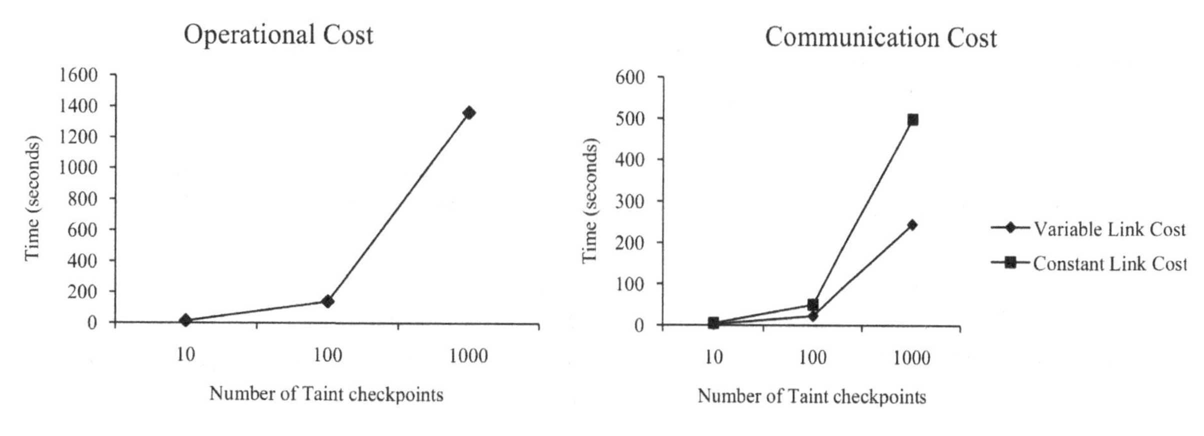

Thus, we explored the verification communication cost as the number of taint checkpoints increases. We observed that the communication cost can grow rapidly in a more realistic scenario where taint checkpoints are at larger variable distances from the check nodes.

11.6 RESPONSES TO COVERT CHANNELS

We enumerate several possible mitigating techniques available to the target organizations when the covert channels are detected in their supply chains. Note that the presence of covert channels in the supply chain can never be completely ruled out, even when privacy-preserving algorithms are used in the underlying RFID technology (Bailey, Boneh, Goh and Juels (2007); Garfinkel, Juels and Pappu (2005)).

11.6.1 Response 1: Authentication

According to the *EPC Gen2* RFID standard, an RFID tag is required to support password protection for read or write access to the tag. The systematic use of passwords can mitigate tag tracking, duplication, and modification attacks. However, this technique requires that the RFID system in the supply chain support and conform to the chosen password schemes. Alternatively, challenge-response based authentication protocols can be used to thwart tag tracking, modification and duplication attacks (Rhee, Kwak, Kim and Won (2005)). Moreover, compromised readers may be authenticated with respect to genuine tags using a technique described in (Paise and Vaudenay (2008)).

11.6.2 Response 2: Pseudonym

An RFID tag using pseudonyms transmits a slightly different ID each time it is queried (Molnar, Soppera and Wagner (2005)). This can prevent the adversary from discovering patterns in a supply chain, but requires the target business to accommodate a pseudonym scheme in its tracking logic. Also, (Burmester and Munilla (2009)) describe an unlinking technique that can be used to prevent tag tracking attacks. Additionally, blocker tags can be utilized to selectively block compromised readers (Juels, Rivest and Szydlo (2003)).

11.6.3 Response 3: Re-Encryption

The use of encryption to conceal the tag's data can still allow an adversary to track a statically encrypted tag over the supply chain. To defeat such an attack, the tags can be re-encrypted after each phase of the supply chain, in order to prevent the adversary from modifying or tracking the tags. Techniques described in (Golle, Jakobsson, Juels and Syverson (2004); Saito, Ryou, and Sakurai (2004)) can be used to anonymize tags via re-encryption.

11.6.4 Response 4: Direct Mitigation

The work of (Reiback, Crispo and Tanenbaum (2005)) describes a device that can be used for sweeping and preventing reader compromise attacks. When a covert channel source is discovered, an operator can physically clear the operating environment while temporarily altering the flow of products. Alternatively, path checking techniques can trace tags that follow altered routes (Oua and Vaudenay (2009)). Furthermore, physically detaching a tag's body from its antenna (temporarily) can serve as an effective (albeit cumbersome) technique to prevent RFID-based attacks en route between supply chain sites (Karjoth and Moskowitz (2005)).

11.6.5 Response 5: Physically Unclonable Functions (PUF)

Physically unclonable functions (PUFs) are easy to evaluate but hard to characterize or duplicate. For example, PUFs may be used to generate random numbers based on the variability inherent in the manufacturing processes of the underlying VLSI circuits (Bolotnyy and Robins (2007)). PUF-based privacy-preserving algorithms can provide a way to build message authentication codes to ensure data integrity and aid in preventing tag modification attacks.

CONCLUSION

We discussed and analyzed vulnerabilities in RFID-enabled supply chains, and enumerated possible attacks that can be perpetrated with relatively modest effort. An adversary can thus surreptitiously learn product flow patterns in the RFID-enabled supply chain of a target organization, which may result in privacy leakage and economical damage. We proposed a simple model for reasoning about supply chain flow and RFID attack mitigation, and demonstrated that attacks can be detected and addressed at selected nodes in a supply chain. We presented a practical heuristic template for the computationally intractable problem of optimal taint checkpoint selection that enables trading off protection coverage against overall cost. We simulated and analyzed verification algorithms, and enumerated several possible responses against detected covert channels.

This chapter represents an important step toward the analysis and mitigation of attacks on RFID-enabled supply chains. We envision that the proposed basic supply chain model can be extended to include additional practical considerations (e.g., multi-product supply chains, adversary cartels, cost-benefit analyses, supply chain topology-specific constraints, fine-grained product flow analyses, market change scenarios, etc.) such as described in (Lee and Whang (2005)). Future heuristics may be fine-tuned to address such additional practical considerations. Finally, it would be interesting to further study the comparative tradeoffs between coverage, cost, and efficiency for different taint checkpoint selection strategies in realistic scenarios.

ACKNOWLEDGMENT

This research was supported by National Science Foundation (NSF) grant CNS-0716635.

REFERENCES

Angeles, R. (2005). RFID technologies: Supply-chain applications and implementation issues. *Information Systems Management*, *22*(1), 51–65. doi:10.1201/1078/44912.22.1.20051201/85739.7

Anylogic Professional 6. (2009). *AB-SD supply chain model simulator*. Retrieved from http://www.xjtek.com

Avoine, G., Lauradoux, C., & Martin, T. (2009). Lecture Notes in Computer Science: *Vol. 5932*. *When compromised readers meet RFID. Information Security Applications* (pp. 36–50). Springer. doi:10.1007/978-3-642-10838-9_4

Bailey, D. V., Boneh, D., Goh, E., & Juels, A. (2007). Covert channels in privacy-preserving identification systems. *14th ACM International Conference on Computer and Communication Security*, (pp. 297-306).

Bolotnyy, L., & Robins, G. (2007). Physically unclonable function-based security and privacy in RFID systems. *5th International Conference on Pervasive Computing and Communications*, (pp. 211-120).

Burmester, M., & Munilla, J. (2009). *A flyweight RFID authentication protocol.* Workshop on RFID Security.

Chawla, K., Robins, G., & Weimer, W. (2010). On mitigating covert channels in RFID-enabled supply chains. *Radio Frequency Identification System Security (RFIDsec), Workshop on RFID Security*, (pp. 135-146).

Chen, J., Kanj, I. A., & Jia, W. (1999). Vertex cover: Further observations and further improvements. *Journal of Algorithms*, *41*, 313–324.

Cormen, T. H., Leiserson, C. E., Rivest, R. L., & Stein, C. (2009). *Introduction to algorithms* (3rd ed.). MIT Press.

EPCGlobal. (2008). *UHF C1 G2 air interface protocol standard*. Retrieved from http://www.epcglobalinc.org/standards/uhfc1g2/uhfc1g2_1_2_0-standard-20080511.pdf

EPCGlobal. (2011). *Tag data standards version 1.6*. Retrieved from http://www.gs1.org/gsmp/kc/epcglobal/tds/tds_1_6-RatifiedStd-20110922.pdf

Finkenzeller, K. (2003). *RFID-handbook: Fundamentals and applications in contactless smart cards and identification* (2nd ed.). Munich, Germany: Wiley and Sons Inc.

Garfinkel, S. L., Juels, A., & Pappu, R. (2005). RFID privacy: An overview of problems and proposed solutions. *IEEE Security and Privacy*, *3*(3), 34–43. doi:10.1109/MSP.2005.78

Golle, P., Jakobsson, M., Juels, A., & Syverson, P. (2004). Lecture Notes in Computer Science: *Vol. 2964. Universal re-encryption for mixnets. Topics in Cryptology - CT-RSA 2004* (pp. 163–178). Springer. doi:10.1007/978-3-540-24660-2_14

Juels, A., & Pappu, R. (2003). Lecture Notes in Computer Science: *Vol. 2742. Squealing Euros: Privacy protection in RFID-enabled banknotes. Financial Cryptography* (pp. 103–121). Springer. doi:10.1007/978-3-540-45126-6_8

Juels, A., Rivest, R. L., & Szydlo, M. (2003). The blocker tag: Selective blocking of RFID tags for consumer privacy. *10th ACM Conference on Computer and Communications Security*, (pp. 103-111).

Karjoth, G., & Moskowitz, P. A. (2005). Disabling RFID tags with visible confirmation: Clipped tags are silenced. *ACM Workshop on Privacy in the Electronic Society*, (pp. 27-30).

Karygiannis, A., Phillips, T., & Tsibertzopoulos, A. (2006). RFID security: A taxonomy of risks. *Conference on Communications and Networking in China*, (pp. 1-8).

Koscher, K., Juels, A., Brajkovic, V., & Tadayoshi, K. (2009). EPC RFID tag security weaknesses and defenses: Passport cards, enhanced drivers licenses, and beyond. *16th ACM Conference on Computer and Communications Security*, (pp. 33-42).

Lee, H. L., & Whang, S. (2005). Higher supply chain security with lower cost: Lessons from total quality management. *International Journal of Production Economics*, *96*, 289–300. doi:10.1016/j.ijpe.2003.06.003

Mandel, J., Roach, A., & Winstein, K. (2004). *MIT proximity card vulnerabilities*. MIT Technical report. Retrieved from http://web.mit.edu/keithw/Public/MIT-Card-Vulnerabilities-March31.pdf

Min, H., & Zhou, G. (2002). Supply chain modeling: Past, present and future. *Journal of Computer and Industrial Engineering*, *43*(1-2), 231–249. doi:10.1016/S0360-8352(02)00066-9

Mitrokotsa, A., Rieback, M. R., & Tanenbaum, A. S. (2008). Classification of RFID attacks. *International Workshop on RFID Technology*, (pp. 73-86).

Molnar, D., Soppera, A., & Wagner, A. (2005). Lecture Notes in Computer Science: *Vol. 3897. A scalable, delegatable pseudonym protocol enabling ownership transfer of RFID tags. Selected Areas in Cryptography* (pp. 276–290). Springer. doi:10.1007/11693383_19

Moskowitz, I. S., & Kang, M. H. (1994). Covert channels here to stay. *9th IEEE International Conference on Computer Assurance*, (pp. 235-243).

Niederman, F., Mathieu, R. G., Morley, R., & Kwon, I. (2005). Examining RFID applications in supply chain management. *Communications of the ACM*, *50*(7), 93–102.

Niedermeir, R., & Rossmanith, P. (1999). Upper bounds for vertex cover further improved. *16th Symposium on Theoretical Aspects in Computer Science, Lecture Notes in Computer Science, Vol. 1563,* (pp. 561-570). Springer.

Orlin, J. B. (1988). A faster strongly polynomial minimum cost flow algorithm. *20th ACM Symposium on Theory of Computing*, (pp. 377-387).

Oua, K., & Vaudenay, S. (2009). *Pathchecker: An RFID application for tracing products in supply-chains.* Workshop on RFID Security.

Paise, R., & Vaudenay, S. (2008). Mutual authentication in RFID. *3rd ACM ASIA Conference on Computer and Communications Security*, (pp. 292-299).

Rhee, K., Kwak, J., Kim, S., & Won, D. (2005). Lecture Notes in Computer Science: *Vol. 3450. Challenge-response based RFID authentication protocol for distributed database environment. Security in Pervasive Computing* (pp. 70–84). Springer. doi:10.1007/978-3-540-32004-3_9

Rieback, M. R., Crispo, B., & Tanenbaum, A. S. (2005). Lecture Notes in Computer Science: *Vol. 3574. RFID guardian: A battery-powered mobile device for RFID privacy management. Information Security and Privacy* (pp. 184–194). Springer.

Saito, J., Ryou, J. C., & Sakurai, K. (2004). Lecture Notes in Computer Science: *Vol. 3207. Enhancing privacy of universal re-encryption scheme for RFID tags. Embedded and Ubiquitous Computing* (pp. 879–890). Springer. doi:10.1007/978-3-540-30121-9_84

Stoer, M., & Wagner, F. (1997). A simple min-cut algorithm. *Journal of the ACM*, *44*(4), 585–591. doi:10.1145/263867.263872

Swaminathan, J. M., Smith, S. F., & Sadeh, N. M. (1998). Modeling supply chain dynamics: A multiagent approach. *Decision Sciences*, *29*(3), 607–632. doi:10.1111/j.1540-5915.1998.tb01356.x

Sweeney, P. J. (2005). *RFID for dummies*. Wiley Publishing, Inc.

Weis, S. A., Sarma, S. E., Rivest, R. L., & Engels, D. W. (2004). Lecture Notes in Computer Science: *Vol. 2802. Security and privacy aspects of low-cost radio frequency identification systems. Security in Pervasive Computing* (pp. 201–212). Springer.

Wilding, R., & Delgado, T. (2004). RFID demystified: Supply chain applications. *Logistics and Transport Focus*, *6*(4), 42–47.

KEY TERMS AND DEFINITIONS

Flow Network: In graph theory, a flow network models the movements of commodities using a weighted graph, where edges have associated maximum capacities, and the total incoming/outgoing flow at each node is conserved.

Market Change Scenario: A change in the underlying dynamics of a market involving supply, demand, consumption, businesses, competition, and consumers.

Maximal Outgoing Flow: A node's property in a flow network where the edge having the maximum flow value among the number of outgoing edges is selected in order to identify the critical path.

NP-Complete: A class of problems in computational complexity theory that are likely to be intractable (i.e. not likely to be solved deterministically within polynomial time).

Pseudonym: An RFID tag ID that changes each time the tag is read.

PUF: Physically Unclonable Function is a function that is easy to evaluate but hard to characterize or duplicate. For example, PUFs may be implemented aboard RFID tags based on the manufacturing variability inherent to the underlying electronic circuits.

Re-Encryption: The process of iteratively composing multiple encryptions of data (onboard a tag), which can vary based on time and location.

RFID: Radio Frequency Identification is a radio-based technology that can be used to track and identify tagged objects.

Supply Chain: A supply chain is a system of processes, organizations, technology, activities, information, resources, and people that move products from producers to consumers.

Taint Check Cover: A minimum set of nodes that cover (i.e. are adjacent to) all the edges of the supply chain flow graph.

Taint Check Point: A node in the taint check cover set.

Chapter 12
Building Scalable, Private RFID Systems

Li Lu
University of Electronic Science and Technology of China, China

ABSTRACT

Due to low cost and easy deployment, RFID has become a promising technology in many applications, such as retailing, medical-patient management, logistics, and supply chain management. Although a number of RFID standards have been issued and widely adopted by many off-the-shelf products, those standards, however, scarcely added privacy concerns because of computing and communication patterns. On the other hand, in RFID systems, RF tags emit their unique serial numbers to RF readers. Without privacy protection, however, any reader can identify a tag ID via the emitted serial number. Indeed, a malicious reader can easily perform bogus authentications with detected tags to retrieve sensitive information within its scanning range. The main obstacle to preserving privacy in RFID systems lies in the capability of tags. Due to the cost consideration, common RFID tags have tight constraints on power, computational capacity, and memory. Therefore, the mature cryptographic tools for bulky PCs are not suitable for RFID devices. In this chapter, the author focuses on the privacy issue to establish scalable and private RFID systems. The chapter first discusses the privacy issue in RFID systems; and then correspondingly introduces privacy preserving techniques including privacy-preserving authentication and secure ownership transfer. Finally, the theoretic formal privacy models for RFID systems are given, in which the author formally defines privacy and the behaviors of adversaries in RFID systems. Based on a formal model, say the weak privacy model, the chapter illustrates the methodology for designing highly efficient privacy-preserving authentication protocols.

12.1 OVERVIEW

Privacy is a basic requirement of human beings. "Privacy is the quality or state of being apart from company or observation; and the freedom from unauthorized intrusion" – (Merriam-Webster (2009)). In the information technology, privacy requires protecting the users' private information from being exposed to unauthorized individuals or organizations.

In RFID systems, RF tags emit their unique serial numbers to RF readers. Without privacy

DOI: 10.4018/978-1-4666-3685-9.ch012

protection, however, any reader can identify a tag ID via the emitted serial number. Indeed, within the scanning range, a malicious reader can easily perform bogus authentications with detected tags to retrieve sensitive information. Today, many companies embed tags in items. Since the tags indicate information of the items, a customer carrying those tags is subject to silent tracking from unauthorized readers. Sensitive personal information might be exposed: the illnesses she suffers from, indicated by the pharmaceutical products; the malls where she shops; the types of items she prefers to buy, and so on. The majority of today's RFID applications lose sight of privacy, which exacerbates serious threats from emerging attacks.

12.2 PRIVACY-PRESERVATION IN RFID SYSTEMS

The basic type of RF tag identification is a "challenge-response" procedure. That is, a reader sends a request as a challenge to a tag, and then receives the corresponding response from the tag, as plotted in Figure 1.

The basic type of RF tag identification leaks the private information of RF tags. To preserve privacy, many Privacy-Preserving Authentication, PPA, Protocols have been proposed to achieve private authentication in RFID systems. Weis (Weis, Sarma, Rivest & Engels (2003)) proposed a hash function based authentication scheme, HashLock, to avoid tags being tracked, as illustrated in Figure 2. In this approach, each tag shares a secret key k with the reader. The reader sends a random number r as the authentication request. To respond to the reader, the tag uses a hash function h to generate a response $h(k, r)$ on the inputs of r and k. The reader then computes $h(k, r)$ of all stored keys until it finds a key to recover r, thereby identifying the tag. The search complexity of HashLock is linear in N, where N is the number of tags in the system. Subsequent approaches in the literature mostly aimed at improving the efficiency of key search. Juels (Juels (2006)) classifies those approaches into three categories.

Synchronization approaches: Such approaches (Ohkubo, Suzuki & Kinoshita (2004), Juels (2004), Dimitriou (2005), Tsudik (2006)) use an incremental counter to record the state of authentication. When an authentication is successfully performed, the tag increases the counter by one. The reader compares the value of a tag's counter with the record in the database. If the difference of the two counter values is in a proper window, the tag is viewed as valid and the reader synchronizes the counter record of the tag. Synchronization schemes are subject to the De-synchronization Attack, in which a malicious reader interrogates a tag so many times such the counter of the tag exceeds the range of the window and the reader fails to recognize a valid tag.

Time-space tradeoff approaches: OSK (Ohkubo, Suzuki & Kinoshita (2004)) and AO (Avoine & Oechslin (2005)) employ Hellman tables to

Figure 1. Challenge-response based RFID authentication

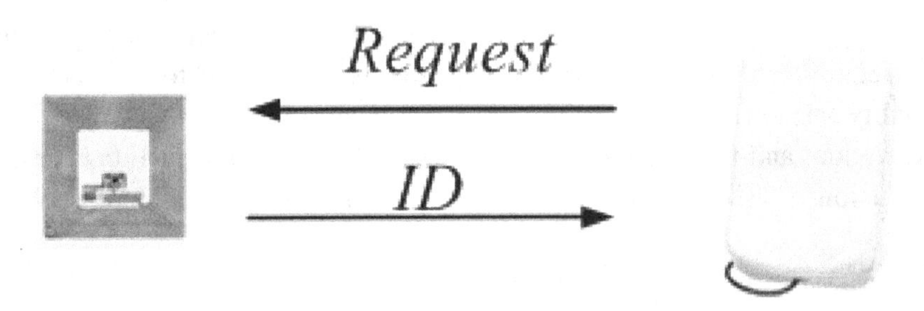

Figure 2. Hashlock authentication

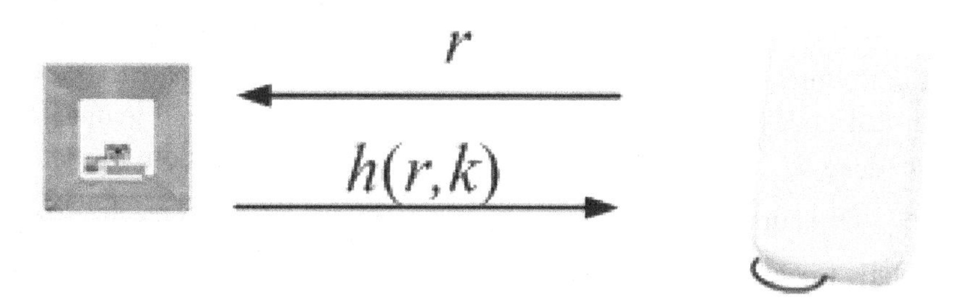

improve key-efficiency. Hellman (Hellman (1980)) studied the problem of breaking symmetric keys and showed that an adversary can pre-compute a Hellman table of size $O(N^{2/3})$ (N is the number of tags in a given RFID system), in which the adversary can search a key with the complexity $O(N^{2/3})$. That means the key-searching efficiency of OSK or AO is also $O(N^{2/3})$. Those approaches are not sufficiently efficient to support a large scale RFID system.

Balanced tree-based approaches: Balanced-tree based approaches (Dimitriou (2006), Molnar & Wagner (2004), Molnar, Soppera & Wagner (2005), Lu, Han, Hu, Liu & Ni (2007)) improve key search efficiency from linear complexity to logarithmic complexity. They employ a balanced-tree to organize and store keys for tags. In a balance tree, each node stores a unique key. Keys in the path from the root to a leaf node are distributed to a tag. Each tag uses these multiple keys to encrypt the identification message. Upon receiving an encrypted message, the reader performs a depth-first search in the key tree with a logarithmic complexity the size of the system.

12.2.1 Tree-Based Privacy-Preserving Authentication Protocols

Current RFID based applications may contain over hundreds of millions tags. They also need to support thousands of authentication requests from tags simultaneously. Both of these requirements demand a prompt key search scheme for PPA based approaches. The most recent attempt is the balanced-tree based schemes (Dimitriou (2006), Molnar & Wagner (2004), Molnar, Soppera & Wagner (2005)), which can improve the search efficiency to $O(logN)$.

Existing balanced-tree based approaches (Dimitriou (2006), Molnar & Wagner (2004), Molnar, Soppera & Wagner (2005)) construct a balanced tree to organize and store the keys for all tags. Each node stores a key and each tag is assigned to a unique leaf node. Thus, there exists a unique path from the root to this leaf node. Correspondingly, those keys on its path are assigned to the tag. For example, tag $\mathcal{T}_1$ obtains keys k_0, $k_{1,1}$, $k_{2,1}$, and $k_{3,1}$, as illustrate in Figure 3. When a reader authenticates a tag, e.g., $\mathcal{T}_1$, it conducts the identification protocol shown in Figure 4. The value $h(k,r)$ denotes the output of a hash function h on two input parameters: a key k and a random number r. The identification procedure is similar to traversing a tree from the root to a leaf. The reader $\mathcal{R}$ first sends a r to tag $\mathcal{T}_1$. $\mathcal{T}_1$ encrypts r with all its keys and includes the ciphertexts in a response. Upon receiving the response from $\mathcal{T}_1$, the reader searches proper keys in the key tree to recover r. This is equal to marking a path from the root to the leaf node of $\mathcal{T}_1$ in the tree.

At the end of identification, if such a path exists, $\mathcal{R}$ regards $\mathcal{T}_1$ as a valid tag. Usually, the encryption is employed by using cryptographic hash functions.

From the above procedure, we see that tags will share some non-leaf nodes in the tree. For example, $\mathcal{T}_1$ and $\mathcal{T}_2$ share $k_{2,1}$, while $\mathcal{T}_1$, $\mathcal{T}_2$, $\mathcal{T}_3$, and $\mathcal{T}_4$ share $k_{1,1}$. Of course all tags share the root k_0. Such a static tree architecture is efficient because the complexity of key search is logarithmic. For the example in Figure 3, any identification of a tag only needs $log_2(8) = 3$ search steps. On the other hand, if the adversary compromises some tags, however, it obtains several paths from the root to these leaf nodes of the corrupt tags, as well as the keys on those paths. Since keys are not changed in the static tree architecture, the captured keys will still be used by those tags which are uncompromised (for simplicity, we denote those tags as *normal tags*). Consequently, the adversary is able to capture the secret of normal tags. Lu (Lu, Han, Hu, Liu & Ni (2007)) evaluates the damage caused by compromising attacks on balanced-tree based approaches: In an RFID system containing 2^{20} tags, and employing a binary tree as the key tree, an adversary, by compromising only 20 tags, has a probability of nearly 100% of being able to track normal tags.

A practical solution is to update keys for a tag after each authentication so that the adversary cannot make use of keys obtained from compromised tags to attack normal ones. However, the static tree architecture makes it highly inflexible and different to provide consistent key-updating. Suppose we update the keys of $\mathcal{T}_1$ in Figure 3, we have to change k_0, $k_{1,1}$, $k_{2,1}$, and $k_{3,1}$ partially or totally. Note that $k_{1,1}$ is also used by $\mathcal{T}_2$, $\mathcal{T}_3$, and $\mathcal{T}_4$, and $k_{2,1}$ is used by $\mathcal{T}_2$. To keep the updating consistent, the keys of all influenced tags must be updated and re-distributed. A challenging issue is that if the position of a key is close to the root, the key-updating would influence more tags. For example, updating $k_{1,1}$ would influence half of all tags in the system ($\mathcal{T}_1$, $\mathcal{T}_2$, $\mathcal{T}_3$, and $\mathcal{T}_4$). One intuitive idea is to periodically

Figure 3. A binary key tree with eight tags

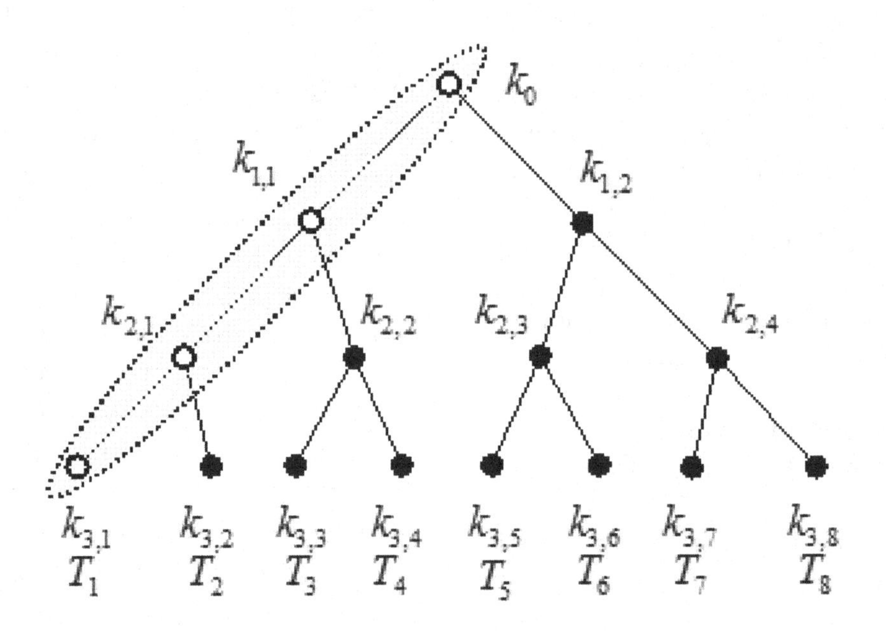

Figure 4. A basic RFID authentication procedure in balanced tree-based approaches

R T_1

$Request, r$

$h(k_0, r), h(k_{1,1}, r), h(k_{2,1}, r), h(k_{3,1}, r)$

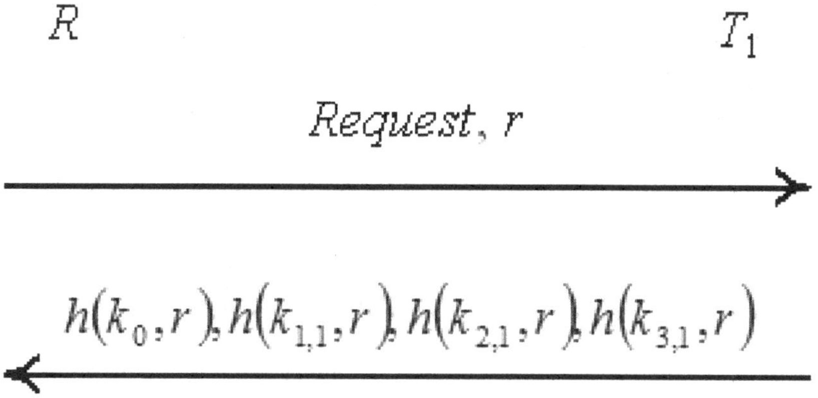

recall all tags and update the keys simultaneously. Unfortunately, such a solution is not practical in large scale systems with millions or even hundreds of millions of tags. Another solution is to collect only those tags subset and update their keys. This is also difficult because we have to collect a large span of tags even though there is only one tag updating its keys.

To address the key-updating issue, Lu et al. propose a PPA scheme SPA (Lu, Han, Hu, Liu & Ni (2007)), for enhancing the balanced-tree base approaches. SPA employs an automatic key-updating after each tag's authentication. For each virtual node in the key tree, SPA introduces a cache and a set of state bits for indicating the key-updating status. Meanwhile, a recursive algorithm is designed for achieving a dynamic key-updating, in which the updated keys of the tag will not incur conflict with other normal tags. As a result, SPA reduces the number of keys shared among compromised and normal tags, and hence lessens the possibility of successful compromising attacks.

SPA is comprised of four components: *system initialization*, *tag identification*, *key-updating*, and *system maintenance*. The first and second components are similar to previous tree-based approaches and perform the basic identification functions. The key-updating is employed after a tag successfully performs its identification with the reader. In this procedure, the tag and the reader update their shared keys. This key-updating procedure will not break the validation of keys used by other tags. SPA achieves this using temporary keys and state bits. A temporary key is used to store the old key for each non-leaf node in the key tree. For each non-leaf node, a number of state bits are used in order to record the key-updating status of nodes in the sub-trees. Based on this design, each non-leaf node will automatically perform key-updating when all its children nodes have updated their keys. Thus, SPA guarantees the validity and consistency of private authentication for all tags.

SPA, however, does not completely eliminate the impact of compromising attacks. Another drawback for balanced-tree based PPAs is the large space needed to store keys in each tag. Balanced-tree based approaches require each tag to hold $O(log_\delta N)$ keys, and the reader to store $\delta \cdot N$ keys, where δ is a branch factor of the key tree. Obviously, due to the limited memory capacity of current RF tags, existing PPAs are difficult to apply in current RFID systems.

To completely eliminate the common key effect and improve the storage efficiency, ACTION (Lu, Han, Xiao & Liu (2009)) was proposed using *sparse trees* as the key organization of structure. Utilizing ACTION, each tag only needs two keys for authentication. After each authentication, a new path is randomly generated for the tag such that the position of the tag's keys is dynamically changed, which removes the common key effect. Such a measure can effectively resist the compromising attack, because the adversary cannot leverage the keys from compromised tags to attack other legitimate tags.

12.2.2 Privacy-Preservation among Heterogeneous RFID Systems

Many RFID applications demand that RF tags can be accepted among heterogeneous systems. For example, e-logistics require that cargos seamlessly and uninterruptedly circulate along the LSCM (the logistic and supply chain management) flow. In those applications, the tags attached to cargos should also support cross-domain circulation, i.e. logistic companies are able to authenticate the tags issued by other companies. In reality, it is impractical to restrict companies to only use their own tags. On the other hand, different companies may deploy different kinds of RFID systems. That means RF tags should be authenticated among different systems. In other words, tags' ownership should be transferred among heterogeneous systems. For the safe of privacy, tag Ownership Transfer protocols, OT, should satisfy three requirements: 1) protect the privacy of each participant in heterogeneous systems; 2) defend against tag impersonation; and 3) achieve high efficiency in communication and computation.

Ownership Transfer protocols (Song (2008), Dimitriou (2008), Fouladgar & Afifi (2007), Saito, Imamoto & Sakurai (2005), Lim & Kwon (2006)) focus on how to transfer a tag between two owners. Before transformation, the tag's original owner updates the tag's secret key and sends the updated key to the tag's new owner, and then the original owner transfers the tag. Upon receiving the tag, the new owner updates the tag's key again. By updating the secret key twice, the original owner cannot track the tag anymore and the new owner cannot retrieve the tag's past information. Obviously these schemes are inefficient when the number of tags is large.

Several designs have been proposed to achieve privacy-preservation in tag ownership transfer. (Molnar, Soppera & Wagner (2005)) propose a pseudonym protocol enabling ownership transfer of RFID tags. In their scheme, all owners trust a trusted centre (TC), and the TC possesses all the tags' secret keys. An owner uses a reader to query its tag and the tag responds with a pseudonym. Then the owner forwards the pseudonym to the TC and authenticates itself to the TC. If the authentication is successful, the TC will reply the tag's information to the owner. To enable time-limited delegation and ownership transfer of tags, all the tag secrets are organized as a tree, and a tag stores the secrets corresponding to the path from the root to the tag. If a reader obtains the secrets for a node in the tree from TC, it can compute the keys of all descendant nodes, and uses these keys to decode tag pseudonyms. By doing this, the reader can read the tag a limited number of times without on-line connectivity to the TC. However, the ownership transfer procedure of this scheme is rather restrictive, in the sense that TC manages all the information of tags to control them all the time and the owner must relay the TC to get a tag's information or limited ability to control it. When a tag is transferred from one owner to another, the TC will forbid the former owner to access the tag and allow the latter one to access it. Thus, strictly speaking, their scheme corresponds to time-limited access delegation rather than ownership transfer.

Lim and Kwon (Lim & Kwon (2006)) proposed an RFID authentication protocol to support tag ownership transfer. They argue that backward and forward untraceability is a fundamentally important security property which should be satisfied during transfer of tag ownership. In their scheme, the tag secret is evolved using a one-way key chain in one of two different ways in every authentication session to achieve both forward and backward untraceability; if the authentication succeeds, then the tag and the server refresh the tag secret probabilistically and if the protocol fails, the tag updates its secret deterministically. To transfer a tag's ownership, the new owner of the tag securely communicates with the old owner to receive all the relevant information for the tag. Then the new owner communicates with the tag outside the reading range of the previous owner to refresh the tag's secret so that no other party can communicate with the tag. Their work only considers backward and forward untraceability under the attack of outside adversaries. Sometimes, however, owners can be malicious. Their scheme cannot protect the old owner's privacy when the new owner is malicious.

Fouladgar and Afifi (Fouladgar & Afifi (2007)) employ specific keys for updating the tag. The old and new owners utilize those keys to refresh the tag before and after the transfer to guarantee the backward and forward untraceability.

The existing Ownership Transfer protocols have some drawbacks: 1) a secure channel is needed to transfer the keys used in tag authentication. A secure channel, however, does not always exist in real applications; 2) an online trusted third party is involved in the ownership transfer, which is costly and inefficient if a large number of tags need to be transferred. Thus, future work will focus on eliminating the secure channel and developing off-line ownership transfer schemes.

12.3 FORMAL PRIVACY MODELS FOR RFID SYSTEMS

To address the issue of privacy, PPA becomes an immediate need for protecting the interactive procedure between the RFID reader and tags (Robinson & Beigl (2004)). Recently, many PPA schemes have been proposed. Nevertheless, the research on RFID is still short of appropriate formal models that can explicitly define privacy as well as maintain the authentication efficiency in a general way. Lacking such models, existing PPA schemes have to employ ad hoc notions of security and privacy (Juels (2006)), and then heuristically analyze security and privacy via those notions. The heuristic analysis, however, only allows those PPA schemes to examine privacy under known attacks using the ad hoc defined notations. It is difficult to explore the potential vulnerabilities and flaws that are vulnerable to newly emerging attacks. To the best knowledge, most existing works, if not all, are subject to one or several attacks.

Juels and Weis (Juels & Weis (2007)) propose a privacy model, named "Strong Privacy", to meet the demands of privacy in RFID systems. Strong Privacy employs *indistinguishability* to represent privacy. Indistinguishability means that RFID tags should not be distinguished from each other according to their output. To achieve indistinguishability, tags randomize their output such that adversaries cannot compromise their privacy. However, to authenticate a tag, the legitimate reader cannot be directly aware of which tag it is interrogating due to the random output of the tags. It has to search all tags in the system to identify the tag instead. Under the Strong Privacy model, such a brute-force search makes authentication efficiency linear in the number of tags in the system. In other words, the PPA scheme might not be practical and applicable in large-scale RFID systems due to the linear-search authentication, although the Strong Privacy model can guarantee privacy.

To be applicative to the large-scale deployments, most PPA schemes focus on high authentication efficiency. Their privacy, however, surely degrades because of the trade-off between privacy and authentication efficiency. They are often vague about the extent of the trade-off: how much privacy degradation is required to improve authentication efficiency to what extent?

Lu *et al.* (Lu, Liu & Li (2010)) propose a weak formal privacy model, termed as Refresh. Different from Strong Privacy, Refresh loosens the strict constraints on the output of tags, such as randomization and unpredictability. Refresh allows a tag to contain a temporally constant field in the output for stating the tag's identity. The field remains unchanged during the time interval between two consecutive interrogations from the legitimate reader. Thus, the field is predictable for the valid reader and used to accelerate authenticating the tag. After a successful authentication with the valid reader, this field will be refreshed (or updated) for the next authentication procedure. The tag encrypts the field using the secret keys shared with the valid reader for preventing adversaries from identifying the tag. By this means, RFID systems can achieve acceptable privacy protection as well as highly efficient authentication.

To learn privacy model of RFID PPAs more, we describe the Strong Privacy model and the weak privacy model Refresh below.

12.3.1 Strong Privacy Model

Generally, an RFID system consists of two types of devices, tags and a reader. A tag $\mathcal{T}$ is a transponder with well-designed circuits that can respond the signals emitted from the RFID reader with its own unique ID. Since there is no battery available on a tag, the communication range is very limited, usually within a meter (Paise & Vaudenay (2008)). Although an RFID tag has constrained capability of computation and storage, it is believed that the next generation of tag is able to perform some cryptographic operations, such as symmetric-key encryption, hash computation, and pseudo-random number generation (Vaudenay (2007), Weis, Sarma, Rivest & Engels (2003)). Strong Privacy also assumes that a tag $\mathcal{T}$ shares a number of secret keys with the reader. In addition, $\mathcal{T}$ cannot defend against tag-compromising (Juels & Weis (2007)).

The RFID reader $\mathcal{R}$ in an RFID system has one or more transceivers. The reader interrogates a tag periodically or at a user's demand, and then reports responses of the tag to the back-end database of the system. The goal of the RFID system is to distinguish legitimate tags (tags which are registered in the back-end database) from unknown tags, and further authenticate them (infer their IDs). RFID systems usually use the back-end database to perform computations and store the information of tags in the system. Generally, the channel between the reader and back-end database is assumed to be secure. Hereafter in this paper, we take the reader device and the back-end database as a whole. Thus, we denote the reader device and back-end database by the "reader" for simplicity.

In an RFID scheme, $\mathcal{T}$ and $\mathcal{R}$ are initiated with a shared secret by the system. There is an interactive protocol describing the manner of message exchanges between $\mathcal{T}$ and $\mathcal{R}$.

Adversaries in RFID systems can be specified by three characteristics: the actions that they can perform (i.e., the *oracles* they can query), the goal of their actions (i.e., the *game* they can play), and the manner in which they can access the system (i.e., the *rules* of the game).

The Privacy Game of Strong Privacy, which describes the methodology of an adversary used to attack an RFID scheme, is derived from the classic INDistinguishability under Chosen-Plaintext Attack (IND-CPA) and under Chosen-Ciphertext Attack (IND-CCA) cryptosystem security games. The idea is that an RFID protocol can be considered private for some parameters if no polynomial-time adversary can win the game with a overwhelming probability. The goal of the adversary $\mathcal{A}$ in the game is to distinguish between two different tags within his computational bounds (i.e., how many times he can query oracles). Due to the page limitation, we do not go through the details of the game, which can be found in Juels's work (Juels & Weis (2007)).

The basic idea of Strong Privacy is that tags randomize their output such that adversaries cannot compromise their privacy. However, to authenticate a tag, the legitimate reader has to search all tags in the system. Such a brute-force search makes the authentication efficiency linear in the number of tags in the system.

In essence, the Strong Privacy model can achieve a strong sense of privacy with an implicitly assumption that both adversaries and the legitimate reader perform algorithms in polynomial time. It is well known that the polynomial time algorithms are considered to be efficient in theory, but not really applicable to practical applications in many protocols due to large hidden constants, or large polynomial degree. For example, to preserve privacy, the output of a tag are encrypted or hashed with a key shared between the reader and the tag. Without the key, an adversary cannot determine which tag is interrogated by the reader. On the other hand, the reader has to search all tags' keys to determine which tag can generate the same output such that it can authenticate this tag. Obviously, authenticating the tag is a brute-force search with the efficiency of authentication being $\mathcal{O}(n)$. For a large-scale RFID system, the authentication efficiency $\mathcal{O}(n)$ may not be acceptable.

12.3.2 Weak Privacy Model

The basic idea of Refresh is that a tag generates independent output each time when being interrogated by the legitimate reader, making it very hard, if not impossible for adversaries to correlate the current output with previous ones. Refresh consists of three components: RFID Scheme, Adversaries, and Privacy Game.

RFID Scheme: In Refresh, an RFID scheme is defined as follows:

An RFID scheme consists of a polynomial-time algorithm $KeyGen(1^s)$ which generates all key materials $k_1, \ldots, k_n$ for the system depending on a security parameter s.

- A setup scheme $SetupTag(ID)$ which allocates a specific secret key k and a distinct ID to a tag. Each legitimate tag should have a pair (ID, k) stored in the back-end database.
- A setup scheme $SetupReader$ which stores all pairs in the reader's back-end database for all legitimate tags'ID s in the system.
- A polynomial-time intracive protocol P between the reader and a tag in which the reader owns the common inputs, the database and the secrets. If the reader fails, it outputs $\perp$; otherwise, it outputs some ID and may update the database.

An RFID scheme has a correct output if the reader executes the protocol P honestly and then infers the ID of a legitimate tag except with a negligible probability. (A function in terms of a security parameter s is called negligible if there exists a constant $x > 0$ such that it is $\mathcal{O}(x^{-s})$). Otherwise, the reader outputs $\perp$ when the tag is not legitimate.

Adversaries: Similar to Strong Privacy, adversaries in Refresh have three characteristics: the oracles they can query, the goal of their actions, and the rules of their actions. According to those characteristics, we define adversaries in RFID systems below.

An adversary $\mathcal{A}$ in an RFID system is a polynomial-time algorithm, which performs attacking behaviors by querying five oracles.

- $Launch \rightarrow \pi$: Execute a new protocol instance π between the reader and a tag.
- $TagQuery(m, \pi, \mathcal{T}) \rightarrow m'$: Send a message m to a given protocol session π on the tag $\mathcal{T}$. The oracle returns a message m' as the output of $\mathcal{T}$.
- $ReaderSend(m, \pi, \mathcal{R}) \rightarrow m'$: Send a message m to a given protocol session π on the reader $\mathcal{R}$. The oracle returns a message m' as the output of $\mathcal{R}$.
- $Corrupt(\mathcal{T})$: Compromise the tag $\mathcal{T}$, and obtain the secret stored in $\mathcal{T}$. The tag $\mathcal{T}$ is no longer used after this oracle call. In this case, we say that the tag $\mathcal{T}$ is destroyed.

- $Result(\pi)$: When π is complete, the oracle returns 1 if the scheme has the correct output; otherwise, it returns 0.

The adversary starts a game by setting up the RFID system and feeding the adversary with the common parameters. The adversary uses the oracles above following a privacy game, which will be described in the next subsection, and produces the output. Depending on the output, the adversary wins or loses the game.

Privacy Game: The game has three phases: Learning, Challenging and Re-learning (see Box 1).

As shown in Box 1, in the Learning phase, $\mathcal{A}$ is able to issue any message and perform any polynomial-time computation (i.e., query oracles in polynomial times). After the Learning phase, $\mathcal{A}$ selects two uncorrupted tags as challenge candidates in the Challenge phase. One of these challenge candidates is then randomly chosen by the system (the *Challenger* $\mathcal{C}$) and presented to the adversary (the oracles of this tag can be queried by the adversary except the *Corrupt* oracle). After that, similar to the Learning phase, $\mathcal{A}$ is offered the oracles of all the tags in the RFID system by $\mathcal{C}$ except the two challenge candidates. This phase is named Re-learning. At the end of Re-learning, $\mathcal{A}$ outputs a guess about which candidate tag is selected by $\mathcal{C}$. If the guess is correct, $\mathcal{A}$ wins the game; otherwise, $\mathcal{A}$ loses.

In addition, there are two requirements for our privacy game to work properly. First, at step (7) in the privacy game, the challenger $\mathcal{C}$ refreshes the private information of the two challenge candidates $\mathcal{T}_0^*$ and $\mathcal{T}_1^*$. Thus, the adversary cannot correlate the output of $\mathcal{T}_b^*$ at the Re-learning phase with the output of $\mathcal{T}_0^*$ and $\mathcal{T}_1^*$ at the Learning phase. Second, if an adversary can corrupt $n-1$ tags and get the keys of these tags, then he can retrieve the output of these corrupted tags. Therefore, any tag in the system can be definitely distinguished from others with the output of the tag. That is why at least two tags need to be uncorrupted.

We denote such a privacy game for an RFID system as $\mathbf{Game}_{\mathcal{A}}^{ref-priv}(s,n,r,t,c)$. Here s is a security parameter, for example, the length of keys, and n, r, t and c are respective parameters for number of tags, number of $ReaderSend$ queries, number of $TagQuery$ queries, and computation steps. An adversary $\mathcal{A}$ with parameters r, t, and c is denoted by $\mathcal{A}[r,t,c]$.

Based on the privacy game above, we define the Refresh Privacy of an RFID scheme as follows:

RFID $(r,t,c) - Refresh - Privacy$: A protocol P of an RFID system achieves $(r,t,c) - Refresh - Privacy$ with parameter s, if for any polynomial-time $\mathcal{A}$, the probability of $\mathcal{A}$ wining under $\mathbf{Game}_{\mathcal{A}}^{ref-priv}(s,n,r,t,c)$ satisfies:

$$\forall \mathcal{A}(r,t,c), \mathbf{Pr}[\mathcal{A} \quad wins] \leq \frac{1}{2} + 1 \,/\, ploy(s)$$

where $poly(s)$ denotes any polynomial function of parameter s.

For a given protocol P in an RFID system, we define the *advantage* of an adversary by:

$$\mathbf{Adv}_P(\mathcal{A}) = \mathbf{Pr}[\mathcal{A} \quad wins] - \frac{1}{2}$$

In $\mathbf{Game}_{\mathcal{A}}^{ref-priv}(s,n,r,t,c)$, the adversary $\mathcal{A}$ can win the game in a trivial way. That is $\mathcal{A}$ picks up a bit b' from $\{0,1\}$ uniformly at random, i.e., $\mathbf{Pr}[b' = b] = \frac{1}{2}$. In this case, $\mathcal{A}$ attacks the system without any knowledge about the tags in the system, the successful attacking probability is the

Box 1. Privacy game of refresh

Game $\mathbf{Game}_{\mathcal{A}}^{ref-priv}(s,n,r,t,c)$:
Setup:
1. KeyGen $KeyGen(1^s) \rightarrow (k_0,\ldots,k_n)$.
2. Initiate $\mathcal{R}$ by $SetupReader$ with $(k_0,\ldots,k_n)$.
3. Initiate each tag $\mathcal{T}_i$ by $SetupTag(\mathcal{T}_i,k_i)$ with the key k_i.
Learning:
4. $\mathcal{A}$ may perform the following actions in any interleaved order.
a. Query $ReaderSend$ oracle without exceeding r times.
b. Query $TagQuery$ oracle without exceeding t times.
c. Query $Corrupt$ oracles of $n-2$ tags arbitrarily selected from n tags.
d. Do computations without exceeding c overall steps. Note that $\mathcal{A}$ may query $Launch$ oracle to initiate an instance of the protocol in communications and computations.
Challenge:
5. $\mathcal{A}$ selects two tags $\mathcal{T}_i$ and $\mathcal{T}_j$ to which he did not query $Corrupt$ oracles, then $\mathcal{A}$ presents these two tags to challenger $\mathcal{C}$.
6. Let $\mathcal{T}_0^* = \mathcal{T}_i$ and $\mathcal{T}_1^* = \mathcal{T}_j$.
7. *Challenger* $\mathcal{C}$ *queries* $TagQuery$ *oracles of* $\mathcal{T}_0^*$ *and* $\mathcal{T}_1^*$ *for one time, respectively.*
8. Challenger $\mathcal{C}$ picks up a bit b from $\{0,1\}$ uniformly at random. Then provide $\mathcal{A}$ access to $\mathcal{T}_b^*$.
Re-learning:
9. $\mathcal{A}$ may perform the following actions in any interleaved order.
a. Query $ReaderSend$ oracle without exceeding r overall queries.
b. Query $TagQuery$ oracle without exceeding t overall queries.
c. Query $Corrupt$ oracles of any tag except $\mathcal{T}_b^*$.
d. Do computations without exceeding c overall steps. Note that $\mathcal{A}$ may query $Launch$ oracle to initiate an instance of the protocol in communications and computations.
10. $\mathcal{A}$ outputs a guess bit b'.
11. $\mathcal{A}$ wins the game if $b' = b$.

lower bound of all attacking activities. Therefore, we define the advantage of any polynomial-time adversary by $\mathbf{Pr}[b' = b] - \frac{1}{2}$.

Unlike the Strong Privacy, Refresh Privacy only requires that tags change their output after being interrogated by the valid reader. Thus, between authentication sessions with the legitimate reader, the output value of a tag is static. Tags are therefore subject to tracking during such intervals of time. After a valid interrogation, the secret keys are refreshed. The output of tags therefore is changed accordingly. Without the secret keys, the adversary cannot predict the output of tags, and thus fail to track tags.

The privacy degradation is acceptable in many applications, which require that the valid reader interrogates tags frequently, such as access control and retailing. With limited privacy degradation, the efficiency of authentication can be improved significantly, since the legitimate reader need not perform brute-force searches to authenticate each tag.

Implications of Refresh: We have a set of important implications of the Refresh model.

First, the $KeyGen$ function should not generate low-entropy or strongly correlated keys. Otherwise, a PPA scheme might suffer the *compromising* attack. As the adversary $\mathcal{A}$ can make $Corrupt$ calls to some tags, $\mathcal{A}$ may infer some information of keys in uncorrupted tags from those corrupted tags. It may gain a significant advantage to win the Privacy Game. The details of the compromising attack can be found in (Lu, Han, Hu, Liu & Ni (2007)).

Second, whether or not $\mathcal{R}$ accepts a certain tag should not be history-dependent; otherwise, a PPA scheme will be vulnerable to the *de-synchronization* attack. For example, we assume that a tag only be interrogated for a maximum m times. After the adversary $\mathcal{A}$ makes m legitimate queries to a certain tag in the Learning phase, $\mathcal{A}$ chooses this tag as one of the two candidates in the challenge phase. Then $\mathcal{A}$ observes whether or not $\mathcal{R}$ accepts the challenge tag $\mathcal{T}_b^*$ in the Challenge phase. Therefore, $\mathcal{A}$ can determine if the challenge tag was the selected one or not. Hence, $\mathcal{A}$ can definitely win $\textbf{Game}_{\mathcal{A}}^{ref-priv}(s, n, r, t, c)$.

Third, the reader $\mathcal{R}$ should not differ significantly in its acceptance rate across tags; otherwise, $\mathcal{A}$ can just pick up one tag which $\mathcal{R}$ accepts frequently and one that $\mathcal{R}$ accepts less frequently as its challenge candidates in the Challenge phase. Thus, $\mathcal{A}$ might have a significant advantage in $\textbf{Game}_{\mathcal{A}}^{ref-priv}(s, n, r, t, c)$. This attack requires that $\mathcal{R}$ must treat all tags in the system equally.

12.3.3 Privacy Analysis of Current Work under Refresh

To show the methodology of formal privacy analysis, we use Refresh to examine several representative RFID PPAs: (1) the OSK scheme (Ohkubo, Suzuki & Kinoshita (2004)) and its variant AO (Avoine & Oechslin (2005)); (2) the YA-TRAP scheme (Tsudik (2006)); (3) and tree-based approaches (Molnar & Wagner (2004), Molnar, Soppera & Wagner (2005), Lu, Han, Hu, Liu & Ni (2007), Dimitriou (2006)). We chose these schemes based on the following observations: first, they are well-known and efficient; second, they lack of formal privacy proofs.

OSK And AO Schemes

In OSK, a tag $\mathcal{T}_i$ is initiated by the reader $\mathcal{R}$ with a secret key k_i. Let f_1 and f_2 be two independent pseudo-random functions. The tag $\mathcal{T}_i$ outputs $M_{i,t} = f_2(f_1^{(t)}(k_i))$ when interrogated by the reader $\mathcal{R}$, where $f_1^{(t)}$ denotes the functional powers of f_1 for natural t. Then $\mathcal{T}_i$ updates its key k_i as $f_1(k_i)$.

Upon receiving $M_{i,t}$, $\mathcal{R}$ performs a brute-force search to authenticate $\mathcal{T}_i$. Note that there is a predefined upper bound on the number of interrogations for each tag, denoted as m. After being accessed m times, a tag will either yield no output or just output random nonce until the reader $\mathcal{R}$ re-initializes the tag, i.e., $1 \le t \le m$. To improve the efficiency of authentication, the reader $\mathcal{R}$ stores all pre-computed output in a giant table $T = \{M_{i,t} \mid 1 \le i \le n, 1 \le t \le m\}$. By looking up $M_{i,t}$ in the table, the reader $\mathcal{R}$ can immediately authenticate the tag $\mathcal{T}_i$.

Avoine et al. propose a variant scheme AO to improve the storage efficiency of OSK by using Hellman tables (Hellman (1980)). The storage efficiency can be improved from $\mathcal{O}(N)$ to $\mathcal{O}(N^{2/3})$, where $N = n \times m$.

The vulnerability of OSK or AO lies in the upper bound m. An adversary can launch the de-synchronization attack on a tag. Interrogating the tag m times, the adversary can exhaust the valid output of the tag. As a result, the valid reader will reject the tag even though the tag is valid. Algorithm 1 shows the attack procedure. Based on the de-synchronization attack, the adversary can always distinguish a tag from others and win the privacy game with probability one. Hence, OSK and AO schemes do not achieve Refresh Privacy.

YA-TRAP Scheme

In YA-TRAP, besides the secret key k_i issued by the reader $\mathcal{R}$, a tag $\mathcal{T}_i$ stores an internal timestamp t_i to record the time of last interrogation of $\mathcal{R}$.

During each interrogating, $\mathcal{R}$ sends the current timestamp t to a tag $\mathcal{T}_i$. Upon receiving t, $\mathcal{T}_i$ compares t with its internal timestamp t_i. If $t > t_i$, then $\mathcal{T}_i$ outputs $M = h(t \,||\, k_i)$ and sets $t_i = t$, where h is a cryptographic hash function and $||$ denotes concatenation; otherwise, $\mathcal{T}_i$ outputs a random response.

To authenticate the tag, $\mathcal{R}$ conducts a brute-force search of its database to find a secret key k_i that can generate a message $h(t \,||\, k_i)$ equal to M. If such a key exists, $\mathcal{R}$ accepts the tag; otherwise, $\mathcal{R}$ rejects the tag.

Similar to OSK/AO, YA-TRAP is also vulnerable to the de-synchronization attack. An adversary can query a tag with a certain future timestamp t_{max} that indicates an extremely distant future time. The tag thereby sets its internal timestamp as t_{max}, and outputs random responses in all subsequent interrogations. Consequently, this tag will always be rejected by the legitimate reader in the following sessions. We show the attack in Algorithm 2. In this scenario, the adversary can win the Privacy Game with probability one. Thus, YA-TRAP is not private under Refresh.

We can see that OSK/AO and YA-TRAP are not private under Refresh because they violate the second and third implications of Refresh. Hence, an adversary can mark a target tag by modifying its inner state (i.e., the upper bound m in OSK/AO, and the timestamp in YA-TRAP) to win the Privacy Game.

Balanced Tree-Based Approaches

In balanced tree-based approaches (Molnar & Wagner (2004),Molnar, Soppera & Wagner

Algorithm 1. De-synchronization attack to OSK/AO

1. In the Learning phase, the adversary $\mathcal{A}$ selects two distinct tags $\mathcal{T}_i$ and $\mathcal{T}_j$ uniformly at random.
2. $\mathcal{A}$ queries the oracle *TagQuery* of $\mathcal{T}_i$ for m times.
3. $\mathcal{A}$ submits $\mathcal{T}_i$ and $\mathcal{T}_j$ as its challenge candidates.
4. In the Re-learning phase, $\mathcal{A}$ first queries the oracle *TagQuery* of $\mathcal{T}_b^*$ and then relays the response to $\mathcal{T}_b^*$'s *ReaderSend* oracle, $\mathcal{A}$ finally queries *Result* oracle.
5. If $\mathcal{T}_b^*$ is valid, the *Result* oracle of $\mathcal{T}_b^*$ returns 1, then $\mathcal{A}$ guesses $b = 1$, i.e., $\mathcal{T}_b^* = \mathcal{T}_i$; otherwise, $\mathcal{A}$ guesses $b = 0$, i.e., $\mathcal{T}_b^* = \mathcal{T}_j$.

Algorithm 2. De-synchronization attack on YA-TRAP

1. In the Learning phase, the adversary $\mathcal{A}$ selects two distinct tags $\mathcal{T}_i$ and $\mathcal{T}_j$ uniformly at random.
2. $\mathcal{A}$ queries the oracle $TagQuery$ of $\mathcal{T}_i$ with t_{max}.
3. $\mathcal{A}$ submits $\mathcal{T}_i$ and $\mathcal{T}_j$ as its challenge candidates.
4. In the Re-learning phase, $\mathcal{A}$ first queries the oracle $TagQuery$ of $\mathcal{T}_b^*$, and then relays the response to $\mathcal{T}_b^*$' $ReaderSend$ oracle, finally makes a query to $Result$ oracle.
5. If $\mathcal{T}_b^*$ is valid, the $Result$ oracle of $\mathcal{T}_b^*$ returns 1, then $\mathcal{A}$ guesses $b = 1$, i.e., $\mathcal{T}_b^* = \mathcal{T}_i$; otherwise $\mathcal{A}$ guesses $b = 0$, i.e., $\mathcal{T}_b^* = \mathcal{T}_j$.

(2005), Dimitriou (2006), Lu, Han, Hu, Liu & Ni (2007)), PPA schemes use a balanced tree to organize and store the keys for all tags. In the key tree structure, we find that each tag shares some inner nodes, more or less, with other tags in the key tree. Therefore, the tree-based approaches are vulnerable to compromising. If the adversary deliberately corrupts some tags, it can obtain several paths related to the corrupt tags, as well as the keys along those paths. Since those keys are shared among tags, some secret keys of uncorrupted tags will be exposed to the adversary. Only via eavesdropping, the adversary may have significant advantages in distinguishing other uncorrupted tags. We define the compromising attack in Algorithm 3. We can see that tree-based approaches are not private under Refresh due to the correlation between the keys of tags, which violate the first implication of Refresh.

Through the formal analysis above, we find that some well-known PPA schemes are not private under our Refresh model, since they do not satisfy the implications of Refresh. Specifically, the output of tags in OSK/AO and YA-TRAP is history-dependent; they are thereby inevitably vulnerable to the de-synchronization attack. On the other hand, due to the correlation of tags' keys, the tree based approaches fail to defend against the compromising attack. Therefore, a challenging issue emerges in designing PPA schemes: achieving the privacy as well as high authentication efficiency.

A Demonstrated Design of Highly Efficient PPA

To show the methodology of efficient PPA design, we raise an example PPA, LAST, from the work of L. Lu et al. (Lu, Liu & Li (2010)).

Scheme: LAST consists of the following components.

- $KeyGen(1^s)$: Generates two sequences of keys, $k_1, \ldots, k_n$ and $Index_1, \ldots, Index_n$, depending on a security parameter s. Each key is generated independently to all others, here $Index_i$ is the index of tag $\mathcal{T}_i$ in the back-end database, and k_i is the key used in the authentication procedure.
- $SetupTag(ID)$: returns a specific secret pair $(Index, k)$ for a particular tag $\mathcal{T}_i$ with identifier *ID*. The tuple $(\mathcal{T}_i, Index_i, k_i)$ exists in the back-end database when the tag $\mathcal{T}_i$ is legitimate.

Algorithm 3. Compromising attack on tree-based approaches

1. In the Learning phase, the adversary $\mathcal{A}$ selects anumber of tags, and queries *Corrupt* oracles of these tags.
2. $\mathcal{A}$ chooses two distinct tags $\mathcal{T}_i$ and $\mathcal{T}_j$ arbitrarily from the set of uncompromised tags.
3. $\mathcal{A}$ submits $\mathcal{T}_i$ and $\mathcal{T}_j$ as its challenge candidates.
4. In the Re-learning phase, $\mathcal{A}$ queries all oracles of $\mathcal{T}_b^*$, except *Corrupt* oracle.
5. $\mathcal{A}$ analyzes the output of $\mathcal{T}_b^*$, then guesses b accordingly.

- $SetupReader$: stores all tuples $(\mathcal{T}_i, Index_i, k_i)$, $1 \leq i \leq n$, in the back-end database.
- P is a polynomial-time interactive protocol between $\mathcal{R}$ and $\mathcal{T}_i$ in which $\mathcal{R}$ uses common parameters and the securet tuples. If the reader failed, it outputs ; otherwise, $\mathcal{R}$ outputs the *ID* of tag $\mathcal{T}_i$ and updates the corresponding tuple in the database. and updates the corresponding tuple in the database.

The protocol P includes three rounds, as illustrated in Figure 5. In the first round, $\mathcal{R}$ generates a random number r_1 (a nonce), and then sends r_1 with the "Request" message to $\mathcal{T}_i$. In the second round, $\mathcal{T}_i$ generates a random number r_2 and computes $h(r_1, r_2, k_i)$, where $h(r_1, r_2, k)$ denotes the output of a cryptographic hash function h on three inputs: a key k of $\mathcal{T}_i$ and two random numbers r_1 and r_2. $\mathcal{T}_i$ computes $V = h(r_1, r_2, k_i)$, and then replies to $\mathcal{R}$ with a message $U = (r_2, Index_i, V)$. $\mathcal{R}$ authenticates $\mathcal{T}_i$ according to U.

Upon receiving U, $\mathcal{R}$ searches for an $Index_i$ in the back-end database, and then output $\perp$ if $Index_i$ does not exist; otherwise, $\mathcal{R}$ picks up the k_i belonging to the tuple $(\mathcal{T}_i, Index_i, k_i)$, and computes $V' = h(r_1, r_2, k_i)$. If $V' = V$, $\mathcal{R}$ accepts the tag and identifies it as $\mathcal{T}_i$; otherwise, $\mathcal{R}$ outputs $\perp$.

After authenticating $\mathcal{T}_i$, $\mathcal{R}$ will update the keys of $\mathcal{T}_i$. $\mathcal{R}$ computes $Index_i' = h(r_1, r_2, Index_i, k_i)$ and $k_i' = h(r_1, r_2, k_i)$, respectively, and then replaces the tuple $(\mathcal{T}_i, Index_i, k_i)$ in the database with $(\mathcal{T}_i, Index_i', k_i')$. After the key-updating procedure, $\mathcal{R}$ replies to $\mathcal{T}_i$ with the message $\sigma = h(r_1, r_2, k_i')$.

Upon receiving σ, $\mathcal{T}_i$ first computes $k_i' = h(r_1, r_2, k_i)$ and $\sigma' = h(r_1, r_2, k_i')$. Then $\mathcal{T}_i$ checks whether σ' equals σ. If yes, $\mathcal{T}_i$ updates the original k_i and $Index_i$ to k_i' and $Index_i' = h(r_1, r_2, Index_i, k_i)$, respectively; otherwise, $\mathcal{T}_i$ discards σ and keeps $Index_i$ and k_i unchanged.

Efficiency: We first investigate the storage efficiency of LAST, and then analyze the authentication efficiency by estimating the number of hash computations in each authentication.

An RFID tag normally has very tiny memory to store users' information as well as keys. The storage efficiency therefore is critical in designing PPA schemes. LAST is efficient in key storage. Specifically, LAST only allocates two keys for each tag, the index and authentication key. On the reader side, the storage is $2n$. The storage efficiency is similar to that of HashLock (one key stored on tag side and n keys on reader side),

Figure 5. Authentication procedure in LAST. Upon receiving U, $\mathcal{R}$'s operations are: 1. authenticating $\mathcal{T}_i$ and key-updating; 2. computing σ and sending it to $\mathcal{T}_i$. $\mathcal{T}_i$ also updates its keys after checking σ.

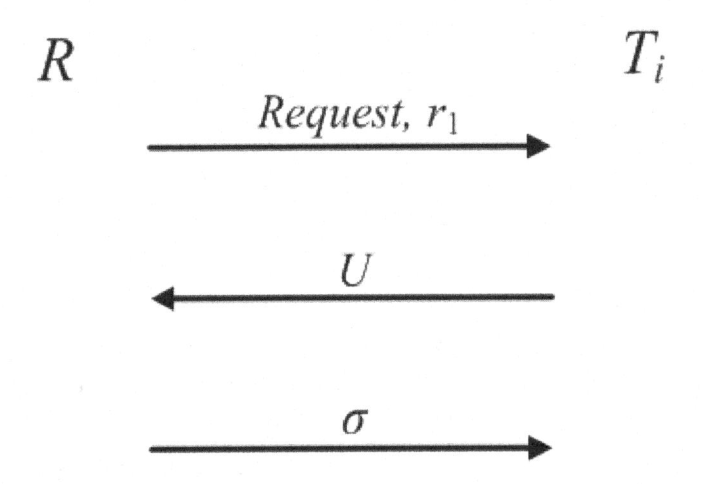

whereas higher than that of balanced-tree based approaches ($\mathcal{O}(\log n)$ on tag side and $\mathcal{O}(n \log n)$ on reader side (Lu, Han, Hu, Liu & Ni (2007))).

The basic operation in LAST is hash computation. Therefore, we can use the number of hash computations to evaluate the authentication efficiency of LAST. According to the authentication procedure, $\mathcal{R}$ just needs four hash computations: one for authentication, three for key-updating. Thus, the authentication of LAST has the complexity $\mathcal{O}(1)$ on the authentication efficiency, which is much more efficient than previous schemes.

Privacy: Based on the Refresh model, LAST can be formally proved as Refresh Privacy. The detailed proof can be found in the work of (Lu, Liu & Li (2010)).

FUTURE RESEARCH AND RECOMMENDATIONS

Existing privacy protection technologies mainly focus on protocol design of PPAs at the application layer in closed RFID systems. In the future, the research for RFID privacy and security will be extended to two major directions. First, technologies to protect privacy at physical or MAC layers will gain more attention. Challenges of this direction include highly efficient implementation of secure hardware, secure encoding algorithms of RF signals, secure protocol design of air-interface, and anti-collision mechanism design under security consideration. The second direction is to protect tags' privacy in interoperable and open systems. A number of technologies, for example secure ownership transfer, will be further developed to establish cross-domain authentication among heterogeneous RFID systems, and to enable seamless, reliable, ubiquitous, and trustworthy services for end users.

CONCLUSION

In this chapter, we introduce the privacy issue, Privacy-Preserving Authentication techniques, PPA, and some representative PPAs in RFID systems. In particular, we explain the inconsistency between scalability and privacy, and illustrate some highly efficient authentication protocols in RFID systems, i.e. tree-based approaches. We also interpret current work on protecting privacy in heterogeneous and interoperable RFID systems, which maintain close contact with real RFID applications. Finally, to deepen the understanding of privacy, we formally define privacy, adversaries and privacy models in RFID systems. Based on formal privacy models, we can not only formally analyze and prove the privacy of a given system, but also discover potential vulnerabilities.

REFERENCES

Avoine, G., & Oechslin, P. (2005). A scalable and provably secure hash based RFID protocol. In *Proceedings of the 3rd IEEE International Conference on Pervasive Computing and Communications (PerCom, Workshop Security)*, Hawaii, USA.

Dimitriou, T. (2005). A lightweight RFID protocol to protect against traceability and cloning attacks. In *Proceedings of the 1st International Conference on Security and Privacy in Communication Networks (SecureComm)*, Athens, Greece.

Dimitriou, T. (2006). A secure and efficient RFID protocol that could make Big Brother (partially) obsolete. In *Proceedings of the 4th Annual IEEE International Conference on Pervasive Computing and Communications(PerCom)*, Pisa, Italy.

Dimitriou, T. (2008). RFIDDOT: RFID delegation and ownership transfer made simple. In *Proceedings of the 4th International Conference on Security and Privacy in Communication Networks (SecureComm)*, Istanbul, Turkey.

Fouladgar, S., & Afifi, H. (2007). An efficient delegation and transfer of ownership protocol for RFID tags. In Proc. *the 1st International EURASIP Workshop on RFID Technology (RFID)*, Vienna, Austria.

Golle, P., Jakobsson, M., Juels, A., & Syverson, P. (2004). Universal re-encryption for mixnets. In *Proceedings of The Cryptographers' Track at the RSA Conference (CT-RSA)*, San Francisco, USA.

Hellman, M. (1980). A cryptanalytic time-memory trade-off. *IEEE Transactions on Information Theory*, *26*(4), 401–406. doi:10.1109/TIT.1980.1056220

Juels, A. (2004). Minimalist cryptography for low-cost RFID tags. In *Proceedings of the 4th International Conference on Security in Communication Networks (SCN)*, Amalfi, Italy.

Juels, A. (2006). RFID security and privacy: A research survey. *IEEE Journal on Selected Areas in Communications*, *24*(2), 381–394. doi:10.1109/JSAC.2005.861395

Juels, A., & Weis, S. (2007). Defining strong privacy for RFID. In *Proceedings of the 5th IEEE International Conference on Pervasive Computing and Communications (PerCom), Workshop Security*, New York, USA.

Lim, C., & Kwon, T. (2006). Strong and robust RFID authentication enabling perfect ownership transfer. In *Proceedings of International Conference on Information and Communications Security (ICICS)*, Raleigh, USA.

Lim, T. L., Li, T., & Gu, T. (2008). Secure RFID identification and authentication with triggered hash chain variants. In *Proceedings the 14th IEEE International Conference on Parallel and Distributed Systems (ICPADS)*, Victoria, Australia.

Lu, L., Han, J., Hu, L., Liu, Y., & Ni, L. (2007). Dynamic key-updating: Privacy-preserving authentication for RFID systems. In *Proceedings of the 5th IEEE International Conference on Pervasive Computing and Communications (PerCom)*, New York, USA.

Lu, L., Han, J., Xiao, R., & Liu, Y. (2009). ACTION: breaking the pravicy barrier for RFID systems. In *Proceedings the 28th Conference on Computer Communications (INFOCOM)*, Rio de Janaro, Brazil.

Lu, L., Liu, Y., & Li, X. (2010). Refresh: Weak privacy model for RFID systems. In *Proceedings of The 29th IEEE Conference on Computer Communications (Infocom)*, San Diego, USA.

Merriam-Webster. (2009). *Privacy*. Retrieved October 12, 2009, from http://www.merriam-webster.com/dictionary/privacy

Molnar, D., Soppera, A., & Wagner, D. (2005). A scalable, delegatable pseudonym protocol enabling owner-ship transfer of RFID tags. In *Proceedings of the 12th International Workshop Selected Areas in Cryptography(SAC)*, Waterloo, Canada.

Molnar, D., & Wagner, D. (2004). Privacy and security in library RFID: issues, practices, and architectures. In *Proceedings of the 11th ACM Conference on Computer and Communications security(CCS)*, Washington DC, USA.

Okubo, M., Suzuki, K., & Kinoshita, S. (2004). Efficient hash-chain based RFID privacy protection scheme. In *Proceedings of the International Conference on Ubiquitous Computing (Ubicomp)*, Nottingham, England.

Paise, R., & Vaudenay, S. (2008). *Mutual authentication in RFID: Security and privacy*. ACM Symposium on Information, Computer and Communications Security (ASIACCS), Tokyo, Japan.

Philip, R., & Beigl, M. (2004). Trust context spaces: An infrastructure for pervasive security in context-aware environments. In *Proceedings the 1st International Conference on Security in Pervasive Computing* (SPC), Boppard, Germany.

Saito, J., Imamoto, K., & Sakurai, K. (2005). Reassignment scheme of an RFID tag's key for owner transfer. In *Proceedings of the International Conference on Embedded and Ubiquitous Computing (EUC Workshops)*, Nagasaki, Japan.

SONG. B. (2008). RFID tag ownership transfer. In *Proceedings of the 4th Workshop on RFID Security (RFIDSec)*, Budapest, Hungary.

Tsudik, G. (2006). YA-TRAP: Yet another trivial RFID authentication protocol. In *Proceedings of the 4th Annual IEEE International Conference on Pervasive Computing and Communications (PerCom)*, Pisa, Italy.

Vaudenay, S. (2007). On privacy models for RFID. In *Proceedings of the 13th Annual International Conference on the Theory and Application of Cryptology and Information Security (Asiacrypt)*, Kuching, Malaysia.

Weis, S., Sarma, S., Rivest, R., & Engels, D. (2003). Security and privacy aspects of low-cost radio frequency identification systems. In *Proceedings of the 1st International Conference on Security in Pervasive Computing (SPC)*, Boppard, Germany.

Weis, S., Sarma, S., Rivest, R., & Engels, D. (2003). Security and privacy aspects of low-cost radio frequency identification systems. In *Proceedings of the 1st International Conference on Security in Pervasive Computing (SPC)*, Boppard, Germany.

Compilation of References

(2010). *Cambridge Dictionaries*. Cambridge University Press.

Abadi, M., Budiu, M., Erlingsson, U., & Ligatti, J. (2005). Control-flow integrity. In *Proceedings of the 12th ACM Conference on Computer and Communications Security*, (pp. 340–353) New York, NY: ACM.

Abawajy, J. (2009). Enhancing RFID tag resistance against cloning attack. In *Third International Conference on Network and System Security*, October 19-October 21, 2009, (pp. 18–23). IEEE Computer Society.

Adams, C., & Lloyd, S. (2002). *Understanding PKI: Concepts, standards, and deployment considerations* (2nd ed.). Addison-Wesley.

Ahmed, E. G., Shaaban, E., & Hashem, M.. (2010). Lightweight mutual authentication protocol for low cost RFID tags. *International Journal of Network Security and its Applications, 1*, 27-37.

Akishita, T., & Takagi, T. (2003). Zero-value point attacks on elliptic curve cryptosystem. In *Information Security Conference - ISC '03, LNCS 2851*, (pp. 218-233). Springer-Verlag.

Anderson, R., Bond, M., Clulow, J., & Skorobogatov, S. (2006, February). Cryptographic processors - A survey. *Proceedings of the IEEE, 94*(2), 357–369. doi:10.1109/JPROC.2005.862423

Angeles, R. (2005). RFID technologies: Supply-chain applications and implementation issues. *Information Systems Management, 22*(1), 51–65. doi:10.1201/1078/44912.22.1.20051201/85739.7

Anylogic Professional 6. (2009). *AB-SD supply chain model simulator*. Retrieved from http://www.xjtek.com

Aura, T. (1997). Strategies against replay attacks. In *The Proceedings of the 10th IEEE Workshop on Computer Security Foundations, CSFW '97*, (pp. 59-68).

Avanzi, R. M. (2005). *Side channel attacks on implementations of curve-based cryptographic primitive*. Retrieved from http://eprint.iacr.org/2005/017.pdf

Avoine, G. (2005). *Adversary model for radio frequency identification* (Tech. Rep. LASEC-REPORT-2005-001). Swiss Federal Institute of Technology, Lausanne, Switzerland.

Avoine, G. (2005). *Cryptography in radio frequency identification and fair exchange protocols*. PhD thesis, Ecole Polytechnique Federale de Lausanne, Lausanne, Switzerland.

Avoine, G., & Oechslin, P. (2005). A scalable and provably secure hash based RFID protocol. In *Proceedings of the 3rd IEEE International Conference on Pervasive Computing and Communications (PerCom, Workshop Security)*, Hawaii, USA.

Avoine, G., Kalach, K., & Quisquater, J.-J. (2008). ePassport: Securing international contacts with contactless chips. In *Financial Cryptography and Data Security, FC 2008*, (pp.141-155). Cozumel, Mexico.

Avoine, G., Lauradoux, C., & Martin, T. (2009). Lecture Notes in Computer Science: *Vol. 5932. When compromised readers meet RFID. Information Security Applications* (pp. 36–50). Springer. doi:10.1007/978-3-642-10838-9_4

Ayoade, J. (2007). Roadmap to solving security and privacy concerns in RFID systems. *Computer Law & Security Report, 23*(6), 555–561. doi:10.1016/j.clsr.2007.09.005

Bailey, D. V., Boneh, D., Goh, E., & Juels, A. (2007). Covert channels in privacy-preserving identification systems. *14th ACM International Conference on Computer and Communication Security*, (pp. 297-306).

Barasz, M., Boros, B., Ligeti, P., Loja, K., & Nagy, D. (2007). *Breaking LMAP*. RFIDSec07.

Bárász, M., Boros, B., Ligeti, P., Lója, K., & Nagy, D. A. (2007). *Passive attack against the M2AP mutual authentication protocol for RFID Tags*. In EURASIP International Workshop on RFID Technology.

Batina, L., Guajardo, J., Kerins, T., Mentens, N., Tuyls, P., & Verbauwhede, I. (2006). *An elliptic curve processor suitable for RFID tags*. 1st Benelux Workshop on Information and System Security (WISSec 2006), Belguim.

Batina, L., Guajardo, J., Kerins, T., Mentens, N., Tuyls, P., & Verbauwhede, I. (2007). *Public-key cryptography for RFID tags*. Fifth Annual International Conference on Pervasive Computing and Communications Workshop. New York.

Batina, L., Mentens, N., Sakiyama, K., Preneel, B., & Verbauwhede, I. (2006). Low-cost elliptic curve cryptography for wireless sensor networks. In *Workshop on Security and Privacy in Ad-Hoc and Sensor Networks: ESAS06, Lecture Notes in Computer Science, vol. 4357*, (pp. 6–17). Springer-Verlag.

Batina, L., Mentens, N., Sakiyama, K., Preneel, B., & Verbauwhede, I. (2007). *Public key cryptography on the top of a needle*. IEEE International Symposium on Circuits and Systems (ISCAS 2007), Special Session: Novel Cryptographic Architectures for Low-Cost RFID. New Orleans.

Batina, L., Guajardo, J., Preneel, B., Tuyls, P., & Verbauwhede, I. (2009). Public key cryptography for RFID tags and applications. In Kitsos, P., & Zhang, Y. (Eds.), *RFID security: Techniques, protocols and system-on-chip design*. Springer-Verlag.

Batina, L., Seys, S., Preneel, B., & Verbauwhede, I. (2008). Public key primitves. In *Wireless sensor network security* (pp. 77–108). IOS Press.

Bellare, M., & Rogaway, P. (1993). Random oracles are practical: A paradigm for designing efficient protocols. In *First ACM Conference on Computer and Communications Security*, (pp. 62-73). ACM Press.

Benoit, C., Canard, S., Girault, M., & Sibert, H. (2006). Low-cost cryptography for privacy in RFID systems. In *The Proceedings of the International Conference on Smart Card Research and Advanced Applications, CARDIS '06.*

Berger, T. P., Cayrel, P.-L., Gaborit, P., & Otmani, A. (2009). Lecture Notes in Computer Science: *Vol. 5580. Reducing key length of the McEliece cryptosystem. Advances in Cryptology (AFRICACRYPT09)* (pp. 77–97). Springer-Verlag.

Bernardi, P., Gandino, F., Lamberti, F., Montrucchio, B., Rebaudengo, M., & Sanchez, E. (2008). An anti-counterfeit mechanism for the application layer in low-cost RFID devices. In *The Proceedings of the 4th European Conference on Circuits and Systems for Communications, ECCSC*, (pp. 227–231).

Blake, I. F., Seroussi, G., & Smart, N. (1999). *Elliptic curves in cryptography*. Cambridge University Press.

Blake, I. F., Seroussi, G., & Smart, N. P. (Eds.). (2005). *Advances in elliptic curve cryptography*. Cambridge University Press. doi:10.1017/CBO9780511546570

Bogdanov, A., Knudsen, L. R., Leander, G., Paar, C., Poschmann, A., & Robshaw, M. J. B. ... Vikkelsoe, C. (2007). PRESENT: An ultra-lighweight block cipher. In P. Paillier & I. Verbauwhede (Eds.), *CHES 2007: LNCS* (pp. 450-466). Santa Barbara.

Bogdanov, A., Leander, G., Paar, C., Poschmann, A., Robshaw, M., & Seurin, Y. (2008). Hash functions and RFID tags: Mind the gap. In Oswald, E., & Rohatgi, P. (Eds.), *CHES 2007, LNCS 5154* (pp. 283–299). doi:10.1007/978-3-540-85053-3_18

Bolotnyy, L., & Robins, G. (2007). Physically unclonable function-based security and privacy in RFID systems. *5th International Conference on Pervasive Computing and Communications*, (pp. 211-120).

Boneh, D., DeMillo, R. A., & Lipton, R. J. (1997). On the importance of checking cryptographic protocols for faults. In *EUROCRYPT' 1997, LNCS 1233* (pp. 37–51). Berlin, Germany: Springer.

Boneh, D., & Franklin, M. (2003). Identity-based encryption from the Weil pairing. *SIAM Journal on Computing*, *32*(3), 586–615. doi:10.1137/S0097539701398521

Bono, S. C., Green, M., Stubblefield, A., Juels, A., Rubin, A. D., & Szydlo, M. (2005). Security analysis of a cryptographically-enabled RFID device. In *The Proceedings of the 14th Conference on USENIX Security Symposium, SSYM '05*, (pp. 1–15).

Borders, K., Weele, E. V., Lau, B., & Prakash, A. (2009). Protecting confidential data on personal computers with storage capsules. In *Proceedings of the 18th USENIX Security Symposium, USENIX Association*, Berkeley, CA, USA, (pp. 367–382).

Bose, A., Hu, X., Shin, K. G., & Park, T. (2008). Behavioral detection of malware on mobile handsets. In *Proceeding of the 6th International Conference on Mobile Systems, Applications, and Services*, (pp. 225–238). New York, NY: ACM.

Brands, S., & Chaum, D. (1993). Distance-bounding protocols. In *Advances in Cryptology EUROCRYPT '93* (pp. 344–359). Berlin, Germany: Springer-Verlag.

Brier, E., Clavier, C., & Olivier, F. (2004). Correlation power analysis with a leakage model. In *Proceedings of 6th International Workshop on Cryptographic Hardware and Embedded Systems,* (pp. 16–29). Berlin, Germany: Springer-Verlag.

Brosch, T., & Morgenstern, M. (2006). *Runtime packers: The hidden problem?* Presented at Black Hat USA, 2006.

Brumley, D., & Boneh, D. (2003). Remote timing attacks are practical. In *Proceedings of the 12th conference on USENIX Security Symposium* - Volume 12, (p. 1).

Burmester, M., & Munilla, J. (2009). *A flyweight RFID authentication protocol.* Workshop on RFID Security.

Burmester, M., van Le, T., & de Medeiros, B. (2006). Provably secure ubiquitous systems: Universally composable RFID authentication protocols. *Conference on Security and Privacy for Emerging Areas in Communication Networks – SecureComm 2006* (pp. 1-10). Baltimore, MD: IEEE Computer Society.

Burmester, M., de Medeiros, B., & Motta, R. (2008). Anonymous RFID authentication supporting constant-cost key-lookup against active adversaries. *International Journal of Applied Cryptography*, *1*(2), 79–90. doi:10.1504/IJACT.2008.021082

Calmels, B., Canard, S., Girault, M., & Sibert, H. (2006). Low-cost cryptography for privacy in RFID systems. *Smart Card Research and Advanced Applications, LNCS, 3928*, 237–251. doi:10.1007/11733447_17

Canetti, R. (2000). *Universally composable security: A new paradigm for cryptographic protocols (Report 2000/067). Cryptology ePrint Archive.* International Association for Cryptologic Research.

Castro, J. C. H., Estevez-Tapiador, J. M., Peris-Lopez, P., & Quisquater, J.-J. (2008), *Cryptanalysis of the SASI ultralightweight RFID authentication protocol with modular rotations.* In CoRR labs.

Chae, H., Yeager, D., Smith, J., & Fu, K. (2007), Maximalist cryptography and computation on the WISP UHF RFID tag. In *The Proceedings of the Conference on RFID Security.*

Charles, D. X., Goren, E. Z., & Lauter, K. E. (2006). *Cryptographic hash functions from expander graphs.* 2nd NIST Hash Functions Workshop.

Chartier, P., Consultants, P., & van den Akker, G. (2008). *GRIFS, global RFID forum for standards, RFID standardisation state of the art report. Technical report.* CEN.

Chaum, D. (1983). Blind signatures for untraceable payments. *Advances in Cryptology- Proceedings of Crypto 82*, (pp. 199-203).

Chawla, K., Robins, G., & Weimer, W. (2010). On mitigating covert channels in RFID-enabled supply chains. *Radio Frequency Identification System Security (RFIDsec), Workshop on RFID Security*, (pp. 135-146).

Chen, J., Kanj, I. A., & Jia, W. (1999). Vertex cover: Further observations and further improvements. *Journal of Algorithms*, *41*, 313–324.

Chevallier-Mames, B., Ciet, M., & Joye, M. (2004). Low-cost solutions for preventing simple side-channel analysis: Side-channel atomicity. *IEEE Transactions on Computers*, *53*, 760–768.

Chien, H.-Y. (2007). SASI: A new Ultralightweight RFID authentication protocol providing strong authentication and strong integrity. *IEEE Transactions on Dependable and Secure Computing*, *4*(4), 337–340. doi:10.1109/TDSC.2007.70226

Chinnappa Gounder Periaswamy, S. (2007). *Fingerprinting RFID tags*. Unpublished master dissertation, University of Arkansas, Arkansas.

Cohen, F. B. (1992). Defense-in-depth against computer viruses. *Computers & Security*, *11*(6), 563–579. doi:10.1016/0167-4048(92)90192-T

Cormen, T. H., Leiserson, C. E., Rivest, R. L., & Stein, C. (2009). *Introduction to algorithms* (3rd ed.). MIT Press.

Coron, J.-S. (1999). Resistance against differential power analysis for elliptic curve cryptosystems. In *CHES' 1999, LNCS 1717* (pp. 292–302). Berlin, Germany: Springer.

Cover, T. M., & King, R. C. (1978). A convergent gambling estimate of the entropy of English. *IEEE Transactions on Information Theory*, *24*(4). doi:10.1109/TIT.1978.1055912

Cox, B., Evans, D., Filipi, A., Rowanhill, J., Hu, W., & Davidson, J. … Hiser, J. (2006). N-variant systems: A secretless framework for security through diversity. In *Proceedings of the 15th conference on USENIX Security Symposium,* (pp. 105–120). Berkeley, CA: USENIX Association.

Cranfield, R., & Pomerance, C. (2001). *Prime numbers: A computational perspective*. Springer-Verlag.

Damgård, I., & Østergaard, M. (2006). *RFID security: Tradeoffs between security and efficiency (Report 2006/234). Cryptology ePrint Archive*. International Association for Cryptologic Research.

Das, R., & Harrop, D. P. (2009). *RFID forecasts, players and opportunities 2009-2019*. Market report, IDTechEx. Retrieved from http://www.idtechex.com/research

Deursen, T., & Radomirović, S. (2009). Algebraic attacks on RFID protocols. In *The Proceedings of the 3rd IFIP WG 11.2 International Workshop on Information Security Theory and Practice. Smart Devices, Pervasive Systems, and Ubiquitous Networks, WISTP'09*, (pp. 38–51). Berlin, Germany: Springer-Verlag.

Dhem, J.-F., Koeune, F., Leroux, P.-A., Mestré, P., Quisquater, J.-J., & Willems, J.-L. (1998). A practical implementation of the timing attack. In *Third Smart Card Research and Advanced Application Conference - CARDIS 98, volume 1820 of Lecture Notes in Computer Science*, (pp. 167-182). Berlin, Germany: Springer-Verlag.

Diffie, W., & Hellman, M. (1976). New directions in cryptography. *IEEE Transactions on Information Theory*, *22*, 644–654. doi:10.1109/TIT.1976.1055638

Dimitriou, T. (2005). A lightweight RFID protocol to protect against traceability and cloning attacks. In *Proceedings of the 1st International Conference on Security and Privacy in Communication Networks (SecureComm)*, Athens, Greece.

Dimitriou, T. (2006). A secure and efficient RFID protocol that could make Big Brother (partially) obsolete. In *Proceedings of the 4th Annual IEEE International Conference on Pervasive Computing and Communications (PerCom)*, Pisa, Italy.

Dimitriou, T. (2008). RFID security and privacy. In P. Kitsos & Y. Zhang (Eds.), *RFID security: Techniques, protocols and system-on-chip design* (pp. 57-79). Springer Science+Business Media LLC.

Dimitriou, T. (2008). RFIDDOT: RFID delegation and ownership transfer made simple. In *Proceedings of the 4th International Conference on Security and Privacy in Communication Networks (SecureComm)*, Istanbul, Turkey.

Dolev, D., & Yao, A. C. (1981). On the security of public key protocols. In *Proceedings of IEEE 22nd Annual Symposium on Foundations of Computer Science*, (pp. 350-357).

Driessen, B., Poschmann, A., & Paar, C. (2008). Comparison of innovative signature algorithms for WSNs. In *The First ACM Conference on Wireless Network Security*, ACM, (pp. 30–35).

Drimer, S., & Murdoch, S. J. (2007), Keep your enemies close: distance bounding against smartcard relay attacks. In *The Proceedings of 16th USENIX Security Symposium on USENIX Security Symposium, SS '07*, (pp. 1–16).

Duc, D. N., Yeun, C. Y., & Kim, K. (2010). Reconsidering Ryu-Takagi RFID authentication protocol. In *the 2nd International Workshop on RFID/USN Security and Cryptography (RISC)*, (pp. 1–6).

Duc, D. N., & Kim, K. (2011). Defending RFID authentication protocols against DoS attacks. *Elsevier's Journal of Computer Communications*, *34*(3), 384–390. doi:10.1016/j.comcom.2010.06.014

Eisenbarth, T., Paar, C., Poschmann, A., Kumar, S., & Uhsadel, L. (2007). A survey of lightweight-cryptography implementations. *IEEE Design & Test of Computers*, *24*, 522–533. doi:10.1109/MDT.2007.178

ElGamal, T. (1985). A public key cryptosystem and a signature scheme based on discrete logarithms. *IEEE Transactions on Information Theory*, *31*, 469–472. doi:10.1109/TIT.1985.1057074

EPCglobal Inc. (n.d.). *Application layer events (ALE) standard*. Retrieved from http://www.epcglobalinc.org/standards/ale/ale_1_1_1-standard-core-20090313.pdf

EPCglobal Inc. (n.d.). *EPC tag data standard* (TDS). Retrieved from http://www.epcglobalinc.org/standards/tds/tds_1_6-RatifiedStd-20110922.pdf

EPCglobal Inc. (n.d.). *Low level reader protocol (LLRP) standard*. Retrieved from http://www.epcglobalinc.org/standards/llrp/llrp_1_1-standard-20101013.pdf

EPCglobal Inc. (n.d.). *Reader protocol (RP) standard*. Retrieved from http://autoid.mit.edu/CS/files/11/download.aspx

EPCglobal Inc. (n.d.). Retrieved from http://www.epcglobalinc.org

EPCglobal. (2007). *EPC information services* (EPCIS) v1.0.1. Technical report, EPCglobal. Retrieved from http://www.epcglobalinc.org/standards

EPCGlobal. (2008). *UHF C1 G2 air interface protocol standard*. Retrieved from http://www.epcglobalinc.org/standards/uhfc1g2/uhfc1g2_1_2_0-standard-20080511.pdf

EPCglobal. (2008). *EPCglobal object naming service (ONS) v1.0.1*. Technical report, EPCglobal. Retrieved from http://www.epcglobalinc.org/standards

EPCglobal. (2008). *UHF class 1 gen 2 standard v1.2.0*. Technical report, EPCglobal. Retrieved from http://www.epcglobalinc.org/standards

EPCGlobal. (2011). *Tag data standards version 1.6*. Retrieved from http://www.gs1.org/gsmp/kc/epcglobal/tds/tds_1_6-RatifiedStd-20110922.pdf

Fabian, B., Guenther, O., & Spiekermann, S. (2005). Security analysis of the object name service for RFID. In *International Workshop on Security, Privacy and Trust in Pervasive and Ubiquitous Computing*, July 2005.

Federal Information Processing Standards Publication 197. (2001). *Advanced encryption standard (AES)*. Retrieved from http://www.itl.nist.gov/fipspubs/

Feldhofer, M., & Wolkerstorfer, J. (2007). Strong crypto for RFID tags – A comparison of low-power hardware implementations. In *IEEE International Symposium on Circuits and Systems (ISCAS07)*, (pp. 1839–1842).

Feldhofer, M., Dominikus, S., & Wolkerstorfer, J. (2004). Strong authentication for RFID systems using the AES algorithm. In *Workshop on Cryptographic Hardware and Embedded Systems (CHES04), Lecture Notes in Computer Science, vol. 3156*, (pp. 357–370). Springer-Verlag.

Feldhofer, M., Wolkerstorfer, J., & Rijmen, V. (2005). AES implementation on a grain of sand. In *IEE Proceedings of Information Security: IS05, Vol. 152*, (pp. 13–20).

Feldhofer, M., & Rechberger, C. (2006). Lecture Notes in Computer Science: *Vol. 4277. A case against currently used hash functions in RFID protocols. OTM 2006 Workshops* (pp. 372–381). Springer-Verlag.

Finkenzeller, K. (2003). *RFID handbook: Fundamentals and applications in contactless smart cards and identification*. John Wiley & Sons, Inc.

Finkenzeller, K. (2003). *RFID-handbook: Fundamentals and applications in contactless smart cards and identification* (2nd ed.). Munich, Germany: Wiley and Sons Inc.

FIPS. (1995). *Secure hash standard. Federal Information Processing Standard, NIST* (pp. 180–181). Washington, D.C.: Department of Commerce.

FIPS. (2003). *Secure hash standard. Federal Information Processing Standard, NIST* (pp. 180–182). Washington, D.C.: Department of Commerce.

FIPS. 186. (2000). *Digital signature algorithm.* Federal Information Processing Standard, NIST, Department of Commerce, Washington, D.C.

FIPS. 197. (2001). *Advanced encryption standard.* Federal Information Processing Standard, NIST, Department of Commerce, Washington, D.C.

FIPS. 46. (1977). *Data encryption standard.* Federal Information Processing Standard. NIST, Department of Commerce, Washington, D.C.

FIPS. 81. (1980). *DES modes of operation.* Federal Information Processing Standard, NIST, Department of Commerce, Washington, D.C.

Floerkemeier, C., & Lampe, M. (2005). RFID middleware design: Addressing application requirements and RFID constraints. In the *Proceedings of the 2005 Joint Conference on Smart Objects and Ambient Intelligence, sOc-EUSAI '05.*

Fouladgar, S., & Afifi, H. (2007). An efficient delegation and transfer of ownership protocol for RFID tags. In Proc. *the 1st International EURASIP Workshop on RFID Technology (RFID)*, Vienna, Austria.

Fouque, P.-A., & Valette, F. (2003). The doubling attack - Why upwards is better than downwards. In *Cryptographic Hardware and Embedded Systems - CHES '03, LNCS 2779*, (pp. 269-280). Springer-Verlag.

Gandino, F., Montrucchio, B., & Rebaudengo, M. (2009). Tampering in RFID: A survey on risks and defenses. [Springer Netherlands.]. *Mobile Networks and Applications*, 1–15.

Gandino, F., Montrucchio, B., & Rebaudengo, M. (2010). Tampering in RFID: A survey on risks and defenses. *Mobile Networks and Applications, 15*(4), 502–516. doi:10.1007/s11036-009-0209-y

Gao, D., Reiter, M. K., & Song, D. (2005). Behavioral distance for intrusion detection. In *Proceedings of the 8th International Symposium on Recent Advances in Intrusion Detection,* (pp. 63–81). Berlin, Germany: Springer-Verlag.

Garfinkel, S. L., Juels, A., & Pappu, R. (2005). RFID privacy: An overview of problems and proposed solutions. [IEEE Computer Society.]. *IEEE Security and Privacy, 3*(3), 34–43. doi:10.1109/MSP.2005.78

Garfinkel, S., & Spafford, G. (1996). *Practical Unix & internet security. O'Reilly.* Associates.

Gennaro, R., Lysyanskaya, A., Malkin, T., Micali, S., & Rabin, T. (2004). Algorithmic tamper-proof (ATP) security: Theoretical foundations for security against hardware tampering. In *TCC 2004*, (pp. 258-277).

Gilbert, H., Robshaw, M. J., & Sibert, H. (2005). An active attack against HB+: A provably secure lightweight authentication protocol. *IEE Electronics Letters, 41*(21), 1169–1170. doi:10.1049/el:20052622

Gilbert, H., Robshaw, M. J., & Seurin, Y. (2008). HB#: Increasing the security and efficiency of HB+. In Advances in Cryptology (Eurocrypt08), Lecture Notes in Computer Science, vol. 4965, (pp. 361–378). Springer.

Girault, M. (1991). Self certified public keys. In Davies, D. (Ed.), *EUROCRYPT '91, LNCS 547* (pp. 490–497). UK.

Girault, M., Poupard, G., & Stern, J. (2006). On the fly authentication and signature schemes based on groups of unknown order. *Journal of Cryptology, 19*, 463–487. doi:10.1007/s00145-006-0224-0

Global, E. (2010). *Class-1 generation-2 UHF air interface protocol standard.* Retrieved from http://www.epcglobalinc.org/standards/.

Goldwasser, S., & Micali, S. (1984). Probabilistic encryption. *Journal of Computer and System Sciences, 28*, 270–299.

Golle, P., Jakobsson, M., Juels, A., & Syverson, P. (2004). Universal re-encryption for mixnets. In *Proceedings of The Cryptographers' Track at the RSA Conference (CT-RSA)*, San Francisco, USA.

Golle, P., Jakobsson, M., Juels, A., & Syverson, P. (2004). Lecture Notes in Computer Science: *Vol. 2964. Universal re-encryption for mixnets. Topics in Cryptology - CT-RSA 2004* (pp. 163–178). Springer. doi:10.1007/978-3-540-24660-2_14

Goodin, D. (2009, 2nd February). Passport RFIDs cloned wholesale by $250 eBay auction spree. *The Register.* Situation Publishing Limited. Retrieved from http://www.theregister.co.uk/2009/02/02/low_cost_rfid_cloner/

Gordon, D. M. (1998). A survey of fast exponentiation methods. *Journal of Algorithms, 27*, 129–146.

Goubin, L. (2003). A refined power-analysis attack on elliptic curve cryptosystems. In *Public Key Cryptography - PKC'03, LNCS 2567*, (pp. 199-210). Springer-Verlag.

Gura, N., Patel, A., Wander, A., Eberle, H., & Shantz, S. C. (2004). *Comparing elliptic curve cryptography and RSA on 8-bit CPUs*. In CHES04.

Hancke, G. (2010). Practical eavesdropping and skimming attacks on high-frequency RFID tokens. *Journal of Computer Security - Special Issue on RFID System Security, 19*(2), 259–288.

Hancke, G. P. (2006). Practical attacks on proximity identification systems. In *IEEE Symposium on Security and Privacy*, (pp. 328-333).

Hancke, G. P., & Kuhn, M. G. (2008). Attacks on time-of-flight distance bounding channels. In *First ACM Conference on Wireless Network Security: WiSec08*, ACM, (pp. 194–202).

Hancke, G. P., & Drimer, S. (2008). Secure proximity identification for RFID. In *Security in RFID and sensor networks* (pp. 170–194). CRC Press.

Hancke, G. P., & Kuhn, M. G. (2005). An RFID distance bounding protocol. In *Security and Privacy for Emerging Areas in Communications Networks (SecureComm05)* (pp. 67–73). IEEE. doi:10.1109/SECURECOMM.2005.56

Hankerson, D., Menezes, A. J., & Vanstone, S. (2004). *Guide to elliptic curve cryptography*. Springer-Verlag.

Hellman, M. (1980). A cryptanalytic time-memory trade-off. *IEEE Transactions on Information Theory, 26*(4), 401–406. doi:10.1109/TIT.1980.1056220

Hellman, M. E. (1977). An extension of the Shannon theory approach to cryptography. *IEEE Transactions on Information Theory, 23*(3). doi:10.1109/TIT.1977.1055709

Hoffstein, J., & Silverman, J. H. & report 012, W. W. N. (2003). *Estimated breaking times for NTRU lattices*. Technical Report 12, NTRU Cryptosystems, Inc.

Hoffstein, J., Pipher, J., & Silverman, J. H. (2001). NSS: An NTRU lattice-based signature scheme. In *Advances in Cryptology (EUROCRYPT01), Vol. 2045*, (pp. 211–228).

Hoffstein, J., Howgrave-Graham, N., Pipher, J., Silverman, J. H., & Whyte, W. (2003). NTRU sign: Digital signatures using the NTRU lattice. In *Topics in Cryptology – The Cryptographers Track at the RSA Conferenc* (*Vol. 2612*, pp. 122–140). Lecture Notes in Computer ScienceSpringer-Verlag.

Hoffstein, J., Pipher, J., & Silverman, J. (2008). *An introduction to mathematical cryptography*. Springer Press Verlag.

Hoffstein, J., Pipher, J., & Silverman, J. H. (1998). NTRU: A ring-based public key cryptosystem. In *Algorithmic Number Theory: ANTS III* (*Vol. 1423*, pp. 267–288). Lecture Notes in Computer ScienceSpringer-Verlag. doi:10.1007/BFb0054868

Hopper, N. J., & Blum, M. (2000). *A secure human-computer authentication scheme* (Report CMU-CS-00-139).

Hu, X., Chiueh, T.-C., & Shin, K. G. (2009), Large-scale malware indexing using function-call graphs. In *Proceedings of the 16th ACM Conference on Computer and Communications Security*, (pp. 611–620). New York, NY: ACM.

Huffman, D. A. (1952). A method for the construction of minimum-redundancy codes. *Institute of Radio Engineers, 40*(9), 1098–1101. doi:10.1109/JRPROC.1952.273898

Hunt, D., Puglia, A., & Puglia, M. (2007). *RFID-A guide to radio frequency identification*. John Wiley & Sons, Inc. doi:10.1002/0470112255

Hutter, M. (2009). *RFID authentication protocols based on elliptic curves: A top-down evaluation survey*. In *International Conference on Security and Cryptography: SECRYPT09*, (pp. 101–110). Retrieved from https://online.tu-graz.ac.at/tug_online/voe_main2.get%volltext?pDocumentNr=106022

ICAO. (n.d.). *Machine readable travel documents: ICAO Doc 9303*. Retrieved from www.icao.int/mrtd/

ISO. IEC 9798-2. (1993). Information technology security techniques entity authentication mechanisms- Part 2: Entity authentication using symmetric techniques. ISO/IEC.

ISO/IEC 14443. (2008). *Identification cards – Contactless integrated circuit cards – Proximity cards*

ISO/IEC 15693. (2006). *Identification cards – Contactless integrated circuit cards – Vicinity cards.*

Juels, A. (2004). Minimalist cryptography for low-cost RFID tags. In *Proceedings of the 4th International Conference on Security in Communication Networks (SCN)*, Amalfi, Italy.

Juels, A. (2005). Strengthening EPC tags against cloning. In M. Jakobsson & R. Poovendran (Eds.), *ACM Workshop on Wireless Security (WiSe)* (pp. 67-76).

Juels, A., & Weis, S. (2007). Defining strong privacy for RFID. In *Proceedings of the 5th IEEE International Conference on Pervasive Computing and Communications (PerCom), Workshop Security*, New York, USA.

Juels, A., Molnar, D., & Wagner, D. (2005). Security and privacy issues in e-passports. In M. Jakobsson & R. Poovendran (Eds.), *Proceedings of the First International Conference on Security and Privacy for Emerging Areas in Communications Networks (SECURECOMM)* (pp. 74-88). Washington, DC: IEEE Computer Society.

Juels, A., Rivest, R. L., & Szydlo, M. (2003). *The blocker tag: Selective blocking of RFID tags for consumer privacy.* Paper presented at the 10th ACM Conference on Computer and Communications Security, CCS 2003, October 27, 2003 - October 31, 2003, Washington, DC.

Juels, A., Syverson, P., & Bailey, D. (2005). *High-power proxies for enhancing RFID privacy and utility. Paper presented at the Privacy Enhancing Technologies.* 5th International Workshop, PET 2005, Revised Selected Papers, 30 May-1 June 2005, Berlin, Germany.

Juels, A. (2006). RFID security and privacy: A research survey. *IEEE Journal on Selected Areas in Communications, 24*(2), 381–395. doi:10.1109/JSAC.2005.861395

Juels, A., & Pappu, R. (2003). Lecture Notes in Computer Science: *Vol. 2742. Squealing Euros: Privacy protection in RFID-enabled banknotes. Financial Cryptography* (pp. 103–121). Springer. doi:10.1007/978-3-540-45126-6_8

Juels, A., & Weis, S. A. (2005). Authenticating pervasive devices with human protocols. In *Advances in Cryptology (Crypto05)* (*Vol. 3126*, pp. 198–293). Lecture Notes in Computer ScienceSpringer.

Kaps, J.-P. (2006). *Cryptography for ultra-low power devices.* PhD thesis, ECE Department of Worcester Polytehcnic Institute.

Karjoth, G., & Moskowitz, P. A. (2005). Disabling RFID tags with visible confirmation: Clipped tags are silenced. *ACM Workshop on Privacy in the Electronic Society*, (pp. 27-30).

Karlof, C., & Wagner, D. (2003). Hidden MARKOV model cryptanalysis. In *Cryptographic Hardware and Embedded Systems - CHES 2003, LNCS 2779*, (pp. 17-34). Springer-Verlag.

Karygiannis, A., Phillips, T., & Tsibertzopoulos, A. (2006). RFID security: A taxonomy of risks. *Conference on Communications and Networking in China*, (pp. 1-8).

Kasper, T., Oswald, D., & Paar, C. (2009). *New methods for cost-effective side-channel attacks on cryptographic RFIDs.* In Workshop on RFID Security.

Katz, J., & Lindell, Y. (2008). *Introduction to modern cryptography. Chapman and Hall.* CRC Press.

Keller, M., & Marnane, W. (2007). Low power elliptic curve cryptography. In N. Azémard & L. J. Svensson (Eds.), *Integrated Circuit and System Design, 17th International Workshop on Power and Timing Modeling, Optimization and Simulation (PATMOS 2007), Lecture Notes in Computer Science, vol. 4644*, (pp. 310–319). Gothenburg, Sweden: Springer-Verlag.

Kfir, Z., & Wool, A. (2005). Picking virtual pockets using relay attacks on contactless smartcard. In *The Proceedings of the First International Conference on Security and Privacy for Emerging Areas in Communications Networks, SECURECOMM '05*, (pp. 47–58).

Kim, C. H., & Avoine, G. (2009). RFID distance bounding protocol with mixed challenges to prevent relay attacks. In *International Conference on Cryptology and Network Security: CANS09, Lecture Notes in Computer Science*, (pp. 119–133). Berlin, Germany: Springer-Verlag.

Kim, H. S., Kim, I. G., Han, K. H., & Choi, J. Y. (2006). Security and privacy analysis of RFID systems using model checking. *High Performance Computing and Communications, Vol. 4208 of Lecture Notes in Computer Science*, (pp. 495–504). Berlin, Germany: Springer.

Kim, H., Smith, J., & Shin, K. G. (2008). Detecting energy-greedy anomalies and mobile malware variants. In *Proceeding of the 6th International Conference on Mobile Systems, Applications, and Services*, (pp. 239–252). New York, NY: ACM.

Kim, S., Kim, Y., & Park, S. (2007). RFID security protocol by lightweight ECC algorithm. In *Sixth International Conference on Advanced Language Processing and Web Information Technology* (ALPIT07), IEEE Computer Society, (pp. 323–328).

Koblitz, N. (1987). Elliptic curve cryptosystems. *Mathematics of Computation*, *48*, 203–209.

Koblitz, N. (1994). *A course in number theory and cryptography*. Springer Verlag. doi:10.1007/978-1-4419-8592-7

Kocher, P. (1996). Timing attacks on implementations of Diffie-Hellman, RSA, DSS, and other systems. In *CRYPTOŸÏ 1996, LNCS 1109* (pp. 104–113). Berlin, Germany: Springer.

Kocher, P., Jaffe, J., & Jun, B. (1999). Differential power analysis. In *CRYPTO 1999, LNCS 1666* (pp. 388–397). Berlin, Germany: Springer.

Kohl, J., & Neuman, C. (1993). The Kerberos network authentication service (v5). In *The Internet Engineering Task Force Request For Comments (IETF RFC) 1510.* Retrieved from http://www.ietf.org/rfc/rfc1510.txt.

Koscher, K., Juels, A., Brajkovic, V., & Tadayoshi, K. (2009). EPC RFID tag security weaknesses and defenses: Passport cards, enhanced drivers licenses, and beyond. *16th ACM Conference on Computer and Communications Security*, (pp. 33-42).

Langheinrich, M. (2009). A survey of RFID privacy approaches. *Personal and Ubiquitous Computing*, *13*(6), 413–421. doi:10.1007/s00779-008-0213-4

Le, T. V., Burnmester, M., & de Medeiros, B. (2007). Universally composable and forward secure RFID authentication and authenticated key exchange. In *the Proceedings of the 2nd ACM Symposium on Information, Computer and Communications Security*, (pp. 242-252).

Lee, H. L., & Whang, S. (2005). Higher supply chain security with lower cost: Lessons from total quality management. *International Journal of Production Economics*, *96*, 289–300. doi:10.1016/j.ijpe.2003.06.003

Lee, Y. K., Sakiyama, K., Batina, L., & Verbauwhede, I. (2008). Elliptic-curve-based security processor for RFID security. *IEEE Transactions on Computers*, *57*, 1514–1527. doi:10.1109/TC.2008.148

Lehtonen, M., Michahelles, F., & Fleisch, E. (2009). How to detect cloned tags in a reliable way from imcomplete RFID traces. In *2009 IEEE International Conference on RFID* (pp. 257-264). Piscataway, NJ: IEEE.

Lenstra, A. K., & Verheul, E. R. (2000). The XTR public key system. In *Advances in Cryptology (CRYPTO 2000), Lecture Notes in Computer Science, vol. 1880*, Springer-Verlag, (pp. 1–19).

Lenstra, A. K., & Verheul, E. R. (2001). Selecting cryptographic key sizes. *Journal of Cryptology: The Journal of the International Association for Cryptologic Research*, *14*(4), 255–293.

Li, T., & Deng, R. (2007). Vulnerability analysis of EMAP - an efficient RFID mutual authentication protocol. In *International Conference on Availability, Reliability and Security*, (pp. 238–245).

Li, T., & Wang, G. (2007). Security analysis of two ultra-lightweight RFID authentication protocols. In *IFIP-SEC07*, (pp. 14–16).

Lim, C., & Kwon, T. (2006). Strong and robust RFID authentication enabling perfect ownership transfer. In P. Ning, S. Qing, & N. Li (Eds.), *International Conference on Information and Communications Security - ICICS'06, Lecture Notes in Computer Science: Vol. 4307*, (pp. 1-20). Springer.

Lim, T. L., Li, T., & Gu, T. (2008). Secure RFID identification and authentication with triggered hash chain variants. In *Proceedings the 14th IEEE International Conference on Parallel and Distributed Systems (ICPADS)*, Victoria, Australia.

Lima, A., Miri, A., & Nevins, M. (2008). RFID relay attacks: System analysis, modelling, and implementation. In *Security in RFID and sensor networks*, (pp. 49–75). Auerbach Publications, Taylor & Francis Group.

Lim, D., Lee, J., Gassend, B., Suh, G., Dijk, M., & Devadas, S. (2005). Extracting secret keys from integrated circuits. *IEEE Transactions on Very Large Scale Integration (VLSI). Systems*, *13*(10), 1200–1205.

Liu, L., Yan, G., Zhang, X., & Chen, S. (2009). VirusMeter: Preventing your cellphone from spies. In *Proceedings of the 12th International Symposium on Recent Advances in Intrusion Detection,* (pp. 244–264). Berlin, Germany: Springer-Verlag.

Lu, L., Han, J., Hu, L., Liu, Y., & Ni, L. (2007). Dynamic key-updating: Privacy-preserving authentication for RFID systems. In *Proceedings of the 5th IEEE International Conference on Pervasive Computing and Communications (PerCom)*, New York, USA.

Lu, L., Han, J., Xiao, R., & Liu, Y. (2009). ACTION: breaking the pravicy barrier for RFID systems. In *Proceedings the 28th Conference on Computer Communications (INFOCOM)*, Rio de Janaro, Brazil.

Lu, L., Liu, Y., & Li, X. (2010). Refresh: Weak privacy model for RFID systems. In *Proceedings ofThe 29th IEEE Conference on Computer Communications (Infocom)*, San Diego, USA.

Lu, S., Park, S., Hu, C., Ma, X., Jiang, W., Li, Z., et al. (2007). MUVI: Automatically inferring multi-variable access correlations and detecting related semantic and concurrency bugs. In *Proceedings of Twenty-First ACM SIGOPS Symposium on Operating Systems Principles*, (pp. 103–116). New York, NY: ACM.

Ma, C., Li, Y., Deng, R. H., & Li, T. (2009). RFID privacy: Relation between two notions, minimal condition, and efficient construction. In *Proceedings of the 16th ACM Conference on Computer and Communications Security,* (pp. 54–65). New York, NY: ACM.

Majzoobi, M., Koushanfar, F., & Potkonjak, M. (2009). Techniques for design and implementation of secure reconfigurable PUFs. *ACM Transactions on Reconfigurable Technology and Systems, 5*(2), 1–33. doi:10.1145/1502781.1502786

Mamiya, H., Miyaji, A., & Morimoto, H. (2004). Efficient countermeasure against RPA, DPA, and SPA. In *CHES' 2004, LNCS 3156* (pp. 343–356). Berlin, Germany: Springer.

Mandel, J., Roach, A., & Winstein, K. (2004). *MIT proximity card vulnerabilities.* MIT Technical report. Retrieved from http://web.mit.edu/keithw/Public/MIT-Card-Vulnerabilities-March31.pdf

Mao, W. B. (2004). *Modern cryptography: Theory and practice.* Prentice Hall.

Martignoni, L., Christodorescu, M., & Jha, S. (2007). OmniUnpack: Fast, generic, and safe unpacking of malware. In *Proceedings of the 23rd Annual Computer Security Applications Conference,* (pp. 431–441). Los Alamitos, CA: IEEE Computer Society.

McClure, S., Scambray, S., & Kurtz, G. (1999). *Hacking exposed: Network security secrets and solutions.* McGraw-Hill.

McLoone, M., & Robshaw, M. J. B. (2007). Public key cryptography and RFID tags. In Abe, M. (Ed.), *CT-RSA 2007 LNCS 4377* (pp. 372–384). San Francisco.

Menezes, A. J., van Oorschot, P. C., & Vanstone, S. A. (1996). *Handbook of applied cryptography.* CRC Press. doi:10.1201/9781439821916

Merkle, R. C. (1990). A fast software one-way hash function. *Journal of Cryptology, 3,* 43–58. doi:10.1007/BF00203968

Merriam-Webster. (2009). *Privacy.* Retrieved October 12, 2009, from http://www.merriam-webster.com/dictionary/privacy

Miller, S., Neuman, B., Schiller, J., & Saltzer, J. (1988). Kerberos authentication and authorization system. In *Section E.2.1, Project Athena Technical Plan, M.I.T. Project Athena.* Cambridge, MA: MIT.

Miller, V. S. (1985). Use of elliptic curves in cryptography. In *Proceedings of Crypto 85, LNCS 218,* (pp. 417-426).

Min, H., & Zhou, G. (2002). Supply chain modeling: Past, present and future. *Journal of Computer and Industrial Engineering, 43*(1-2), 231–249. doi:10.1016/S0360-8352(02)00066-9

Mitrokotsa, A., Rieback, M. R., & Tanenbaum, A. S. (2008). Classification of RFID attacks. *International Workshop on RFID Technology,* (pp. 73-86).

Molnar, D., & Wagner, D. (2004). Privacy and security in library RFID: Issues, practices, and architectures. In *Conference on Computer and Communications Security* (CCS04), ACM, (pp. 210–219).

Molnar, D., Soppera, A., & Wagner, D. (2005). A scalable, delegatable pseudonym protocol enabling owner-ship transfer of RFID tags. In *Proceedings of the 12th International Workshop Selected Areas in Cryptography (SAC)*, Waterloo, Canada.

Molnar, D., Soppera, A., & Wagner, A. (2005). Lecture Notes in Computer Science: *Vol. 3897. A scalable, delegatable pseudonym protocol enabling ownership transfer of RFID tags. Selected Areas in Cryptography* (pp. 276–290). Springer. doi:10.1007/11693383_19

Moskowitz, I. S., & Kang, M. H. (1994). Covert channels here to stay. *9th IEEE International Conference on Computer Assurance*, (pp. 235-243).

Moskowitz, P., & Karjoth, G. (2005, November 7). IBM proposes privacy-protecting tag. *RFID Journal*.

Moskowitz, P., Lauris, A., & Morris, S. S. (2007). A privacy-enhancing radio frequency identification tag: implementation of the clipped tag. In *Fifth IEEE International Conference on Pervasive Computing and Communications Workshops (PerComW'07)*, March 19-March 23 2007, White Plains, New York.

Munilla, J., & Peinado, A. (2010). Enhanced low-cost RFID protocol to detect relay attacks. In *Wireless Communications and Mobile Computing*, Vol. 10, (pp. 361–371).

Munilla, J., & Peinado, A. (2008). Distance bounding protocols for RFID enhanced by using void-challenges and analysis in noisy channels. *Wireless Communication and Mobile Computing*, *8*(9), 1227–1232. doi:10.1002/wcm.590

Muthukumaran, D., Sawani, A., Schiffman, J., Jung, B. M., & Jaeger, T. (2008). Measuring integrity on mobile phone systems. In *Proceedings of the 13th ACM Symposium on Access Control Models and Technologies*, (pp. 155–164). New York, NY: ACM.

Newsome, J., & Song, D. (2005). Dynamic taint analysis for automatic detection, analysis, and signature generation of exploits on commodity software. In *Proceedings of the Network and Distributed System Security Symposium*, Internet Society, Reston, Virginia, USA.

Ng, C., Susilo, W., Mu, Y., & Safavi-Naini, R. (2008). RFID privacy models revisited. *13th European Symposium on Research in Computer Security – ESORICS 2008, Lecture Notes in Computer Science, 5283/2008*, 251-266. Springer.

Niederman, F., Mathieu, R. G., Morley, R., & Kwon, I. (2005). Examining RFID applications in supply chain management. *Communications of the ACM*, *50*(7), 93–102.

Niedermeir, R., & Rossmanith, P. (1999). Upper bounds for vertex cover further improved. *16th Symposium on Theoretical Aspects in Computer Science, Lecture Notes in Computer Science, Vol. 1563*, (pp. 561-570). Springer.

NSA. (n.d.). *Defense in depth: A practical strategy for achieving information assurance in today's highly networked environments*. Technical report, National Security Agency. Retrieved from http://www.nsa.gov/ia/_files/support/defenseindepth.pdf

ODonnell. A. J., & Sethu, H. (2004). On achieving software diversity for improved network security using distributed coloring algorithms. In *Proceedings of the 11th ACM Conference on Computer and Communications Security*, (pp. 121–131). New York, NY: ACM.

Ohkubo, M., Suzuki, K., & Kinoshita, S. (2003). *Cryptographic approach to "privacy-friendly" tags*. Paper presented at RFID Privacy Workshop, MIT, Massachusetts, USA.

Ohkubo, M., Suzuki, K., & Kinoshita, S. (2004). Efficient hash-chain based RFID privacy protection scheme. In *the Proceedings of International Conference on Ubiquitous Computing, Workshop Privacy*.

Ohkubo, M., Suzuki, K., & Kinoshita, S. (2005). RFID privacy issues and technical challenges. *Communications of the ACM*, *48*(9), 66–71. doi:10.1145/1081992.1082022

Ohta, K., & Okamoto, T. (1988). Practical extension of Fiat-Shamir scheme. *Electronics Letters*, *24*, 955–956. doi:10.1049/el:19880650

Okamoto, T. (1992). Provably secure and practical identification schemes and corresponding signature schemes. In *Advances in Cryptology (CRYPTO92)* (*Vol. 740*, pp. 31–53). Lecture Notes in Computer ScienceSpringer.

Okubo, M., Suzuki, K., & Kinoshita, S. (2004). Efficient hash-chain based RFID privacy protection scheme. In *Proceedings of the International Conference on Ubiquitous Computing (Ubicomp)*, Nottingham, England.

Oren, Y., & Shamir, A. (2006). *Power analysis of RFID tags*. Advances in Cryptology - CRYPTO 2006. Retrieved from http://www.wisdom.weizmann.ac.il/yossio/rfid/

Oren, Y., & Shamir, A. (2007). Remote password extraction from RFID tags. *IEEE Transactions on Computers*, *56*(9), 1292–1296.

Orlin, J. B. (1988). A faster strongly polynomial minimum cost flow algorithm. *20th ACM Symposium on Theory of Computing*, (pp. 377-387).

Oua, K., & Vaudenay, S. (2009). *Pathchecker: An RFID application for tracing products in supply-chains*. Workshop on RFID Security.

Padioleau, Y., Tan, L., & Zhou, Y. (2009). Listening to programmers - Taxonomies and characteristics of comments in operating system code. In *Proceedings of the 31st International Conference on Software Engineering*, (pp. 331–341). Washington, DC: IEEE Computer Society.

Paise, R., & Vaudenay, S. (2008). Mutual authentication in RFID. *3rd ACM ASIA Conference on Computer and Communications Security*, (pp. 292-299).

Peris-Lopez, P., Castro, J. C. H., Estévez-Tapiador, J. M., & Ribagorda, A. (2009). Advances in ultralightweight cryptography for low-cost RFID tags: Gossamer protocol. In *International Workshop Information Security Applications: WISA08, Lecture Notes in Computer Science, vol. 5379*, (pp. 56–68). Springer.

Peris-Lopez, P., Hernandez-Castro, J. C., Estevez-Tapiador, J. M., & Ribagorda, A. (2006). EMAP: An efficient mutual authentication protocol for low-cost RFID tags. In *The Proceedings of the OTM Federated Conferences and Workshop: IS Workshop, IS '06, Vol. 4277*, (pp. 352–361).

Peris-Lopez, P., Hernandez-Castro, J. C., Estevez-Tapiador, J., & Ribagorda, A. (2006). M2AP: A minimalist mutual-authentication protocol for low-cost RFID tags. In *The Proceedings of the International Conference on Ubiquitous Intelligence and Computing, UIC '06, Vol. 4159*, (pp. 912–923).

Peris-Lopez, P., Hernandez-Castro, J., Tapiador, J. M. E., & Ribagorda, A. (2006). *LMAP: A real lightweight mutual authentication protocol for low-cost RFID tags*. Workshop on RFID Security (RFIDSec'06), Austria.

Pescatore, J., Young, G., Allan, A., Girard, J., Feiman, J., & MacDonald, N. (2008). *Gartner 2008 IT security threat projection timeline*. Technical report, Gartner. Retrieved from http://www.gartner.com

Philip, R., & Beigl, M. (2004). Trust context spaces: An infrastructure for pervasive security in context-aware environments. In *Proceedings the 1st International Conference on Security in Pervasive Computing* (SPC), Boppard, Germany.

Phillips, T., Karygiannis, T., & Kuhn, R. (2005). Security standards for the RFID market. *IEEE Security Privacy*, *3*(6), 85–89. doi:10.1109/MSP.2005.157

Poschmann, A. (2009). *Lightweight cryptography: Cryptographic engineering for a pervasive world*. Dissertation, Doktor-Ingenieur, Ruhr-University Bochum.

Poschmann, A., Leander, G., Schramm, K., & Paar, C. (2006). *New lightweight crypto algorithms for RFID*. Workshop on RFID Security (RFIDSec'06), Austria.

Potdar, V., & Chang, E. (2006). Tamper detection in RFID tags using fragile watermarking. *IEEE International Conference on Industrial Technology, ICIT'06* (pp. 2846–2852).

Pramstaller, N., Mangard, S., Dominikus, S., & Wolkerstorfer, J. (2005). Efficient AES implementations on ASICs and FPGAs. In *Advanced Encryption Standard (AES05)* (*Vol. 3373*, pp. 98–112). Lecture Notes in Computer ScienceSpringer. doi:10.1007/11506447_9

Preneel, B. (2010). The first 30 years of cryptographic hash functions and the NIST SHA-3 competition. In Pieprzyk, J. (Ed.), *CT-RSA 2010, LNCS 5985* (pp. 1–14). doi:10.1007/978-3-642-11925-5_1

Rec, I. T. U.-T. X.509 (revised). (1993). *The directory-authentication framework*. Geneva, Switzerland: International Telecommunication Union.

Reid, J., Nieto, J. M. G., Tang, T., & Senadji, B. (2007). Detecting relay attacks with timing-based protocols. In *2nd ACM Symposium on Information, Computer and Communications Security: ASIACCS07*, (pp. 204 – 213).

Rescorla, E., Ray, M., Dispensa, S., & Oskov, N. (2010). Transport layer security (TLS) renegotiation indication extension. In *The Internet Engineering Task Force Request For Comments (IETF RFC) 5746*. Retrieved from http://tools.ietf.org/html/rfc5746

Research, C. (2000). *Standards for efficient cryptography - SEC2: Recommended elliptic curve domain parameters*. Retrieved from http://www.secg.org/download/aid-386/sec2_final.pdf

RFIDJournal. (2005). *Attack on a cryptographic RFID device*. Retrieved from http://www.rfidjournal.com/article/view/1415/1/82

Rhee, K., Kwak, J., Kim, S., & Won, D. (2005). Lecture Notes in Computer Science: *Vol. 3450. Challenge-response based RFID authentication protocol for distributed database environment. Security in Pervasive Computing* (pp. 70–84). Springer. doi:10.1007/978-3-540-32004-3_9

Rieback, M. R. (2006). *A hackers toolkit for RFID emulation and jamming*. Presented at 23rd Chaos Communication Congress.

Rieback, M. R. (2008). *Security and privacy of radio frequency identification*. Unpublished doctoral dissertation, Vrije Universiteit, Amsterdam.

Rieback, M. R., Crispo, B., & Tanenbaum, A. S. (2005). *RFID guardian: A battery-powered mobile device for RFID privacy management*. Paper presented at the Information Security and Privacy 10th Australasian Conference, ACISP 2005, 4-6 July, Berlin, Germany.

Rieback, M. R., Crispo, B., & Tanenbaum, A. S. (2006). *Is your cat infected with a computer virus?* Paper presented at the 4th Annual IEEE International Conference on Pervasive Computing and Communications, PerCom 2006, March 13, 2006 - March 17, 2006, Pisa, Italy.

Rieback, M. R., Gaydadjiev, G. N., Crispo, B., Hofman, R. F. H., & Tanenbaum, A. S. (2006). A platform for RFID security and privacy administration. In *Proceedings of the 20th conference on Large Installation System Administration*, (pp. 89–102). Berkeley, CA: USENIX Association.

Rieback, M. R., Crispo, B., & Tanenbaum, A. S. (2005). Lecture Notes in Computer Science: *Vol. 3574. RFID guardian: A battery-powered mobile device for RFID privacy management. Information Security and Privacy* (pp. 184–194). Springer.

Rieback, M. R., Simpson, P. N., Crispo, B., & Tanenbaum, A. S. (2006). RFID malware: Design principles and examples. *Pervasive and Mobile Computing*, *2*(4), 405–426. doi:10.1016/j.pmcj.2006.07.008

Rivest, R. L., Shamir, A., & Adleman, L. (1978). A method for obtaining digital signature and public-key cryptosystems. *Communications of the ACM*, *21*(2), 120–126. doi:10.1145/359340.359342

Rolfes, C., Poschmann, A., Leander, G., & Paar, C. (2008). Ultra-lightweight implementations for smart devices - Security for 1000 gate equivalents. In Grimaud, G., & Standaert, F. X. (Eds.), *CARDIS 2008 LNCS 5189* (pp. 89–103). doi:10.1007/978-3-540-85893-5_7

Ryu, E.-K., & Takagi, T. (2009). A hybrid approach for privacy-preserving RFID tags. *Journal of Computer Standards and Interfaces*, *31*, 812–815. doi:10.1016/j.csi.2008.09.001

Saito, J., Imamoto, K., & Sakurai, K. (2005). Reassignment scheme of an RFID tag's key for owner transfer. In *Proceedings of the International Conference on Embedded and Ubiquitous Computing (EUC Workshops)*, Nagasaki, Japan.

Saito, J., Ryou, J. C., & Sakurai, K. (2004). Lecture Notes in Computer Science: *Vol. 3207. Enhancing privacy of universal re-encryption scheme for RFID tags. Embedded and Ubiquitous Computing* (pp. 879–890). Springer. doi:10.1007/978-3-540-30121-9_84

Sawyer, S., & Tapia, A. (2005). The sociotechnical nature of mobile computing work: Evidence from a study of policing in the United States. *International Journal of Technology and Human Interaction*, *1*(3), 1–14. doi:10.4018/jthi.2005070101

Schnorr, C. (1989). Efficient identification and signatures for smart cards. In *Advances in Cryptology (CRYPTO89)* (*Vol. 435*, pp. 239–252). Lecture Notes in Computer ScienceSpringer.

Schnorr, C. (1990). Efficient identification and signatures for smart cards. *Advances in Cryptology, Crypto '89. LNCS*, *435*, 239–252.

Shamir, A. (2008). SQUASH - A new MAC with provable security properties for highly constrained devices such as RFID tags. *In the Proceedings of Fast Software Encryption 2008* [Springer-Verlag.]. *LNCS*, *5086*, 144–157.

Shankarapani, M., Sulaiman, A., & Mukkamala, S. (2009). Fragmented malware through RFID and its defenses. *Journal in Computer Virology*, *5*(3), 187–198. doi:10.1007/s11416-008-0106-0

Shannon, C. E. (1948). A mathematical theory of communication. *The Bell System Technical Journal*, *27*(3), 379–423.

Shannon, C. E. (1948). Communication theory of secrecy. *The Bell System Technical Journal*, *28*, 656–715.

Shannon, C. E. (1949). Communications theory of secrecy systems. *The Bell System Technical Journal*, *28*, 656–715.

Shor, P. W. (1994). Algorithms for quantum computation: Discrete log and factoring. In S. Goldwasser (Ed.), *Proceedings of the 35th Annual Symposium on the Foundations of Computer Science*, (pp. 124-134).

Singelée, D., & Preneel, B. (2007). Distance bounding in noisy environments. In *European Workshop on Security and Privacy in Ad-Hoc and Sensor Networks: ESAS07, Lecture Notes in Computer Science, vol. 4572*, (pp. 101–115). Springer-Verlag.

Smart, N. (2003). *Cryptography: An introduction*. McGraw-Hill.

Song, B. (2008). *RFID tag ownership transfer*. Workshop on RFID Security - RFIDSec'08, Budapest, Hungary.

Song, B., & Mitchell, C. J. (2008). RFID authentication protocol for low-cost tags. In V. D. Gligor, J. Hubaux, & R. Poovendran (Eds.), *ACM Conference on Wireless Network Security – WiSec '08*, (pp. 140-147). Alexandria, VA: ACM Press.

Song, B., & Mitchell, C. J. (2009). Scalable RFID pseudonym protocol. *The 3rd International Conference on Network & System Security - NSS 2009*, (pp. 216-224). Gold Coast, Australia: IEEE Computer Society.

Song, B., & Mitchell, C. J. (2011). Scalable RFID security protocols supporting tag ownership transfer. [Elsevier.]. *International Journal Computer Communications*, *34*(4), 556–566. doi:10.1016/j.comcom.2010.02.027

Spinellis, D. (2003). Reliable identification of bounded-length viruses is NP complete. *IEEE Transactions on Information Theory*, *49*(1), 280–284. doi:10.1109/TIT.2002.806137

Stallings, W. (2006). *Cryptography and network security: Principles and practice* (4th ed.). New Jersey: Prentice-Hall.

Stinson, D. (1995). *Cryptography: Theory and practice*. CRC Press.

Stinson, D. R. (2002). *Cryptography, theory and practice* (2nd ed., pp. 73–116). Chapman and Hall/CRC.

Stinson, D. R. (2005). *Cryptography: Theory and practice* (3rd ed.). Chapman & Hall/CRC.

Stoer, M., & Wagner, F. (1997). A simple min-cut algorithm. *Journal of the ACM*, *44*(4), 585–591. doi:10.1145/263867.263872

Sulaiman, A., Mukkamala, S., & Sung, A. (2008). SQL infections through RFID. *Journal in Computer Virology*, *4*(4), 347–356. doi:10.1007/s11416-007-0075-8

Swaminathan, J. M., Smith, S. F., & Sadeh, N. M. (1998). Modeling supply chain dynamics: A multiagent approach. *Decision Sciences*, *29*(3), 607–632. doi:10.1111/j.1540-5915.1998.tb01356.x

Sweeney, P. J. (2005). *RFID for dummies*. Wiley Publishing, Inc.

Tan, L., Zhang, X., Ma, X., Xiong, W., & Zhou, Y. (2008). AutoISES: Automatically inferring security specifications and detecting violations. In *Proceedings of the 17th USENIX Security Symposium,* (pp. 379–394). Berkeley, CA: USENIX Association.

TCG. (n.d.). *Trusted computing group*. Retrieved from http://www.trustedcomputinggroup.org

Tiri, K., Akmal, M., & Verbauwhede, I. (2002). A dynamic and differential CMOS logic with signal independent power consumption to withstand differential power analysis on smart cards. In *28th European Solid-State Circuits Conference(ESSCIRC 2002)*, (pp. 403-406).

Traub, K., et al. (2010). *The EPCglobal architecture framework*. Retrieved from http://www.gs1.org/gsmp/kc/epcglobal/architecture/architecture_1_4-framework-20101215.pdf

Tsudik, G. (2006). YA-TRAP: Yet another trivial RFID authentication protocol. In *Proceedings of the 4th Annual IEEE International Conference on Pervasive Computing and Communications (PerCom)*, Pisa, Italy.

Tsudik, G. (2007). A family of dunces: Trivial RFID identification and authentication protocols. In N. Borisov & P. Golle (Eds.), *PET 2007, LNCS 4776* (pp. 45-61).

Vajda, I., & Buttyan, L. (2003). *Lightweight authentication protocols for low-cost RFID tags*. 2nd Workshop on Security in Ubiquitous Computing.

Vaudenay, S. (2007). On privacy models for RFID. In *Proceedings of the 13th Annual International Conference on the Theory and Application of Cryptology and Information Security (Asiacrypt)*, Kuching, Malaysia.

Vaudenay, S. (2007). On privacy models for RFID. In Kurosawa, K. (Ed.), *Advances in Cryptology – Asiacrypt 2007* (*Vol. 4833*, pp. 68–87). Lecture Notes in Computer ScienceSpringer. doi:10.1007/978-3-540-76900-2_5

Wang, X., Li, Z., Xu, J., Reiter, M. K., Kil, C., & Choi, J. Y. (2006), Packet vaccine: Black-box exploit detection and signature generation. In *Proceedings of the 13th ACM Conference on Computer and Communications Security,* (pp. 37–46). New York, NY: ACM.

Washington, L. C. (2003). *Elliptic curves: Number theory and cryptography*. Chapman and Hall, CRC Press.

Weis, S., Sarma, S., Rivest, R., & Engels, D. (2003). Security and privacy aspects of low-cost radio frequency identification systems. In *Proceedings of the 1st International Conference on Security in Pervasive Computing (SPC)*, Boppard, Germany.

Weise, J. (2001). Public key infrastructure overview. In *Global Security Practice, Sun BluePrint™ OnLine*. Retrieved from http://www.sun.com/blueprints/0801/publickey.pdf

Weis, S. A., Sarma, S. E., Rivest, R. L., & Engels, D. W. (2004). Lecture Notes in Computer Science: *Vol. 2802. Security and privacy aspects of low-cost radio frequency identification systems. Security in Pervasive Computing* (pp. 201–212). Springer.

Weis, S. A., Sarma, S. E., Rivest, R. L., & Engels, D. W. (2004). Security and privacy aspects of low-cost radio frequency identification systems. In Goos, G., Hartmanis, J., & van Leeuwen, J. (Eds.), *Security in Pervasive Computing, LNCS* (*Vol. 2802*, pp. 55–59). Berlin, Germany: Springer. doi:10.1007/978-3-540-39881-3_18

Wilding, R., & Delgado, T. (2004). RFID demystified: Supply chain applications. *Logistics and Transport Focus*, *6*(4), 42–47.

Witteman, M. (2005). Attacks on digital passports. *What the Hack*. Retrieved from http://wiki.whatthehack.org/index.php/Track:Attacks_on_Digital_Passports

Xie, Y., & Kim, H.-A. OHallaron, D., Reiter, M. K., & Zhang, H. (2004). Seurat: A pointillist approach to anomaly detection. In *Proceedings of the 7th International Symposium on Recent Advances in Intrusion Detection,* (pp. 238–257). Berlin, Germany: Springer-Verlag.

Yamamoto, A., Suzuki, S., Hada, H., Mitsugi, J., Teraoka, F., & Nakamura, O. (2008). A tamper detection method for RFID tag data. In *The Proceedings of the IEEE International Conference on RFID*, (pp. 51–57).

Yee, B., Sehr, D., Dardyk, G., Chen, J. B., Muth, R., & Ormandy, T. … Fullagar, N. (2009). Native client: A sandbox for portable, untrusted x86 native code. In *Proceedings of the 2009 30th IEEE Symposium on Security and Privacy*, (pp. 79–93). Washington, DC: IEEE Computer Society.

Yuksel, K. (2004). *Universal hashing for ultra-low-power cryptographic hardware applications*. Master's thesis, Worcester Polytechnic Institute, Worcester, Massachusetts, USA.

Zhang, X., Aciicmez, O., & Seifert, J.-P. (2009). Building efficient integrity measurement and attestation for mobile phone platforms. In *Proceedings of the First International Conference on Security and Privacy in Mobile Information and Communication Systems*, (pp. 71–82). Berlin, Germany: Springer-Verlag.

Zhou, Z., & Huang, D. (n.d.). RFID keeper: An RFID data access control mechanism. In *The Proceedings of the IEEE Global Telecommunications Conference, GLOBECOM '07*, (pp. 4570 –4574).

About the Contributors

Ali Miri has been a Full Professor at School of Computer Science, Ryerson University, Toronto since 2009. He has also been with the School of Information Technology and Engineering and Department of Mathematics and Statistics since 2001 as an Assistant Professor, and later as an Associate Professor in 2005, a Full Professor at 2008, and an Adjunct Professor since 2009. He has held visiting positions at the Fields Institute for Research in Mathematical Sciences, Toronto in 2006, and Universite de Cergy-Pontoise, France in 2007, and Alicante and Albecete Universities in Spain in 2008. His research interests include computer networks, digital communication, and security and privacy technologies and their applications, in which he has authored and co-authored more than 140 referred articles, 4 books, and 2 patents. He has served as a guest editor for *Journal of Ad Hoc and Sensor Wireless Networks, Journal of Telecommunications Systems, Security and Communication Networks,* and on the editorial board of *International Journal on Advances in Internet Technology*. He currently serves as an Associate Editor for the *Canadian Journal of Electrical and Computer Engineering*. He is a member of Professional Engineers Ontario, and a senior member of IEEE.

* * *

Ian F. Blake received his undergraduate education at Queen's University in Kingston, Ontario and his Ph.D. at Princeton University in New Jersey. From 1967 to 1969 he was a Research Associate with the Jet Propulsion Laboratories in Pasadena, California. From 1969 to 1996 he was with the Department of Electrical and Computer Engineering at the University of Waterloo, in Waterloo, Ontario. He is currently with the Department of Electrical and Computer Engineering at the University of Toronto. He has spent sabbatical leaves with the IBM Thomas J. Watson Research Center, the IBM Research Laboratories in Switzerland and M/A-Com Linkabit in San Diego, California. From 1996-1999 he was with the Hewlett-Packard Labs in Palo Alto, California. His research interests are in the areas of cryptography, algebraic coding theory, digital communications, and spread spectrum systems. He is a Fellow of the IEEE, the Royal Society of Canada and the Canadian Academy of Engineers. He was awarded an IEEE Millenium Medal.

Robert H. Deng is currently a Professor with the School of Information Systems, Singapore Management University. He has 26 patents and more than 200 technical publications in international conferences and journals in the areas of computer networks, network security and information security. He has served as general chair, program committee chair, and program committee member of numerous international conferences. He is an Associate Editor of the *IEEE Transactions on Information Forensics and Security*, Associate Editor of *Security and Communication Networks Journal*, and member of Editorial Board of *Journal of Computer Science and Technology*. He received the University Outstanding Researcher Award from the National University of Singapore in 1999 and the Lee Kuan Yew Fellow for Research Excellence from the Singapore Management University in 2006. He was named Community Service Star and Showcased Senior Information Security Professional by $(ISC)^2$ under its Asia-Pacific Information Security Leadership Achievements program in 2010.

Dang Nguyen Duc received his Bachelor's degree from Hanoi University of Technology and Science, Vietnam in 2000. In 2003, he received his Master's in Computer Science from Information and Communications University (now part of KAIST), Republic of Korea. He then completed his PhD in Cryptography and Information Security at KAIST, Republic of Korea in 2010. From March, 2010 to October, 2010, he was Post-Doctoral researcher at Department of Computer Science, KAIST, Republic of Korea. Dr. Duc is currently a Senior Engineer at System LSI, Semiconductor, Samsung Electronics. His research interests include design and analysis of cryptographic protocols, efficient implementation of cryptographic primitives, and countermeasures to side-channel attacks.

Kwangjo Kim has received the BS and MS degrees of Electronic Engineering from Yonsei University, Korea in 1981 and 1983, respectively. He has also finished his Ph.D in Div. of Electrical and Computer Engineering in Yokohama National University, Japan in 1991. He was Section Head of Coding Tech. #1 in ETRI (Electronics and Telecommunications Research Institute) (1983~ 1997). He was Professor at School of Engineering in ICU (1998 ~ 2009). Currently, he is Professor at Computer Science Department in KAIST, Korea and Director of IRIS. He has published more than 60 papers in SCI-ranked journals and more than 1,000 papers in international and domestic conference proceedings. He was elected to serve Board of Director Member of IACR (1999 ~ 2004) and Asiacrypt Steering Committee Chair (2005 ~ 2008). He is currently Member of IEEE, IACR, and IEICE. He is also Honorable President of KIISC and Member of ASIACRYPT Steering Committee. He is an editor of JMC. He has been served as Program Committee or Organizing Member in a variety of the international conferences on information security and cryptology more than 10 times per each year.

Fanyu Kong received his BS, MS, and PhD degrees in Computer Science from Shandong University, China, in 2000, 2003, and 2006, *respectively*. He joined Institute of Network Security of Shandong University and became a Lecturer in 2006. Since 2008, he has been an Associate Professor at Shandong University. He is also a member of Key Laboratory of *Cryptologic* Technology and Information Security, Ministry of Education, China. His research interests include public key cryptology, computational number theory, information security, and computer algorithms. He taught several graduate courses such as Public Key Cryptography, Analysis and Design of Algorithms. He is a member of China Computer Federation and Chinese Association for Cryptologic Research.

Ming Li was born in 1982. He received the BS degree in Computer Science from School of Computer Science and Technology in Shandong University in 2004 and the MS degree from Institute of Network Security in Shandong University in 2007. He now is a PhD candidate in School of Computer Science and Technology in Shandong University. He taught in a graduate course Network Security. He was working in Computational Laboratory in Coding & Cryptography at University of Ottawa with Dr. Ali Miri as a visiting scholar in 2009. His research interests include elliptic curve cryptography, network security, design and analysis of algorithms, and the theory of computational complexity.

Yingjiu Li is currently an Associate Professor in the School of Information Systems at Singapore Management University. He received his PhD degree in Information Technology from George Mason University in 2003. His research interests include RFID security and privacy, applied cryptography, and data applications security and privacy. He has published over 80 technical papers in international conferences and journals. He has served in the program committees for over 50 international conferences and workshops. Yingjiu Li is a Senior Member of the ACM and a member of the IEEE.

Li Lu received his BS and MS degrees from Zhejiang University, China, in 2000 and 2003, respectively, all in Electrical Engineering. He received PhD degree in Computer Science and Engineering at Chinese Academy of Science in 2007. He is now an Associate Professor in the School of Computer Science and Engineering at University of Electronic Science and Technology of China. His research interests include RFID system, security in wireless networks, and applied cryptography. He is a member of the IEEE Computer Society, and a member of ACM.

Behzad Malek received his Master's degree in Electrical Engineering and his PhD degree in Electrical and Computer Engineering from University of Ottawa, Canada in 2005 and 2010, respectively. He currently holds a postdoctoral fellow position at Ryerson University in Toronto, Canada. His research interests are communications security and privacy. His specializations are in applied cryptography and design of cryptographic security/privacy systems.

Zhang Ning received the BSc degree in Communication Engineering from the Xidian University, Xian, China, in 2002, the MSc and the PhD degree in Cryptography from Xidian University in 2005 and 2008, respectively. From 2005, she was employed as a researcher and teacher in Xidian University. She is a member of IEEE and Chinese Association of Cryptology Research (CACR). She is doing or has completed projects including the National Natural Science Foundation of China, the National High Technology Research and Development Program (863 Program) of China, the National Basic Research Program (973 Program) of China in area of information security and cryptography. Her research interests are broadly in the area of information security. The areas of special interests include: cryptography, wireless network security, RFID system security, elliptic curve cryptosystem, cryptographic protocol, digital signature, and provable security.

Ilker Onat received a Bachelor degree in Electrical and Electronics Engineering from Middle East Technical University, Turkey, a Master's degree in Electrical and Computer Engineering from New Jersey Institute Technology, USA, and a PhD degree in Electrical and Computer Engineering from University of Ottawa, Canada. His research interests are mainly focused on the energy-efficient cross-layer design algorithms for low-power wireless embedded devices. He is currently working as a senior technical specialist at Turkish Aerospace Industries (TAI) Space Systems Group.

Boyeon Song received her B.Eng. degree in Communications and Information Engineering from Korea Aerospace University in 1998, her M.Sc. degree in Crypto and Information Security Group from Information and Communications University in Korea in 2001, and her Ph.D. degree in Information Security Group from Royal Holloway, University of London in the UK in 2010. She was a researcher in Cryptographic Technology Team of Korea Information Security Agency, which she joined in 2001. Between 2002 and 2004, she was a Lecturer in Department of Information Technology at Mongolia

International University in Mongolia. In 2010, she was a Research Professor in Center for Information Security Technologies of Korea University. From February 2011, she is working as a Researcher in Cryptographic Technical Team of National Institute for Mathematical Sciences. Her research interests mainly relate to information security, specifically privacy and security in wireless communications.

Ehsan Vahedi is a PhD candidate at Electrical and Computer Engineering Department, the University of British Columbia (UBC), Vancouver, BC, Canada. He received his BSc degree from K.N.T. University of Technology, Tehran, Iran (2004) and his MSc degree from University of Tehran, Tehran, Iran (2007), both in Electrical Engineering. His research interests are in mathematical modeling and security analysis of RFID systems and wireless networks, watermarking, and digital image and video processing.

Vincent W.S. Wong received the B.Sc. degree from the University of Manitoba, Winnipeg, MB, Canada, in 1994, the MSc. degree from the University of Waterloo, Waterloo, ON, Canada, in 1996, and the PhD degree from the University of British Columbia (UBC), Vancouver, BC, Canada, in 2000. From 2000 to 2001, he worked as a systems engineer at PMC-Sierra Inc. He joined the Department of Electrical and Computer Engineering at UBC in 2002 and is currently an Associate Professor. His research areas include protocol design, optimization, and resource management of communication networks, with applications to the Internet, wireless networks, smart grid, RFID systems, and intelligent transportation systems. Dr. Wong is an Associate Editor of the *IEEE Transactions on Vehicular Technology* and an Editor of *KICS/IEEE Journal of Communications and Networks*. He is the Symposium Co-chair of IEEE Globecom '11, Wireless Communications Symposium. He is a senior member of the IEEE.

Qiang Yan is currently a PhD student in the School of Information Systems at Singapore Management University. He received his BEng and MSc in Software Engineering from Fudan University, China, in 2006 and 2009. His current research areas include RFID security, mobile security, and system security.

Index

U

V

W

X

Y